EMPIRES OF TIME

EMPIRES OF TIME

Calendars, Clocks, and Cultures

Anthony F. Aveni

KODANSHA INTERNATIONAL
New York • Tokyo • London

Kodansha America, Inc.
114 Fifth Avenue, New York, New York 10011, U.S.A.

Kodansha International Ltd.
17-14 Otowa 1-chome, Bunkyo-ku, Tokyo 112, Japan

Published in 1995 by Kodansha America, Inc.
by arrangement with the author.

First published in 1989 by Basic Books, Inc.

This is a Kodansha Globe book.

Library of Congress Cataloging-in-Publication Data

Aveni, Anthony F.
Empires of time : calendars, clocks, and cultures / Anthony F. Aveni.
p. cm.
Originally published: New York : Basic Books, c1989.
Includes bibliographical references and index.
ISBN 1-56836-073-8
1. Time. I. Title.
QB209.A94 1995
529—dc20 95-15531

Book design by Vincent Torre

Printed in the United States of America

95 96 97 98 99 RRD/H 10 9 8 7 6 5 4 3 2 1

CONTENTS

PART III

Their Time: Following the Order of the Skies

PART IV

A World of Time

LIST OF ILLUSTRATIONS

PART TITLE ILLUSTRATIONS

FIGURES

ACKNOWLEDGMENTS

I OWE MY GRATITUDE to many people in the making of this book. Among them, Roger Hoffman for providing materials and stimulating discussions on biorhythmic activities; Robert Garland for interesting conversations on time and timekeeping in the Classical World; Liz Meryman for advice on the selection of illustrative materials; Warren Wheeler for his superior photographic work; Phoebe Hoss for her expert editing; Gary Urton, Michael Peletz, Mary Moran, Margaret Maurer, Albert and Rebecca Ammerman, Curtis Hinsley, Horst Hartung, Wilbur Albrecht, Robert Linsley, and Tom Zuidema, along with the Colgate students who attended my seminars on comparative cosmology, for numerous discussions on Western and non-Western world views; and Jim McCoy for technical assistance in the preparation of the manuscript. Though I do cite them in context, I want to thank collectively all of my colleagues and all of the institutions that have consented to allow me to reproduce their illustrations and diagrams. The most gratitude in concrete terms I attribute to my wife, Lorraine, both editor and supportive critic, who has unrelentingly challenged and questioned in a helpful, constructive way every idea I put forward that could be re-examined or more clearly expressed.

Finally, retreating all the way back on my own personal time line as far as I can go, I thank my mother and my father, who ultimately emerge as the very wise people who started me out on the road that has led to this inquiry. They unselfishly viewed as their highest priority the task of providing me with a formal education—in their day an exalted privilege rather than a right, and a luxury that, unfortunately, they were denied from pursuing formally to the fullest extent. To them, I dedicate this work.

EMPIRES OF TIME

Four Years Before Lunch. Wendell Castle. Courtesy of Alexander F. Milliken Gallery, Inc., New York. Photo by Bruce Miller.

INTRODUCTION: OUR TIME— AND THEIRS

I WAS BORN. This I am told for I cannot remember it. But looking backward, I *think* that by talking to myself I can remember being very young, barely able to walk, incapable of tying my shoes or riding a tricycle. Fairly early on, I remember sensing a pattern, a set of interlocking harmonies that defined a rhythm to my life. I remember night following day, the moon replacing the sun, and winter following summer as sledding on the ice took the place of long hours at the beach.

As I grew older, the rhythm became more fascinating. Harmonic overtones and subtle subbeats that I had never heard before were added to the music of my life. My day and my year became ever more finely divided and subdivided—school and summer vacation, play and work, home and away from home. Every moment of my existence seemed to take on increased significance. As I look back among the pages of my life's score, it seems that some portions of the rhythm, particularly the opening stanzas, had a sort of largo tempo. But the last several pages have been played at a much quicker beat. Have more musical notes been crammed onto these pages? I know this for sure: the music has never flowed evenly. Sometimes it is more dense and compact, and at other times it becomes attenuated. Often it is free flowing, but then it is bounded and restricted—yesterday's busy slate of meetings, a late afternoon class, some household chores, then social visits in the evening.

There is a difference between what the clock says and the way I conceive of the past in this inner dialogue I've been having with myself. I attach no numbers or dates to my happenings. There is no 1945, 1963, or 1969—only the end of a war, the assassination of a president, astronauts landing on the surface of the moon; or my graduation from high school, my marriage, and the death of a friend. The absolute chronology called history is thrust upon me by the ways of modern, industrialized, scientific society. It conflicts with the way I naturally want to think about the past—purely as a sequence of events that happened in a particular order, like knots on a long string that run from my origin to the present. Who cares how long the string is or how many inches between the knots? Does it matter whether the musical tempo is doubled or halved, so long as the notes are kept in recognizable order?

What about the future? I can be sure that as soon as I put my pen down, about two pages from now, I will have lunch. Farther forward on the event-line string of anticipation, I can already imagine painting the kitchen ceiling. That will happen this weekend, for some things are, after all, inevitable. Yes, I can see myself, even the clothes I am wearing. My hand swishes the brush back and forth. I am concerned about paint dripping onto the floor, onto myself—especially my glasses. Of course, this anticipation imagery of the future is not real; it is concocted in my mind, fashioned out of a memory of the past when I last performed the same act, teetering on that ladder several years ago. What I recollect in my mind's eye concerning that last painting episode surely is not the same as the real event when it happened. I must be hauling out of storage that old "re-collected" set of images of the past to fabricate this weekend's preview of coming attractions.

I cannot hear the tune far ahead along that empty musical staff that connects present to future. It is still too distant, though some parts of it seem to be getting close enough at least to begin to decipher. I can see my grandchild—faceless, nameless, a life still eventless, a string of undetermined length containing no knots. Though still a month short of birth, that child (like me) will live and learn and love. I have confidence he or she, too, will early on sense the rhythm I sensed, and see the pattern I saw.

Does it not make sense to think of time as a line or an arrow that points in one direction, tying past to future, with an infinitely narrow plane cutting it to represent the instantaneous present? Don't we all think of our lives this way in linear fashion—like a road, a river upon which we travel? Didn't our ancestors think of it this way? Where did this metaphor of life's road or life's music come from? Is our own consciousness responsible for thinking of sequences of events, or is the pattern already there waiting for us to fit ourselves into?

I know, too, that someday the music will come to an end for me. Will that be it—or will they strike up the band again? I really don't know. Unlike my encounters with the kitchen ceiling, I have no previous experience by which to anticipate what, if anything, will happen next. There are no images to recollect. Is life final, terminal, and irreversible? Or, does it go on eternally as we advance to another state of existence or perhaps through an interminable series of states? Maybe life repeats itself. Sooner or later we all get preoccupied with the issue of our own mortality, perhaps the more so the older we get.

All of these questions, and the thoughts or experiences that prompted their phrasing, can be embraced by a single word, one that evokes all conceivable ideas about how we relate to one another, to the universe around us, and to that all-pervasive spiritual element of which each of us grasps the essence in different ways: that word is *time*. This is the music, the rhythm, the road, the line cut by the plane; this is the framework in which we recollect the past and anticipate the future.

Time. It gets more room in Webster's unabridged dictionary[1] than nouns as general as *thing* and *god*, more than basic adjectives like *good* and *evil*, much more space than *space*, its archetypal counterpart. In fact, poring through the 2,006 pages of my own copy of the great lexicon, I am hard-pressed to find a word that has more descriptions or conflicting and confusing meanings than this innocuous member of the family of four-letter words.

Time is, first of all, an *idea*—the idea that an ordered sequence can be recognized in our states of consciousness.[2] As the philosopher C. D. Broad puts it, "All the events in the history of the world fall into their places in a single series of moments";[3] or as Ecclesiastes states, "For everything there is a season, and a time for every matter under heaven" (3:1). But time is also a measurement, the measure of duration between events. As a metric quantity, we think of it as unbounded, continuous, homogeneous, unchangeable, indivisible, and endless!

Before the turn of this century, we would have added the term *simultaneous* to the adjectives that describe what we mean by time; but the high-speed, rapidly changing world of the 1900s has forced us to redefine it. In physics, the principle of the special theory of relativity has led us to think of time and space as inseparable. Since information simply cannot get conveyed to the observer instantaneously, whether two events be conceived as having occurred at the same time depends upon where you are in the universe when you witness them. While a commuter waiting by the side of the rail tracks may see lightning strike opposite ends of a speeding train at the same time, a passenger situated at one end of the moving train might say that one strike occurred shortly before the other. And both observers would be correct.

We divide time into years, months, weeks, days, hours, minutes, and seconds—divisions that make it seem absolute, like something that exists apart from everything else in the world including even my own consciousness. Maxwell or Einstein—indeed, any respectable physicist—would say that temporal order is already there. We only need to recognize it and deal with it. But time is also relative, as when we measure lapsed portions of it between events. And when we reckon it by our senses rather than some independent, objective yardstick, we get different durations.

We assign various verbs and adjectives to time so that it can be likened to more tangible things. Time gets spent, wasted, killed, kept, and lost. We have leisure time, quality time, good times, bad times, hard times, and even hot times.

On the other hand, time has qualities of its own, often expressed through simile. For the ancient Greeks, there were changes in the day from dawn to dusk as well as in the year from summer to winter. When we say "*tempus fugit*" ("time flies"), we may not realize that the Romans, who coined that phrase, were equating time with weather. Time not only blows by us like a Roman wind, it also flies like a bird, walks like an old man carrying a scythe, flows like a river, and gets out of joint like our bones. In team sports, we can get out of it and back into it—a much more difficult prospect in most of life's other affairs. In science fiction, we travel through its many warps. Punctual businessmen are on it, old people are beyond it, prodigies before or ahead of it, and great comics and musicians are said to have a marvelous sense of it. How can a single concept hold to such a varied field of descriptions? How dare anyone even address such an enigma in a book of less than a thousand pages?

Time is also related to motion. Somehow we all have faith that *real* time ought to be determined by some sort of physical model, a moving body that passes repeatedly over equal spaces in the same way, like the sun passing across the sky and casting a moving shadow, sand falling through an hourglass, a pendulum bobbing back and forth over its arc, the mainspring of a clock unwinding, or the vibrating microscopic crystal in a digital watch.

Every technical device we have created to keep track of time has its moving parts. Every human-made example we can cite has as its heart an analogue device that tries to mimic the way we think nature actually behaves. All oscillate from a starting point: the pendulum at its greatest amplitude in one direction; the crystal at maximum expansion; the balance wheel wound to its highest tension, twisted all the way to one extreme. Then all three let loose, just as a child kicks up its feet after being pulled all the way back on a playground swing. Each mechanism passes through a continuous series of states or to the other extreme: the pendulum swings completely to the other side; the balance wheel twists taut in the opposite direction; the crystal shrinks to minimal expansion. Then they reverse themselves once more and proceed in backward order through each phase of the oscillation, returning to their original state—the pendulum in a second, the balance wheel in half a second, the crystal in a millionth of a second. In all three cases, there is motion—endless, repetitive, oscillatory motion, an eternal return from beginning back to beginning.

Some say that countless generations of watching the heavens turn led our ancestors to make temporal models that oscillate. After all, all celestial rhythms are basically periodic. Day endlessly follows night; the moon, after waxing to full, wanes in backward order to its new phase and returns once again to display its thin crescent phase, so prominently visible low in the west after sunset. Longer intervals behave the same way. Over the year, the setting sun oscillates back and forth between extremes along the horizon from solstice to solstice, exactly like a pendulum, only with a much slower beat. The earth completes a rotation on its axis and moves through a full revolution about the sun. As the seasons follow upon one another, biorhythms portray celestial rhythms, and we feel the extremes of swing in nature's pendulum in a more direct way. When cold winters replace hot summers, animals hibernate; they become inactive only to become active once again when their habitat warms up. Plants flourish but then wither and atrophy, only to bloom anew. All of time, regardless of whether we become aware of it biologically or astronomically, seems to be embraced by this cyclic quality of reversible recurrence.

Just how universal is the mechanical-motive way of seeing the world change? Is it peculiar to us, to you and me, to all Americans, to

Europeans? Or, does it extend to India and China, to Polynesia, to aboriginal Australia? Did the ancestors of the Chinese and the Polynesians also fabricate mechanical models for understanding time as our ancestors did? Do, indeed, all civilizations on earth, past and present, think about, much less measure out in exactly the same way, this intimate, personal entity we call time?

While many of us, in speaking of the seasonal cycle, automatically attribute to other cultures our spatial metaphor of the *circle* of time, cycle and circle do not necessarily go together. Indeed, this question lies at the root of a scholarly dispute that has been raging for some time. Those who believe that knowledge is determined by culture argue that time is so closely tied to social behavior and organization that the way a people perceive it can be revealed only by studying, in detail, the way they reckon its passage—that is, the phenomena they see, the units they define to relate events to one another, and most important of all, the *images* and *metaphors* they use to conceptualize time.

Others believe that all human knowledge is absolute in nature, that it is separate from culture. They argue that if other cultures really did have different ideas than we do about time, if they were not interested in using any of the imagery we employ to describe its nature, then we could not even begin to communicate with them. The absolutists suggest that all human beings think of time as duration and of duration as either cyclic or linear. Some go further and suggest that of the two kinds of duration, the cyclic is static (in the sense that all things return to the same point in time) and associated with ritual behavior, while linear time is dynamic and has more to do with practical behavior. Now is not the time to deal with the issue of what human knowledge really is, for we are only just getting our feet wet; but in the course of describing and comparing both how time is sensed and how it is marked or reckoned by us as well as by others, we shall be forced to wade up to our necks.

Hundreds of books have been written about time and timekeeping in different cultures of the world. Thousands of volumes exist that deal with time as it is conceived and measured in the Western world, and as it is comprehended from the perspective of the humanities, the sciences, and the social sciences. The term *Western world* usually denotes Western Europe, the Anglo-Franco-Hispanic industrialized Americas, and the background cultures out of which have sprung present ideas, attitudes, customs, and patterns of behavior—those traits that lead us to think of all of these aforementioned people as essentially "the same." One historical common denominator of these cultures is the European Renaissance, when the great artistic styles and scientific methods and models that have developed into today's forms of creative expression were originally and unmistakably forged. If we look further into the past to seek our identity, then we also must travel eastward into the Greco-Babylonian world, for the Greeks and Babylonians together devised the logical-mathematical way of comprehending nature that endures today. Also, we need to look to the east for our borrowings from China, which may have planted the seeds that led to the technological upbringing we all share today. We must extend ourselves southward to the north coast of Africa in order to grasp the significance of Islam's contribution to our own quantitative mathematical development. And so in speaking of "our culture," we draw various elements of our recently acquired identity from diverse places and times.

Then, there are the Others, both present and past, the ones who are not like Us. We can sense the contemporary otherness of the Arab world, of the people of Central Africa, and remote regions such as highland Peru, the jungles of Yucatan, or Arctic Alaska, Canada or Siberia—the way they look, the way they talk or worship, their dress, relationships between their men and their women. But that sense of otherness is fast disappearing. We can see the difference in China—even in the past ten years. There are not many people left on earth who are all that different from Us. Since the time of Columbus, through the age of satellite television, we have been unleashing our civilization upon them. It is possible to see, in less than a lifetime, Them becoming Us.

Bronislaw Malinowski, the early twentieth-century anthropologist, once said that his task was not simply to idly hunt curios or amuse himself over superstitions and barbarous customs but rather to enter into the soul of the savage and, through his eyes, to look at the outer world and feel ourselves what it must feel to him to be himself.[4] Though this was a typical way of addressing the extreme Other sixty years ago, today's anthropologists are beginning to realize that Malinowski's was a difficult, if not impossible, assignment.

The trouble with the past is that we have no direct dialogue with it. We are forced to rely on textual evidence in its several forms: books filled with writing and pictures, ceramics and other archeological remains, art and sculpture. Oddly enough, little has been written about the ways in which other societies, present or past, deal with time. Where the *present-Other* is concerned, matters seem to be left to the domain of the ethnologist, while ethnohistorians and archaeologists usually focus upon the *past-Other.* Such is the compartmentalization of knowledge in our specialized world.

Can it be that our society really does not value the concepts and ideas of the Other? Why study them anyway? What knowledge is to be gained? This is what many students enrolled in anthropology courses ask about the *present-Other.* The *past-Others* seem even less relevant. Not only are they not *Us,* but also they are dead. Even when they were alive, they were not Us. People in general have a tendency to deny the Other a history, and we in the West often perceive other cultures as static and unchanging. We lump all of them together and refer to them as the "traditional cultures." So, why study them? Malinowski's arresting answer went beyond just being tolerant of other people in the world, of simply appreciating their values and customs; it is also, he said, "to better understand our own nature and to make it finer, intellectually and artistically."[5] Other societies become mirrors to gaze into, so that we can see ourselves more clearly—as but one culture in the context of manifold possibilities.

More important than being basically "Western Europeans" in outlook (even though America is conceived as a melting pot of many cultures), we are, above all, human beings and therefore ought to be concerned about the religions, the sciences, the politics of all the people on earth as they converge inexorably toward One World. Because time is a universal concept, its study serves as an excellent way of attempting to get inside the heads of these other people to see what makes them tick, to better understand the world as they see it, even to risk the prospect of wrenching ourselves out of our acquired point of view by glimpsing something admirable we either might have lost or failed to acquire in our cultural past.

I am interested in how *we* think about and measure time as contrasted with how *they* do it. The goal is a comparative cross-cultural look at time. My fascination with *our* time emanates from my training as an astronomer. Official time, precise time, time marked by the rolling of stars and planets around the sky—this has been my preoccupation. But, over the past twenty years, that interest has expanded into the domain of those peculiar Others. I have become as interested in them as in ourselves. What time did they keep? Did they and do they have months and years? How did they reckon them? With what precision? Did they use technology? Was their time scientific, like ours? Malinowski was correct. The last three questions especially reflect back upon us: in exploring them, we are really inquiring about ourselves.

To approach these questions, we must raise even deeper-seated ones. Why did they have timekeeping? Who used the knowledge of time, and for what purpose? Is there only one kind of knowledge about astronomy, time, and the calendar in this world, or are there many kinds as different as the colors of lights on a Christmas tree? Did they think about time so differently that none of Webster's "Western" definitions fits them?

My inquiry into time is not a complete compendium. It does not include everybody's ideas about time in the history of the world. Nor is it a series of explanations about how every civilization of the world, then and now, has measured time. I have chosen those societies with which I have become the most familiar. I offer not the final word, but rather a sketch, a perspective designed to raise more questions than answers in the comparative study of time.

I begin with *our* time, by dissecting the modern Western calendar, largely a product of imperial Rome, but having pastoral roots embedded in the ancient Greek poetry of Hesiod. I chose this strategy because our ordinary desk or wall calendar is part of our daily experience—something tangible to which we can all relate. The calendar we live by is loaded with hidden meaning; and though we use it unceasingly, we seem to have lost all contact with its roots. Yet we go right along taking our midmorning break, collecting our weekly paycheck, planning our monthly agenda and our yearly vacation. In the first part of the book, I want to demonstrate the complex and manifold attributes of our calendar, the many components that comprise it and where they came from. In respect to events in Western social, economic, and political history, today's way of keeping and thinking about time is the result of a series of accidents, of wrong as well as right turns we took in the past. Christ had as much to do with the calendar as did Einstein. Julius Caesar, Charles Darwin, Pope Gregory XIII, and Aristotle had a hand in it, too.

Not only individuals but also great social movements have

affected it. Our notion of time has been molded by the free-thinking spirit of the Renaissance, by the rise of a medieval merchant class, by the theories of Marxism, evolution, and existentialism, and by the development of experimental science. We can trace it all the way back to the great stories of creation: Genesis, the Greek *Theogony*, and the Babylonian *Enuma Elish*. Not just science, but also politics, economics, and, above all, religion, have played a role in the fabrication of today's calendar. To understand our time is to chart the course of Western Judeo-Christendom, to be made aware of the many ways the future could have been different had the past's elements combined otherwise.

My discussion of *their* time is intended to be more than just a description of other timekeeping systems. I aim to reflect back and forth continually, between them and us, and ideas about time and methods of keeping track of it. On the one hand, I deal with time in tribal societies or chiefdoms, like the Nuer of Africa and the Pacific Trobrianders, those who are said to have been organized in less complicated ways than we, at least in terms of fixity of territory, specialization, and social stratification. Somewhat paradoxically, these societies, which early anthropologists tended to think of as primitive or savage, often have certain attributes, such as systems of kinship, that are far more complex than our own. On the other hand, I also examine time in societies that can be characterized as states, those with bureaucratically organized forms of government wherein central authority is strong and often possesses many ancillary branches. Some of these states are empires that create their own time. We categorize them by such terms as *imperial*, like the ancient Inca and the Aztec civilizations of the pre-Columbian New World. Others are called *city-states*, a term that has been applied to the ancient Maya civilization of Central America as a way of connoting a lesser degree of political hegemony. I chose these New World civilizations for two reasons: first of all, I have been especially interested in the preoccupation with time as evidenced by many Native American cultures; and, secondly, by good fortune these cultures were totally isolated from the colonial tentacles of Western Europe until the arrival of such Spanish explorers as Hernán Cortés and Francisco Pizarro early in the sixteenth century. Thus, we can study their timekeeping almost the way a chemist conducts an experiment, in a kind of hermetically sealed cultural laboratory. Still, we must be aware of our ever-present biases as well as of those of the sixteenth- and seventeenth-century historical chroniclers who were responsible for having passed much of the indigenous calendrical information down to us today.

I have made room in this book for Eastern cultures like India and especially China. I think real differences about the meaning of time emerge from these highly organized societies compared with those less rigidly organized. But I also find some similarities that are not superficial. So vivid are these that they move me to wonder whether we are all mentally wired the same way, simply because we are human.

PART I

Sensing and Marking Time

Lemons and Oysters. Georges Braque. The Phillips Collection, Washington, D.C.

1

THE BASIC RHYTHMS

> First the tide rushes in, plants a
> kiss on the shore,
> Then rolls out to sea and the sea
> is very still once more.
>
> —"Ebb Tide"

METAPHOR is a figure of speech by which the meaning of one word is transferred to another. Our century is dominated by computer and mechanical metaphors. We use terms like *network, input, feedback, gridlock,* and *information flow* to describe social situations and interactions as if to imply that society is a system that functions according to the laws of physics. But in the lyrics above from a popular song of the 1950s, the tide becomes the metaphor to describe the love one person experiences for another.* Like our emotions, the tide waxes and wanes from one extreme to another. Or, is it like the feeling of being torn apart by circumstances after coming together? For artists and poets, the changeability of human feelings finds expression in the way the waters rise and fall, not just over a day, but even from moment to moment, for the beat of each wave that breaks upon the shore imitates in microcosm the slower rhythm of ebb and flow that makes up the eternal tidal harmony.

*Actually *time* and *tide* are derived from the same Anglo-Saxon root word, *tīd*, meaning "season," "hour."

Tracking Down the Sense of Time

I am engaging in this romantic talk to show that we use nature's behavior as a model to describe something we feel. In the passage from that old song, we feel time not only as an endless flow of metronomic beats but also as a kind of rhythmic surge, a recurring pattern we can trace to our very roots, to an age before we could even call ourselves human beings—when we came out of the sea.

It is well known that the life cycles of marine organisms respond to the ebb and flow of the tides. The periodic inundation and exposure that results from tidal flow controls changes in temperature, pressure, agitation, salinity, and feeding conditions. Take oysters. When the sea is high, they open their shells for a longer period of time than when it is low: not much longer—it is too dangerous; only three or four minutes more per hour. Just enough longer to take safe advantage of the fresh source of nourishment brought in by the turning of the tide.

In the early 1950s, biologists pulled about a dozen oysters from New Haven harbor and shipped them to Northwestern University in Illinois for study.[1] The oysters were submerged in their original harbor water and kept in total darkness. To explore their feeding patterns, the researchers tied to the shells fine threads that could activate recording pens every time the oysters' muscular movements caused the hinged shells to part or come together. Just as expected, the oysters continued to open and shut their shells as if they still were snug on the bottom of their home harbor, even though they had been displaced to another time zone more than a thousand miles to the west. Then, after about two weeks, something strange happened. Gradually the hour of maximal opening of the shells began to drift from day to day. Now, anyone who lives near the shore knows that the high- and low-water marks shift gradually from day to day. Tides are synchronized not with the place of the sun in the sky; rather it is the moon's schedule of appearance that matters, and the moon runs about 50 minutes, or eight tenths of an hour, later than the sun's cycle. On the average, successive high and low tides occur nearly an hour later each day. We would expect all oysters to open and shut on a 24.8-hour schedule. But, the biologists in Illinois were witnessing a daily drift that corresponded to a different beat. After four weeks of recording and analyzing the data, they had determined beyond any doubt that the oysters had restabilized the rhythmic opening and closing of their shells to the tidal cycle that would occur in Evanston, Illinois, had there been an ocean in that location. For the rest of the time the oysters were observed, they continued to maintain this new cycle (figure 1.1). It was as if they had gradually adjusted their life's pace to correspond to the time when the moon was overhead as seen from Northwestern University rather than from New Haven harbor. Could this lowly form of life actually feel the moon's presence through the sealed walls

FIGURE 1.1 Oysters adapt to a new time standard. The average number of minutes per hour their shells are open is plotted (*vertically*) against the hour of the lunar day (*horizontally*) for fifteen oysters transported from New Haven, Conn., to Evanston, Ill., and maintained in constant environmental conditions: (*A*) for the first two weeks in Evanston (arrows indicate time of high tide in the oysters' home waters); (*B*) for the next two weeks; and (*C*) for the last two weeks of the study. The time of maximum shell opening gradually shifts to, and then stabilizes at, the times of upper and lower transits of the moon across the local meridian, as the dotted line suggests. SOURCE: F. A. Brown et al., *The Biological Clock* (Englewood, N.J.: Heath, 1962), p. 25.

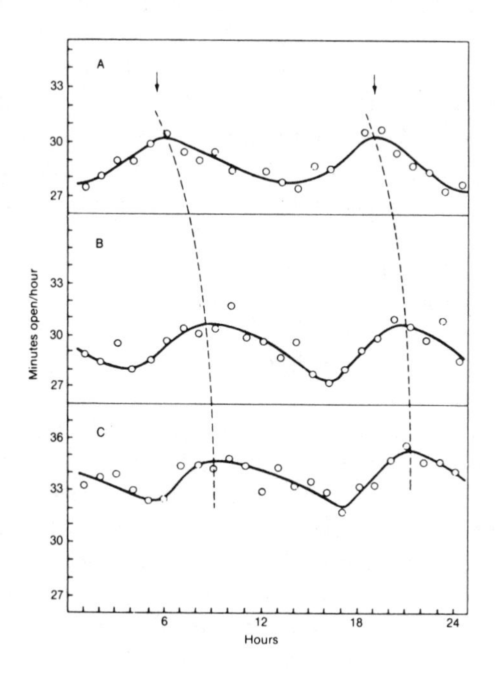

of the laboratory? (Think of that the next time you start to douse a plate of oysters with the stinging pungency of a few drops of Louisiana hot sauce!)

If you are impressed with the sagacity of a brood of New England mollusks, consider the even lowlier potato. Experimental biologists have charted its metabolism by measuring the rate at which it uses oxygen.[2] They removed the sprouting eyes of potatoes and placed them in hermetically sealed containers shielded from outside fluctuations in temperature, pressure, humidity, and light intensity. But the deprived spuds kept the same rhythmic cycles they had before they were snatched out of their natural environment. Peak consumption of oxygen occurred at 7 A.M., noon, and 6 P.M. every day. And when the unseen sun was gone from the sky, oxygen consumption fell to the standard nighttime low. There were annual changes, too. When it was summertime outside the container, the noontime peak was lessened; and in wintertime, it became enhanced.

As insignificant a link as it may seem in the great chain of being, a blindfolded potato still knows not only the time of day but also the season of the year! Furthermore, the rates of potato metabolism were found to correspond to the barometric pressure outside the containers in a most unusual way: they indicated what the barometer read, both yesterday and the day after tomorrow. While we can understand that yesterday's pressure changes might make for alterations in humidity and temperature that can be connected in turn to today's oxygen consumption, it is difficult to comprehend just how information can be conveyed to an imprisoned potato which enables it to adjust the height of its noontime oxygen consumption peak by just the correct amount to correspond to the barometric pressure two days later. A meteorologist removed from his usual post in front of a TV weather map and sealed away in a container could hardly be expected to do as well in predicting the weather.

Of all the rhythmic time cycles in biological organisms, those of the honeybee probably are the most well known and certainly among the most elaborate. Like many other scientists, the German biologist Karl von Frisch initiated his studies, back in the 1930s, quite by accident. One of his colleagues told him that whenever he breakfasted on his terrace, he noticed that the bees seemed always to be there at the right time to anticipate the savory taste of the jams and marmalades that often were set out. They showed up whether or not sweets were at the table. How could a bee acquire this punctuality? Was there within itself, a sort of built-in clock that goes off at the right hour, or did the insect receive clues from the outside environment?

To try to find the answer, von Frisch set up a series of experiments.[3] Food sources were set out in different directions at different times of the day. When bees coming straight out of the hive darted immediately to the food source, it became clear that somehow information about the location of that source was being transferred back to the hive by incoming members of the bee community. Peering inside the hive, von Frisch discovered that the transfer mechanism was a kind of round dance performed by a foraging bee shortly after it arrived back at the hive to dump a load of nectar. It would dance rapidly in a narrow circle, completing half a course in a clockwise direction, then would run along the diameter, and finally complete the other half loop in a counterclockwise sense—a sort of figure eight (figure 1.2). As the bee danced, spectators tagged along after it in great excitement, touching the tip of its wiggling abdomen with their feelers as they went. Cleaning itself off, the forager-communicator then left the hive and returned directly to the food source, only to be joined moments later by attentive comrades.

What information had been conveyed through this curious round dance? How did the bees watching the dance know where to proceed? These questions continued to puzzle von Frisch and his co-workers for a long time. Surely, it was not the sense of smell that caused the bees to arrive at their target, for they showed up regardless of whether the experimenters laid out honey or sugar water. And the bees certainly could not *see* the dancer perform, for the comb on which all the action took place was sealed away in darkness. Any information about orientation must have been picked up by other means—for example, by touch, inside the hive. The experiments continued.

When von Frisch varied the distance to the feeding place, he discovered that the farther away the source, the longer the dancing bee took to complete a dance loop—from forty runs a minute for a source just a few hundred meters away, down to just a few turns a minute for a feeding place situated several kilometers from the hive. And, if the source was nearby, the communicator-bee seemed to wag its abdomen much more rapidly during the straight diametrical portion of the dance. Did this action reflect that the dancer had consumed more energy when running a long way from the hive? Perhaps a tired bee

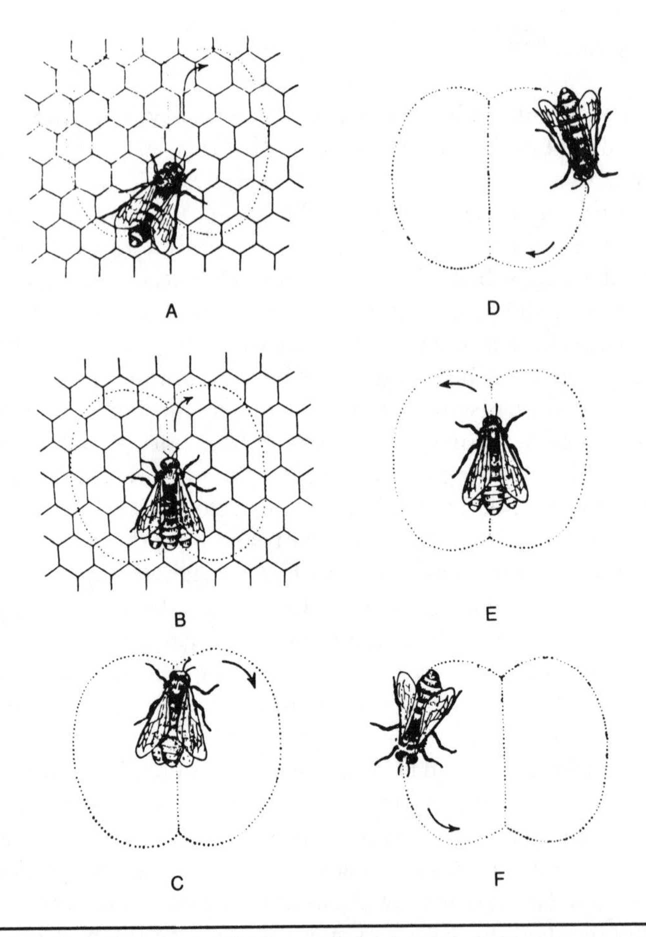

FIGURE 1.2 The wagging dance of the forager bee that has just re-entered the hive. The long axis of the kidney-shaped course gives the direction of the food source, while the intensity of abdominal wagging on the straight run (*b* and *c*) tells the distance. There is little question that the bee's sense of timing comes from the sun. SOURCE: K. von Frisch, *The Dancing Bees* (New York: Harcourt Brace, 1933), fig. 43, p. 117.

automatically conveys information about the distance to a food source because it has less energy in reserve and consequently performs a much slower dance. But what about direction? If the companion bees know how far to fly, how do they know which way to take off?

This information is conveyed in a most remarkable way. Actually the bees were navigating by the sun. Von Frisch discovered they had developed a brilliant way of measuring the angle between the food source and the sun as projected from the hive. On a horizontal comb surface, the angle between the straight portion of the dance and the bearing of the sun is the same as the direction between the sun and the source as seen from the hive. In other words, if a food source lies 40 degrees to the left of the sun as viewed from the hive, then the bee conducts the straight portion of its dance at a 40-degree angle to the left of the direction of the sun, which, it will be remembered, cannot actually be seen from the dance floor. What if the comb surface is vertical? Then the forager bee performs the dance so that the angle to the food source is measured from the vertical instead of from the direction to the sun. In our example, the straight portion of the bee's dance would be 40 degrees east of vertical. Von Frisch could not fool the bees by changing the orientation of either the nectar or the hive. They always made a "beeline" directly to the feeding place once informed about direction and distance via the wagging dance.

If bees know when to turn up at a feeding place, they must possess some sense of time. And, because they convey information by marking off the angle of the sun, it would appear likely that they are cued in to this timing sense directly by the sun. The bee learns that when the rays of the sun arrive from certain portions in the sky, it is time to forage. Von Frisch even trained his bees to become attuned to several different foraging times each day. But this idea of environmental cueing has its problems, for bees, like potatoes, when totally isolated from all outside periodic changes, continued to behave the same way. They appeared at a feeding spot at the time of day for which they had been trained, even when they were exposed to constant artificial light, temperature, and humidity. After years of experimentation with time sensing in the honeybee, von Frisch concluded that "we are dealing here with beings who, seemingly without needing a clock, possess a memory for time, dependent neither on a feeling of hunger nor an appreciation of the sun's position, and which, like our own appreciation of time, seems to defy any further analysis."[4]

We can say the same of potatoes that adjust to the forthcoming barometric pressure and of oysters that manage to convince themselves that they live in a nonexistent ocean just west of Chicago. But where is the elusive rhythmic clock for which von Frisch was searching?

Cycles of about a day's length possessed by nearly all living organisms are called *circadian rhythms* (from the Latin *circa*, "about," and *dies*, "day") and have been recognized in Western culture at least since the time of Aristotle. Twenty-four centuries ago, he observed that certain plants raised and lowered their leaves on a regular day-night schedule. But it was not until the age of controlled scientific experimentation (twenty-two centuries later) that we began to probe the detailed nature of this biological clockwork. In 1729, the French scientist Jean de Mairan conducted the first controlled light-dark experiment on plants in a laboratory.[5] He found that the daily periodic oscillations of plant leaves persisted even when the plants were isolated from the natural environment. This so-called de Mairan phenomenon has been observed in practically all living forms from humans down to one-celled organisms. In all cases, the cycle is close to, but never precisely the same as the earth's rotation period (generally it varies between 23 and 27 hours for different kinds of organism). Furthermore, most subjects can be trained after several days to adapt to a new artificial period through environmental control.

Mice are particularly cooperative. Biologists have weaned them away from their normal morning activity of running on a wheel by subjecting them to a single hour of artificial daylight and then keeping them in the dark for the remaining 23 hours of the day.[6] As the graph in figure 1.3 shows, after several days of such exposure, the mice adopted a routine: they would always run the wheel immediately after the hour of exposure to the light. After several weeks, the mice were allowed to "run free": that is, they were taken out of the artificial periodic environment and kept in darkness all the time. Under these conditions, wheel-running activity often persisted for several days and occurred at the same time of day to which the mice had become "entrained," to use the biologists' terminology; but then the mice began to become "disentrained," and their activity period drifted slowly toward times slightly earlier in the day. This free-running rhythm had a shortened but astonishingly precise period of 23.5 hours. (Notice the straight sloped line in figure 1.3 that fits the time of initiation of wheel activity from the 60th day forward.) Also, the length of time the mice spent running in the wheel became a little bit longer during this stage of the experiment. After 90 days, the biologists tried to switch the mice to a new period by exposing them to light 18 out of 24 hours per day. Released to run free after the 130th day, the mice were active for longer intervals; nevertheless, they kept to a rhythmic cycle that once again corresponded precisely to something less than a day. These experiments not only demonstrate the complicated nature of circadian rhythms, they also raise a fundamental question: Is a sense of time built into all organisms, or are we really being driven by the earth clock outside us?

In our world there is little to be gained by marching to the tune of

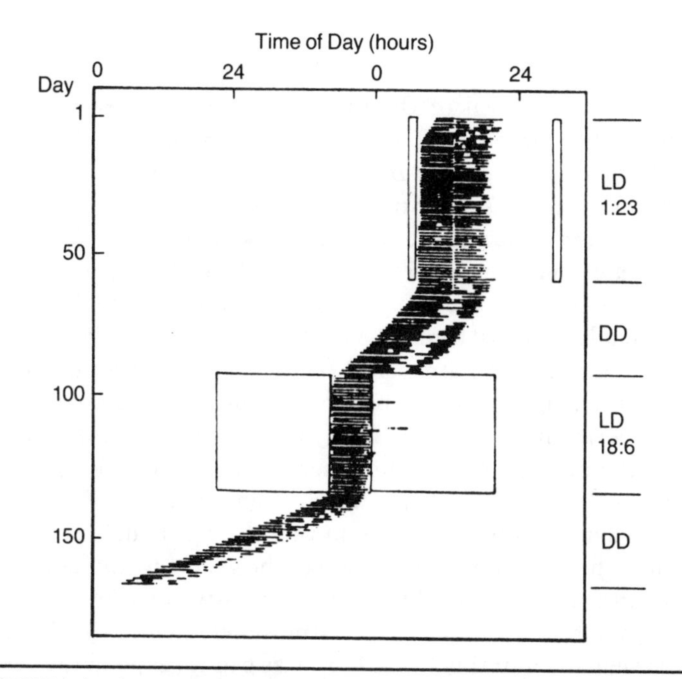

FIGURE 1.3 A mouse's circadian clock. The free period of daily wheel running (*dark horizontal streaks*) by a mouse is followed over several months. For the first 60 days, the mouse is subjected to 1 hour of light and 23 hours of dark each day (*long vertical bars*), with the result that it runs on the wheel soon after being exposed to light. When kept in constant darkness (between the 60th and the 90th day), the mouse's activity period seems to slip backward through the 24-hour day. Following entrainment to a regimen of 18 hours of light and 6 of darkness (90th to 130th day), the mouse is once again left in total darkness; as a consequence, wheel running experiences another backward phase shift (after the 130th day). SOURCE: M. Moore-Ede et al., *The Clocks That Time Us* (Cambridge, Mass.: Harvard University Press, 1982), fig. 2.8, p. 48.

a different drummer. Little wonder that most biological rhythms are virtual duplicates of nature's basic periods—the day, month, and year, all of which have their celestial origins in the rotation of the earth and the movement of the sun and the moon. To rephrase my question: How is the connection between life cycles and celestial rhythm established? One theory, the hypothesis of the internal timer (or the endogenous hypothesis) suggests that because these rhythms persist when the organism is deprived of functioning within the natural environment, every piece of living matter must be its own timer. In other words, every living thing has the capacity to develop its own internal, chemically based timing system. This idea makes good sense because the theory of evolution teaches that the mechanisms of natural selection favor the survival of the organism that achieves an adaptive advantage. Having evolved over millions of years, the oysters that "know" when to keep their mouths shut (or open) get fed; the rest become losers. Every successful class of organism needs to inherit and further develop an accurate biological clock so that it can know when to anticipate environmental change better than its competitors.

In stark contrast, the exogenous hypothesis of time sensing proposes that all biological timing depends on cosmic stimuli. Organisms oscillate with natural geophysical frequencies because they respond directly to changes in the forces of an all-pervasive environment. While experimenters may *think* they have totally isolated their potatoes and their oysters by shielding them from light and changes in barometric or thermal conditions, these organisms have subtle ways of sensing what is really happening in the world outside. For example, they might be responding to changes in the geomagnetic field or electrical charges in the air, to shifts in the intensity of background radiation, or even to tides in the earth's atmosphere that rise and fall on a daily as well as a monthly basis. How else can we explain the way the New Haven oysters adjusted themselves to the invisible moon over the nonexistent harbor in the middle of the Great Plains? Exogenists believe that oysters have a way of sensing the difference between the gravitational pull of the moon when it is directly above and below them, thus causing high tide, as opposed to when it lies on either horizon and consequently produces a lower watermark.

We already know that honeybees have good gravitational sense because they make use of it in their navigational system. How else could the forager who dances on the vertical comb convey the location of food to its hivemates unless it were able to translate the angle between food source and sun into the angle between food source and the vertical? The bee substitutes the direction of the pull of gravity for the direction to the sun. To mark the direction of the pull of the moon, the principle is the same: one need only be more sensitive, more attuned to the environment. Time, the giver of all rhythm, must originate from the outside: time sensing is hidden away in that secret communication between animal and nature that we dismiss casually as "instinct." So say the exogenists.

But consider the punctual fruit flies, or *Drosophila*, who, unlike the oysters, seem to behave as if the whole time-sensing scheme operated internally rather than externally. All of the adults emerge from the pupal case at the same time of day—close to dawn. When reared under normal alternations of light and dark for several generations and then suddenly switched to a constant light-and-temperature situation, succeeding generations of *Drosophila* continue to burst out of the pupa stage and go on to the mating stage precisely when it is dawn outside. Furthermore, the fifteenth generation of flies is just as punctual as the first, none of them ever having seen the natural light of day. Given this sort of experimental evidence, it is difficult to get away from the idea that the clock lies in the fruit fly, that the sense of timing is inherited genetically. Indeed, some experiments on this species even support the idea that there might exist a biochemical clock-gene built into the molecular structure of all living tissue. But how does it work?

Erwin Bünning, a German biologist who conducted many experiments on *Drosophila,* is responsible for framing the hypothesis of how the internal timing mechanism operates in a way that seems to be capable of embracing the photo-periodic responses of a wide variety of living organisms from bee feeding to plant-leaf movements (though he originally intended that it be applied only to leaf movements in certain plants).[7] The mechanism consists of two phases of about one-half day each that alternate. In the "light-loving" phase, the energy falling on a body will enhance growth and activity; while, during the "dark-loving" phase, it will inhibit it. True, organisms draw energy from outside sources, but they convert it by strictly internal means into useful activity (or inactivity). This idea sounds bizarre, for it seems to treat living organisms as if they were self-operating electrical appliances like coffeemakers or clock radios that switch themselves on and off and operate by alternating currents.

I shall not venture into a discussion of all the experiments that led Bünning to his hypothesis, except to say that it had become clear to him in the lab that all circadian rhythms could be broken down into two opposing half-cycles. There seemed to be definite periods when Bünning's plants would respond only to light treatment, after which they would react only to dark treatment. This Jekyll–Hyde pattern of behavior can have but one origin. Had the organism, over its millennia of evolution, internally adapted itself to the alternating day-night cycle of nature?

The ultimate test of the internal-external controversy about time sensing would be to remove a rhythmic organism from all conceivable geophysical influences—say, by taking it into space and putting it into solar orbit far away from the earth—and see what happens to it. Surprisingly, given all the money spent on the exploration of space, no such experiment has ever been conducted in detail. But biologists have taken the next best alternative: they have conducted experiments in an environment where most day-dependent variables can be eliminated—at the South Pole and on a rotating turntable that compensates for the earth's rotation. Hamsters and cockroaches were run there, plant leaf movements were timed, even fruit flies were bred in a turntable laboratory. The results of these experiments weighed heavily in favor of the internal clock hypothesis. The organisms went right on functioning as if the earth's rotation were still present in the environment. But the exogenists, as might be expected, still argue that the test organisms could have harbored a subtle sense of the diurnal variations in spite of all the biologists' attempts to shield them. For example, could not the organisms have taken their cues from the 24-hour movement of the geomagnetic pole around the geographic pole? Endogenists claim to have blocked that escape route by conducting separate tests in which they demonstrated that plants reared in the vicinity of powerful magnets differ in no way from those cultivated under normal conditions. But experiments with laboratory magnets do not prove the case to everyone, and the controversy rages on.

Today even the most stubborn internalist would concede that there are some exogenous effects on biorhythmic activity; and the tenacious externalists, who in today's world of genetically based research have been relegated to an active, vocal minority, have been forced to admit that some sort of biological clock does indeed exist within all of us. Still, the environment is the ultimate source, the giver of time—or *zeitgeber*, as biologists call it. It is the environment that entrains all biological organisms to oscillate in the first place. Tide, temperature, season, and moonlight—all represent the primal input that drives us to execute our clockwork, both chemical and genetic.

In the present context, I am not terribly concerned with the debate about whether life's sense of time is better explained internally or externally; whether it is a direct response to nature or has been nurtured over countless generations of evolution. Rather, I have used oysters and bees as a way of attempting to establish that all living things, regardless of where they stand on the ladder of biological complexity, respond to environmental change in surprisingly sophisticated and detailed ways. And though I write in an artificially lighted room, and you likely will read what I write in a heated or an air-conditioned enclosure, we both are tied directly to the environment outside. Yet we never sense time directly. There is no single organ to monitor time the way the eye detects light, the ear responds to sound, or the tongue to taste. Nevertheless, all living organisms sense time by responding to phenomena that change.

A Multitude of Inner Rhythms

This undeniable sense of timing over which we have been puzzling is as inherent in humans as it is in all the lower organisms; but, paradoxically, knowing ourselves appears to be more difficult than understanding how other life forms express their time sense. The extent to which human time is either *biological* or *psychological* seems to be the root issue.

More than a hundred functions are known to oscillate within our bodies each day, and we share most of them in different degrees of variability with all mammals. At the short end of the time spectrum, there is the 1/10-second oscillation of brain waves on an electroencephalogram, the 1-second basic cardiac rhythm, the 6-second respiratory cycle, and the various sleep stages leading up to the 24-hour sleep-wake period. Our body temperatures also fluctuate on a twenty-four-hour cycle, being highest toward evening and lowest toward morning. At the longer end of the time scale lies the 28-day menstrual cycle and the vestige of a 365-day cycle of hibernation.

Physiologists have their own language for dealing with these phenomena, and its electronic metaphors (italicized in this and the next paragraph) suggest how they think our bodies and brains operate. A *pacemaker* is some entity within the body which is somehow capable of sustaining oscillations and entraining them to other bodily rhythms. Initially, the forced oscillations, or *zeitgebers*, are received through the *auditory or visual* channel before being picked up by *transducers* or *receptors* within the body. Though these transducers acquire the time-based information from nature's cues, the pacemaker is capable of measuring time in the absence of these cues. It acts as a kind of alternating-current device which sends out measured pulses of information. Biologists have adopted this information-processing model as a way of explaining biorhythms in the higher life forms. They formulated it on the basis of experiments with whole animals, deprived of normal environmental conditions. The tactic consists of manipulating environmental parameters and then recording the resulting behavior of successive generations of animals. But when biologists try to pinpoint the location of the pacemaker within the organism, things get really complicated. Here the strategy, like that of most scientists, is to divide and conquer. When they examine the rhythm of isolated body parts (the heart taken out of body, liver cells, or blood cells), they discover that the rhythms of the whole animal are really comprised of multiple oscillations, all running together harmoniously in a perfectly *tuned* condition in the normal specimen. But are all the rhythms in an organism *driven* by a single pacemaker, or is there a hierarchical organization of pacemakers? And, if the latter, where is the *command center* in this metaphor that says, "My body is a computer"?

Most experimental biologists believe the major pacemakers in mammals are found in the hypothalamus area of the brain—at the top of the brain stem, directly behind the eyes, to be exact. The site is a tiny bundle of nerve cells known as the suprachiasmatic nuclei (SCN). In certain instances, the information channeled from *entryway* to *command post* even can be traced—from the eye, which receives the zeitgeber, along a *neural network* to the hypothalamic area—by the injection into the retina of substances that can then be followed by radiographic detection techniques.

By studying the free-running behavior of rats after subjecting them to every conceivable manner of sensory deprivation—such as blinding; removal of adrenal gland, pituitary gland, gonads—biologists find no disruption of normal activity. Only when one begins to make knife cuts in the selected areas of the brain in the hypothalamic area does a rat begin to alter its feeding and drinking rhythm drastically from the normal oscillatory cycle. Lesions that destroy the SCN appear to affect the rhythmic discharge of certain chemicals from the adrenal glands, which control eating and drinking behavior as well as the wake-sleep cycle.

In humans there is no circadian cycle more dramatic than the alternation between being awake and being asleep. We are so conditioned to passing through that unremembered instant between the drastically opposed states of consciousness and unconsciousness that any attempt to deprive us of the ability to do so is doomed to failure. On the average, we spend one third of each earth rotation asleep (by contrast, a donkey spends one seventh; an opossum, four fifths); and, of course, we do it during the dark period. On the other hand, anyone who forgets to "put out the cat" becomes painfully aware that felines undergo alternating bouts of sleep and wakefulness, each one to two hours long. The animal "catnaps," paying no heed whatever to the day-night cycle, a condition to which it may have adapted as a result of its predatory nature. The opossum, by contrast, is awake only during morning and evening hours.

It is extremely difficult for a human being to remain awake continuously for more than two or three days. The extreme case recorded in the *Guinness Book of World Records* is eleven days for a seventeen-year-old boy who achieved his pinnacle only with constant prodding from the outside and with a special motive in mind—to set a world record.

Our strict entrainment to the earth's diurnal period is far-reaching. When you pass a night without sleep, though you are feeling extremely fatigued during the working hours, you quickly get back on track with the normal daily cycle. By midmorning, even though you may be struggling through the twenty-seventh consecutive hour of wakefulness, somehow you feel better than you did while getting through that miserable twenty-first hour during the pre-dawn or the dark part of the diurnal cycle. However, by the time you see the sun go down for the second time, you pay the price twice over by falling into a stupor. Like the experimental animals I discussed earlier, when we depart from the normal conditions of the environment, we struggle to compensate internally.

Just like the free-running rats, people deprived of the normal light-dark cycle by living in caves, or by being confined to sealed rooms under conditions of constant light, become disentrained after a couple of weeks. They alter their biological sleep-wake cycle to a period somewhat different from the usual 8- to 16-hour cycle. Psychologically speaking, they lose all track of time, often believing they have been awake far longer than has actually been the case. They even wonder whether they have slept long enough; but these concerns usually vanish after an extended period of isolation. One subject who was kept in a closed room for several weeks stated, "After a great curiosity about 'true' time, during the first two days of bunker life, I lost all interest in this matter and felt perfectly comfortable to live 'timeless.'"[8]*

*One inexplicable phenomenon resulting from nature deprivation is what I call the *six-hour phase slide* in the sleep-work cycle of practically all college students: rarely do they rise before the sun hits the meridian, and they can be seen prowling around from

We know that the percentage of time we sleep decreases with age; it begins as a prolonged affair but is nonetheless fitful in infants (as any parent knows, it takes about two years before a child is fully fixed into the nocturnal sleep cycle). In old age, somnolence becomes fitful again.

Sleep is not just a single state. Rather, it seems to be made up of subsidiary rhythms. From the rapid eye movements that occur during the time that we dream, we call this period "REM sleep." We spend about one quarter of our sleep time at it. During REM sleep, there is no muscle tone—and fortunately we cannot act out our dreams as we dream them. During non-REM sleep, we do most of our tossing and turning. Muscles are active, but the frequency of brain waves slows down, passing through various substages. These two cycles, REM and non-REM, alternate on approximately a 90-minute cycle throughout the night in all human beings.

This "ultra-dian" oscillation cannot be connected with any known period in the environment; but like all other behavior during the sleep cycle, it seems to be controlled by the interaction among clusters of nerve cells at several sites within the brain, most of them within the brain stem. One such cluster seems to play a role in REM, and another in non-REM sleep. Motor inhibition, or loss of muscular control, is operated by a third area. We know of these connections because the stimulation of these areas results in one or another reaction in the sleeping patient. But still unanswered is the fundamental question of exactly how changes in the activity of different brain-stem nuclei alters the overall activity in the entire central nervous system.

While the human sleep cycle is circadian, the reproductive cycle is predominantly *circa-lunar* and *circ-annual.* Indeed, the very word *menstrual* is derived from the *menses,* or month (a contraction of *moon-th*). At 28 days' length, the menstruation cycle lies, probably not by coincidence, close to the moon's period. In practically all primates, the production of estrogen, variations in bodily temperature, the time of ovulation—all are controlled by a period that spans the interval 25 to 35 days. In southern California, the running of the grunion is a major event during the period of the year when the highest tides occur. At either the dark or the full-moon phase of the monthly

library to dormitory (and other places) long after the sun has left the sky—hardly photoperiodic creatures.

cycle, millions of females of this slender species (about the size of a big sardine) suddenly turn up on the beaches. They perform a vertical wiggling dance, burrowing their tails and half their bodies in the sand where they deposit their eggs. The males coil round them and eject sperm to fertilize the eggs. The hole is then covered with wet sand. Two weeks (or half a lunar cycle) later, the eggs hatch, just in time for the young offspring to be helped by the advancing tide back into the ocean. These moon-based cycles may be a reflection of our original ascent from the sea. Remember the oysters? Many feedings and reproductive cycles of organisms that inhabit the ocean receive their input signals from the tidal period which, in turn, is regulated by the position and appearance of the moon.

Half a billion years ago, the moon lay closer to the earth in its orbit. It has been receding ever since; and as a result, the month, measured by the period of revolution of the moon about the earth, has become longer. There is some evidence that when the lunar period was measurably different, some 350 million years ago, living organisms were entrained to that period. During that time, the so-called Middle Devonian period, the month was about a day and a half shorter, making for more months in a year—closer to thirteen, rather than the present twelve. The evidence shows up in fossil coral embedded in sandstone laid down during the Middle Devonian period. Paleontologists have examined the growth ridges in coral which are deposited on both a lunar and an annual "breeding" cycle. During the summertime, when the waters are warmer, the thickness of the chalky deposit is larger than during the winter, when the organism lies in a more dormant state. From a statistical point of view, the data suggest that the coral time clock was ticking according to a different beat than today, one that reflects the natural environment that once existed but has since undergone gradual change (see figure 1.4).

We are a long way from relating the behavior of 350-million-year-old fossil corals directly to the estrous cycle of twenty-first-century women, but the moon's rhythm could have some bearing. For early humans, the time around the full moon offered an extended period of light during which limited hunting and gathering could take place. And under the difficult conditions in which they must have lived, our prehistoric ancestors needed all the light-time they could get. It is reasonable to suppose that the dark-of-the-moon period might have been given over to sedentary activity, which included mating. To use the

FIGURE 1.4 An archaic month calendar frozen in time. A count of the horizontal growth ridges deposited during full-moon phases by this coral from the Devonian period (350 million years ago) indicates that our months once were shorter and there were 13 of them in a year instead of 12. SOURCE: Courtesy Colgate University Department of Geology. Photo by Warren Wheeler.

physiologist's lingo, this kind of lunar entrainment could have led to the development of bodily pacemakers that controlled the secretion of chemicals related to the breeding cycle. In other words, our brains constructed a lunar-analogue time clock.

The year-clock is evident in mammals that seem to have developed an annual reproductive cycle that ensures that the young will be born at the most promising time for survival; this is particularly true of species residing in temperate climates, where the change of seasons is more dramatic than in the tropics. Animals precisely measure changes in the daily light period. For example, the testes of hamsters remain in regression during the long winter period leading up to the spring equinox. Then, all of a sudden, the weight of the testes is multiplied by 10 in just a few days. This dramatic change seems to take place just when a light period of between 12 to 12¼ hours per day is reached. It is as if the critical day length around the equinox triggers a switch within the animal whereby it anticipates the mating season and rapidly begins to prepare for it.

Now, all of this talk sounds distinctly exogenous, though most animal biologists believe that the time clock is switched on by an endogenous circadian system, something like that advocated in Bünning's two-phase hypothesis. Thus, on long days light would tend to fall on a greater portion of the cycle that would stimulate a physiological response. In fact, experiments conducted in the laboratory showed that when a light pulse fell on one of the "light-loving" phases of the 24-hour cycle, the testes became fully functional.[9]

The circadian clock is blended in with the lunar and the solar reproductive cycles in many other ways. For example, spontaneous births in humans increase dramatically between 3 and 4 A.M.; and the estrous cycle and the time of ovulation follow a 24-hour subperiod within the menstrual cycle. But what coordinates these periods?

When freed of all time cues, humans, as well as animals, seem to go through a series of stages or adjustments before becoming totally desynchronized. They behave as if two coupled pacemakers that once provided information to one another had become totally decoupled. There seems to be an attempt on the part of the organism for the first few weeks to imitate the normal period. Then, when a certain phase between two different bodily cycles, such as body temperature and sleep-wake rhythms, is reached, the organism becomes "trapped," and a new oscillation period is settled upon and followed for both. Finally, after a few more weeks, as if a switch were pulled, the pacemakers become totally desynchronized, and the two periods start to drift rapidly away from one another. In each instance, the coupled pacemakers appear to be contained within or close to the SCN neural clusters. Thus, the human clock, at least as we conceive it, seems to behave much like a machine.

We've got rhythm! There's no doubt about it. And our desire to memorize, mark, and record it—the subject of the next chapter—seems to have been bred within us right from the start, before we could even call ourselves human. Like all living organisms, we march in time to the dependable beat impressed upon us by nature's background music. For us today, that environmental combination of tunes happens to consist of a 24-hour alternating light-dark point-counterpoint, backed up by the song of the moon, sung both in different time and in a different key. The sun chimes in with its second chorus; the music of the seasons sounds with a much lower pitch that conflates with the lunar beat to produce a kind of disharmony. All creatures of the earth automatically tap their extremities to keep time with its tunes. Ours is a beat that could have been different had the earth's past history been otherwise (see figure 1.4). Why not a 5-hour day, a 100-day year, and maybe 3 moons in the sky instead of 1? Different planet, different music. But this is the way it is here and now, at our place.

Mechanism as Metaphor

Am I a man or a violin? When my archaic ancestors finally became rational and began to become interested in the problem of accounting for their own behavior, they chose to see themselves as well-tuned instruments—machines, in fact. When we are in good working order, we happily hum away at the melody. Our pacemakers sustain and sing out every one of nature's vibrations. They sense the downbeat, capture it, and make it fit with the other beats already encoded into the body instrument.

As I have noted, the language used to describe the way we in the West think the coupled circadian oscillation works consists of neural networks, transducers, pacemakers, command posts, control mechanisms, cues, signals, and oscillators—that familiar language of information retrieval and processing in which the twentieth century is embedded. At least since the Industrial Revolution, if not from the Renaissance, when we first began to manipulate and experiment with nature, we have come to have faith in the machine, to believe in mechanism as a way of understanding. Machinery is, for us, the power tool of metaphor. After all, machine technology has helped the Western world prosper. Why not stick with a good thing and use its principles to create new ideas about how all things function? And so we try to account for natural events by inventing mental models that parallel the way things behave: the brain as a sponge, the nervous system as a public transit network, the atom as a miniature solar system, the solar system as a set of revolving spheres all linked together by a gear train—as in a planetarium.

Of course, the physicist will say that the atom is not really made up of little round balls in orbits, and the biologist that the computer is just a contemporary body metaphor. Certainly neither scientist has any intention of portraying these mental fabrications as reality. The body is *really* not a machine—no more than do atoms have a "glue" that bonds them together as molecules. The tenuous relationship between metaphor and reality arises again in the next chapter on how natural time sensing led, in the higher forms of life, to deliberate time reckoning; that is, the conscious attempts by rational human beings to mark time, to remove it from its natural rhythmic casing, to mold and fashion it, to develop and use it. Let me, then, turn to how, when, and why the very time we as humans—along with the corals, the oysters, and the potatoes—had been feeling within ourselves throughout evolutionary history, whether by internal or external impulses, became metamorphosed in the West into something outside ourselves, an entity so strange and novel that today we must strain our senses even to recognize it.

2

EARLY TIME RECKONING

> Keep all these warnings I give you, as the year is completed
> and the days become equal with the nights again, when once more
> the earth, mother of us all, bears yield in all variety.
>
> —HESIOD, *Works and Days*

MAN THE THINKER, man the tool maker, man the symbol maker—these are some of the ways we differ from other animals that have backbones. True, some other animals do use tools. A burrowing wasp uses a small pebble to tap down the soil over the nest where she has deposited her eggs, and a finch employs a cactus spine to pick insects out of a narrow opening in the bark of a tree. A monkey carefully manipulates an edible object before bashing it open on a rock. When animals display skilled behavior in this manner, we say they possess *instinct.* But when humans do it, we call it *intelligence.* The two are not the same.

We think of *instinct* as a certain power or disposition that—without instruction, experience, or deliberation—directs an animal to do whatever is necessary to preserve itself and its kind; while *intelligence* is intended to characterize behavior that is more reflective, learned on the basis of experience. We can train hamsters to run on a wheel or a dog to beg for a bone. In either case, the animal knows that the outcome may be a reward. But most psychologists would argue that a groveling dog is not behaving intelligently because it lacks the deliberate element, the collective knowledge and wisdom, the weighing of alternatives, that goes into a rational decision.

While the distinction between animal and human intelligence is subtle and complicated,[1] our general assumption is that animals do not think. They have no ideas rolling about their heads. They neither judge nor decide, meditate nor recollect, presume nor venture. Scientists use these carefully chosen words to separate human behavior from that of all other living creatures, as if we were deliberately choosing to differentiate ourselves from the rest of the living world.

Take language, for instance, which is often said to be unique to human beings. We think of it as communication through the use of symbols cast in the form of sounds that have meaning only to one who knows the code. Yet the vocalized utterances taken by themselves have nothing directly to do with the acts and objects they represent. True, some of the sounds rolling off our tongues betray what they represent, like *buzz, hiss,* and *whippoorwill.* A school of philological thought even proposes that all our word making began this way—by the childlike imitations of the sounds of objects or actions, such as *bow-wow, quack-quack,* and *ka-boom.**

With but a handful of utterances, we are able to transmit and receive a vast constellation of images and ideas, not only because our tongues and vocal chords can waggle, and our eardrums can vibrate, but also because our brain possesses the ability to coordinate word function, patterned sentence structure, and syntax. The spoken word, the tool held in the hand: regardless of why and how they developed, language and technology emerge as the two indispensable, learned skills that must have preceded any attempts by our ancestors to deliberately and willfully keep track of the flow of events in the human environment, to reckon time, to set it all down in a logical order—to make a calendar.

*Precisely how words ultimately came to stand for concepts and ideas remains a mystery. We do know, however, that animals have languages of their own, and that we can force some of them to learn our language. Sarah, a famous laboratory chimpanzee of the last generation, was able to acquire a vocabulary of over one hundred words.[2] Sarah's words took the form of colored plastic cutouts, each of which was associated with a different object or action. Eventually she could learn to pattern these forms into sentences by arranging them on a table. For example, when given the visual message "Sarah insert apple pail banana dish" and then confronted with each of the real representative objects, she was able to execute the correct command by placing the pail on the dish while putting the two pieces of fruit into the pail. Some chimpanzees have even learned to mouth words like *mama, papa,* and *cup.* To perform her task, Sarah's trainers contend that she needed to learn a language—that is, to identify each object with one of those particular plastic cutout symbols. She was required to learn the name of each thing. Then she needed to deliberate, to discern and decide. Should she put the banana onto the dish or into the pail? Or should the dish go into the pail? In a sense, she needed to think.

Some say that to make sense of time, to gain control of it, one must be literate; and to be literate, one must be able to write. Our version of writing consists of transforming the voice symbols, the code of *bow-wows* and *quack-quacks,* into a permanently retainable medium through some sort of developed tools, such as chisel and stone, stylus and tablet, pen and parchment. Perhaps even a cluster of knotted cords, a woven, multicolored textile, or a pattern of seashells arranged on the shore would suffice.

I find these restrictions confining, a result of our civilization's dependence on technology. Few of us have any experience with what it takes to retain, analyze, and disseminate information with minimal technical skills and without a material medium in which to express ourselves. Yet I believe that people once were able to develop intricate, detailed, and sophisticated time-reckoning schemes by oral means alone: that is, before the hunter could represent a time sequence by making marks on the ground, he could count a kill on his fingers or tell his companions how many days the food would need to last until the next expedition under the full moon's light.

I shall thus in this chapter discuss first the oral mode of time reckoning and then the development of the written. I begin with early Greek poetry and some of the material on creation mythology which came down to us in written form in the Old Testament. The Greek and Babylonian stories of creation, originally also recited without a written text, give a wealth of information concerning how our predecessors thought about time as well as how they marked it. Survivals of the earliest forms of written texts that deal with time come down to us in the form of carved bones from the paleolithic era. Standing stones, such as those at Stonehenge, may represent another concrete way of indelibly preserving the mark of time—a kind of unwritten text.

The Oral Mode

Oral myth making was a powerful vehicle for conveying ideas about what time is as well as about how to reckon it. Indeed, in ancient Greece and medieval Europe, the singer of songs, the oral poet, could chant for hours on end, having committed to memory every part of a story—27,000 lines between the *Iliad* and the *Odyssey* alone—by devising formulas, mnemonics, and epithets to hold it all together. The technique consisted mainly in fitting thought to a rhythmic pattern of words and then linking this rhythm to a melody by programming the fingering of a musical instrument to the words. Once a line of rhythm is repeated over and over, the oral poet is permitted some variation in melody as well as in the timing of syllables within the rhyme—and in parts of the Balkans such chants survive today.[3]

The Farmer's Year: Hesiod's *Works and Days*

We in this day and age, with so many ways of keeping track of time—from the wall calendar to the digital watch—find it hard to imagine that one could reckon time usefully by memory alone. But seven centuries before Christ, the Greek poet Hesiod laid out a time-reckoning code as detailed, precise, and full of predictive power as any basic written calendar; and yet the whole of it resided in his mind, as in the minds of farmers generations before him. In Hesiod's ancient song-of-the year cycle, *Works and Days*, there is a season and a time for everything.

Though the *Works and Days* we know was written down in the fifth century B.C. in classical Greece, originally it was belted out like a ballad, accompanied by the string music of the lyre, the way the *Iliad*, the *Odyssey*, and the Nordic sagas were told—to an audience of avid listeners most of whom were probably already well familiar with the poem's basic ideas and images. Hesiod was a Greek poet of the early seventh century B.C. Unlike Homer, the wanderer, Hesiod was a farmer who spent all of his life in northern Greece, where he became accustomed to the harshness and knowledgeable about the vagaries of a deceptive and unpredictable climate. He must have held a low opinion of women, for he often parallels the fickleness of Mother Nature to that of women in general. His *Works and Days* reads like an archaic self-help text (see table 2.1), in which the poet tells how to lead an orderly, structured life in the difficult agricultural environment of the mountainous Peloponnese. (His advice on how to run the farm is directed toward his younger brother, apparently a wayward rascal in need of a good deal of discipline and some careful instruction, the kind only an older brother can provide.) The oral epic's captivating metronomic meter can still be tapped out even as we read today's written

TABLE 2.1

Annual Calendar of Celestial, Natural, and Human Activities Drawn from the "Works" section of Hesiod's Works and Days*

(line numbers are in parentheses)

		Oct.	*Nov.*	*Dec.*	*Jan.*	*Feb.*	*Mar.*	*Apr.*	*May*	*June*	*July*	*Aug.*	*Sep.*	
Celestial Activity	Sun			Winter solstice (479)		Winter solstice +60 days (565)		Spring equinox (562)				Summer solstice +50 days (663)	Autumn equinox	
	Stars	Pleiades hel.set (384) Pleiades-Hyades-Orion hel.set (615) Pleiades-Orion heliacal set (618)				Arcturus heliacal rise (566)		Pleiades hidden 40 days (383)	Pleiades heliacal rise (383; 571)	Orion heliacal rise (597)	Sirius heliacal rise (585)	Arcturus heliacal rise when Orion and Sirius on meridian (609; 414)	+15 days (613)	
Natural Activity	Birds	crane migrates (448)				Swallows appear (568)		Cuckoo sings (487)						
	Plants	leaves shed (421)					all variety-in growth (563)	fig leaf-(680)		artichoke in flower (581)				

	Weather	rains of autumn (417)	winter rains begin (440)	Boreas, snow (535)				3 days rain (486)		Zephyros (594)			
	Seasons	agricultural cycle; start and end season of new wine (674)		winter (Lenaion) (493)	spring starts (568)					summer (581)		end of summer (663)	
Human				keep busy (504)	make clothes (563, etc.)					sit in shade (587)			
Activity	Cereals	start plowing; start planting? (384; 615)		late plowing (479)	(448)	plow fallow (462)		harvest (383; 573)	winnow, store (598; 600)				
	Grapes					prune and dig vines by (571; 572)						harvest grapes (610)	press grapes (613)
	Animals			keep oxen indoors (462)							bring in hay and fodder (606)		
	Sailing	do not sail (618)	haul ship on land (624)				spring sailing (680)		summer sailing (663)				
	Other												cut wood (419)

*A schematic calendar from pre-Classical Greece. Derived strictly from information passed by word of mouth rather than in writing, this "year clock," from the "Works" section of Hesiod's *Works and Days*, indicates when certain human activities ought to take place based upon happenings in the coordinate world of nature. Sky events (*listed at the top of the table*) served as the main pointer on Hesiod's mental clock, simply because they are the most dependable. After M. L. West, *Hesiod Works and Days* (Oxford: Clarendon Press, 1978), p. 381.

version. With practice, one would find the long passages relatively easy to memorize.

Hesiod's *Works and Days* is filled with information about the Greek immersion in a world of time ten to fifteen generations before they built the Parthenon and Plato and Aristotle roamed the agora and wrote the texts that ultimately became the cornerstone of Western logic and reason. As we read it, we become aware of the tension-filled life of the farmer, the outcome of whose labors is subject to the whim of the elements. Time unfolds as a pattern that begins with mounting anxiety: will the rains destroy the crops, will the harvest be terminated by an early unscheduled frost? Then comes relaxation and relief, with accompanying celebration, when things work out. The bends and turns along life's bumpy road of tension and resolution are marked by signposts in the natural environment—happenings that show the way and portend the future. And it is the sky events that are the most reliable.

While, in *Works and Days*, Hesiod aimed to tell how to run a farm and not how to mark time, its basic framework is still a calendar which dwells deliberately on the notion of time as a progression of orderly events.

Hesiod places his own present into the wider temporal framework of a cosmological model by dealing first with the previous ages of humanity on earth. The ages of past history seem to be related to a perceived hierarchial decline in value as he proceeds to name them according to the list of precious metals. He gives these ages in chronological order, beginning with the Golden Age when everyone lived a blessed life under the god Kronos, the personification of time itself. But from then on, things got worse. The Silver Age was a degenerated version of the Golden, and the people who lived in the Brass Age were even more decadent than their predecessors. Respect for age seems important to Hesiod: he conveys the distinct impression that this cosmic degeneration was still manifest in his day in the lack of respect people had been paying their godly ancestors. Hesiod believed his race, though originally connected to a godly origin, had already descended below the Brass Age and into the age of Iron, the basest of all metals. Only the goodness emanating from human sweat and hard work could forestall, though probably never halt, the ultimate decay of all humanity on its downward slide through history. Perhaps there even would come a time when all justice would disappear from the face of the earth, when babies would be born with gray hair as punishment for human beings having paid no heed to the advice of their elders:

> When the father no longer agrees with his children,
> nor children with the father,
> when guest no longer at one with host,
> nor companion to companion,
> when your brother is no longer your friend,
> as he was in the old days.
> Sad times!
>
> (182–84)[4]

Hesiod's prescription for the action present-day human beings must undertake to hold their declining world together is laid out in the "Works," the first part of the *Works and Days.* A kind of morally based farmer's clock, it tells not only when to work but also how to work. Nature's signals for the time to harvest grain or to plow take the form of a series of periodic time-marking events that can be seen in the sky:

> At the time when the Pleiades, the daughters of Atlas, are rising,
> begin your harvest, and plow again when they are setting.
> The Pleiades are hidden for forty nights and forty days,
> and then, as the turn of the year reaches that point
> they show again, at the time you first sharpen your iron.
>
> (383–87)

The Pleiades is a conspicuous star group located in the constellation of Taurus in the zodiac—the busy celestial highway transited by sun, moon, and planets. Popularly the Pleiades are known as the "Seven Sisters" (the daughters of Atlas in later Greek mythology). By "rising," Hesiod probably means the *heliacal rising,* the time when these stars make their first annual appearance in the fleeting moments of morning twilight after having been obscured for several weeks by the light of the rising sun, which had previously passed on its annual round through the region of the zodiac demarcated by Taurus. Likewise, "setting" means *heliacal setting,* or the last annual disappearance in the west following sunset.

In the seventh century B.C. in northern Greece, these two sky events were separated by a period slightly in excess of forty days, from late March to early May (see table 2.1).

The first morning rise of the bright stars of Orion (which occurred about 20 July) was used to determine when to separate the chaff from the grain once it had been harvested and dried out:

> Rouse up your slaves to winnow the sacred yield of Demeter
> at the time when powerful Orion first shows himself.
>
> (597–98)

One of the major events in viticulture was timed by three different celestial references that provided a kind of crosscheck on one another. In Greece one could not always expect to have clear skies:

> Then, when Orion and Seirios are come to the middle of the sky,
> and the rosy-fingered Dawn confronts Arcturus,
> then, Perses, cut off all your grapes, and bring them home
> with you.
> Show your grapes to the sun for ten days and for ten nights,
> cover them with shade for five, and on the sixth day
> press out the gifts of bountiful Dionysos into jars.
>
> (609–14)

Rosy-fingered dawn is one of the many *epithets*, convenient sayings or word formulas, the oral poet could easily memorize as the equivalent for the twilight period. Thus, the colorful image of the orange star Arcturus confronting the rosy-fingered dawn is a poetic way of saying "at the heliacal rising of Arcturus." In the age and latitude of pre-Classical Greece, this event occurred early in September and at the same time that both the constellation of Orion and Sirius, the brightest star in the sky, reached their highest elevations in the heavens. Thus, the farmer who is serious about producing the best wine ought to follow Hesiod's simple time-based formula: after harvesting the grapes according to the appearance of the stars, he should then count in tens and fives, using the convenient units of his fingers and hands to discover that the fall equinox is the time to pour the wine into jars for aging, a process wine makers still follow there today at the same season of the year.

The task of planting grape vines utilizes another celestial crosscheck, this one involving astronomical and biological time signals working together:

> But when House-on-Back, the snail crawls from the ground up
> the plants, escaping the Pleiades, it's no longer
> time for vine-digging;
> time rather to put an edge to your sickles,
> and rout out your helpers.
>
> (571–73)

In another image-laden passage, meteorological and astronomical events seem to conspire together as Hesiod offers advice not to the farmer but to sailors and fishermen about when not to put to sea. We can almost see the stars chasing each other across the sky, one fleeing the other in endless pursuit:

> But if the desire for stormy seagoing seizes upon you:
> why, when the Pleiades, running to escape from Orion's
> grim bulk, duck themselves under the misty face of the water,
> at that time the blasts of the winds are blowing
> from every direction,
> then is no time to keep your ships on the wine-blue water.
>
> (618–22)

Other passages in the "Works" suggest that the disappearance of these particular stars was associated with the deterioration of the weather, with consequent danger particularly to sailors.

There is even a right time and a wrong time in Hesiod to have sex:

> then is when goats are at their fattest, when the wine tastes best,
> women are most lascivious, but the men's strength fails them
> most, for the star Seirios shrivels them, knees and heads alike.
>
> (585–87)

This period corresponds to early July (note that when women are most desirous, men are in their most dried-up condition!). Here the pessimistic Hesiod takes the star Sirius to be the cause of sexual impotence. He even connects this anatomical imagery of what happens inside a man's body to the length of time Sirius spends in the sky overhead. In an earlier passage:

> At the time when the force of the cruel sun diminishes,
> and the sultriness and the heat, when powerful Zeus brings on
> the rains of autumn, and the feel of a man's body changes,
> and he goes much lighter, for this time the star Seirios
> goes only a little over the heads of hard-fated mankind
> in the daytime, and takes a greater part of the evening.
>
> (414–19)

The event being timed would have occurred about 12 September in our calendar. That date divides the year into periods when the star is visible in the sky during either more or less than one half the interval of total darkness, the Greeks apparently even then—three thousand years ago—attempting to divide up the time of day and possibly foreshadowing our system of hours.

The unwritten farmer's almanac continues all the way throughout the year cycle—when to geld horses, when to hunt birds, when the north wind will blow. All along—as in the epigraph to this chapter—Hesiod admonishes his inattentive brother about the necessity of following the strict order of nature.

Hesiod's calendar codifies the association of celestial rhythms with the biorhythms present in all living things since their beginning. His scheme is intricate and rich in detail in its predictive power, a device expressly fashioned to gain a foothold on nature, to assist a rugged Peloponnesian farmer in his desperate quest to exercise a measure of control over his environment. And his astronomic timings and crosschecks are accurate, as accurate as any time-marking scheme, using these events, that can be written down.

By contrast, the "Days" portion of the *Works and Days* is relatively brief—only sixty lines long. In an abrupt shift, the poet moves from a rational view of time to a superstitious one. Unlike the commonsense, pragmatic "Works," which is based on the annual solar cycle, the "Days" is a purely ritualistic calendar laid out in the framework of time as marked by the phases of the moon (see table 2.2). Hesiod sets forth not only the holy days but also those days supposed to be good or bad for such events as the birth of children and the performance of specific activities: for example, "Avoid the thirteenth of the waxing month for the commencing of sowing. But it is a good day for planting plants" (781–82). In contrast with the down-to-earth treatment of the various "Works" any self-respecting farmer ought to carry out during the course of the agricultural year, the activities in the omen-bearing "Days" strike the modern listener as irrational, metaphysical, or mystical.

Hesiod says something about most days of the month, and the careful reader can discern three different methods of counting them, all cleverly interwoven in the oral text. First is a straightforward counting through the days of the month: for example, "on the twenty-seventh day, [such-and-such should be done]." Second, Hesiod counts in relation to visible events in the moon's phase cycle measured from new moon, as in "the thirteenth of the waxing month" in the previous quotation. Essentially he means when the moon is short a day or two of full: when it rises late in the afternoon with a distinctly visible chunk shaved off its lower left side. And third, the month is divided into three units of ten called *decades*; thus, a day may be indicated as the "middle ninth" (the nineteenth day of the month) or the "first sixth" (the sixth day of the month).

Throughout, the intensity of activities seems to swell and diminish with the lunar phases. Note in table 2.2 that more activity takes place during the waxing phases, especially around full moon. The "Days" passages recall the more animate ways of life our ancestors must have experienced. Birthing, sowing, feeding: all wax and wane just like the tides. In the "Days," Hesiod brings us a little closer to those oysters that open and shut their shells to the lunar beat—at least closer than is the hard-bitten farmer in the "Works" who, though he must follow Nature, nevertheless lashes out at her, curses her, and begins to feel the need to manipulate and control her.

The stark contrast between the two ways of experiencing time in the "Works" and the "Days" makes one wonder whether the "Days" refers to an archaic calendar system that became attached to the main body of the poem at a later time. Or, were the "Works" and the "Days" two systems the Greeks employed simultaneously but for different purposes, like our fiscal and seasonal calendars? Perhaps we are witnessing the conversion of an old lunar calendar system into a newer solar one—a conversion that did not become fully realized until the advent of the even more precise, detailed written calendar of classical Greece.

TABLE 2.2
Monthly Calendar from the "Days" Section of Hesiod's Works and Days*

Day Number	*Kind of Day*	*Good for:*	*Not good for:*	*Observed Lunar Phase*
1.	holy day	—	birth of girls; wedding day	
2.	—	—	—	
3.	—	—	—	
4.	holy day	bring bride home; start a ship; open wine jar	—	
5.	hard day	—	—	
6.	—	geld kid and sheep; build sheep pens; birth of boys	birth of girls	FIRST QUARTER
7.	holy day	—	—	
8.	—	geld boar and bull	—	WAXING
9.	harmless day	birth of boys and girls	—	
10.	—	birth of boys	—	
11.	holy day	shearing; reaping	—	
12.	—	shearing; reaping; geld mule; woman at loom	—	
13.	—	planting plants	sowing	FULL
14.	holy day (above all)	birth of girls	—	
15.	hard day	—	—	
16.	—	birth of boys	plants	
17.	holy day	cutting wood; cast grain on threshing floor	—	WANING
18.	—	—	—	
19.	—	—	—	
20.	—	—	—	LAST QUARTER
21.	holy day	—	—	
22.	—	—	—	
23.	—	—	—	
24.	holy day	—	—	
25.	hard day	—	—	
26.	—	—	—	
27.	holy day	open wine jar; yoke oxen; launch ship	—	
28.	—	—	—	
29.	—	—	—	NEW
30.	—	inspect works; divide rations	—	

*By contrast to the year clock in table 2.1, the calendar laid out in the "Days" section reflects a more metaphysical animate quality: it is less like a railroad timetable, more a farmer's almanac. After M. L. West, *Hesiod Works and Days* (Oxford: Clarendon Press, 1978), p. 381.

The ancient Greeks may well have understood as complementary the sun's rational cycle and the moon's irrational appearances. For within the larger annual cycle of farming—events that Hesiod presents in essentially a "natural" manner, exactly as these events occur within the seasons of the year—there lies embedded a recurrent monthly cycle whose days are richly textured with ritual practices and beliefs, days that alternate in cycles of good and bad, favorable and unfavorable.

Hesiod's word imagery shows us that the early Greeks did not think of time as some abstract phenomenon rated on a clock, the way we think of it today. For them, time was the ordered cycle of sensible natural events to which human beings were meant to relate the events in everyday life from tilling the soil to worshiping the gods. For Hesiod the true essence of time lay in a dialogue continually going on between nature and culture: we must live within nature's bounds; we must endeavor to make the best use of the gift of the gods which comes to us in the form of nature's secrets, and which we can discover if we observe nature closely and carefully. This is the message of this everyman-poet. Nature speaks to the farmer through a repetitive, oscillatory chain of events that can be associated with one another—the appearance of a star, the migration of a bird (see figure 2.1), the blossoming of a flower, the arrival of a gust of wind. Neither taxonomist nor scientist, Hesiod seems unconcerned with *compartmentalizing* nature's happenings into astronomical, ornithological, horticultural, or meteorological categories (though I did just that in table 2.1). When we listen to his calendar song, we can be sure that we are hearing about entities that sing along together harmoniously; they are part of a holistic world view. The message seems to be that Nature, though fickle and often misunderstood, is really here to nurture us; and if we pay close attention to the collective order, we can use her behavior to help structure a better society. We have a role to play. Our response to Nature should be to use her subtle time signals to the best of our ability, to maximize our wealth and our success by creating order in the social and political environment we administer.

Creation Myths

While we think of myths as fabulous conjectural narrative, stories that appeal to the imagination by personifying things and people as

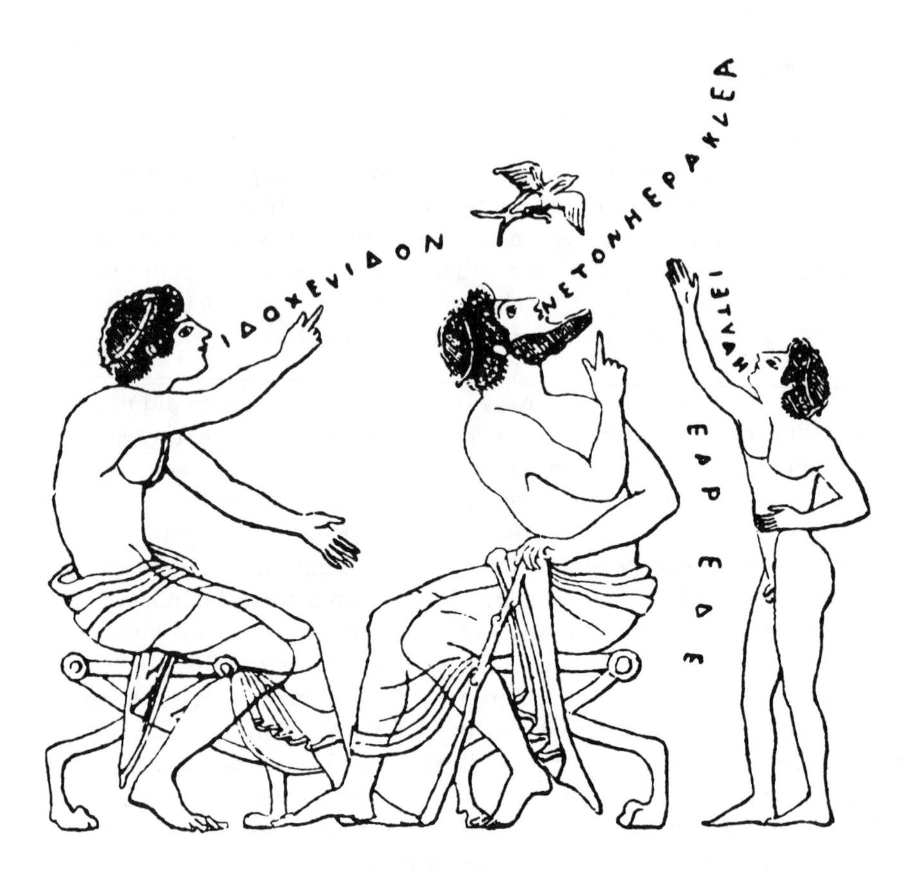

FIGURE 2.1 Early Greeks used nature's direct signs to construct a calendar. Here a group of men sight a swallow and exclaim: "Look at the swallow," "So it is, by Herakles," "There it goes," "Spring is here!" SOURCE: Vatican Museum drawing after J. Harrison, *Epilegomena and Themis* (New Hyde Park, N.Y.: University Books, 1962), p. 98.

spirits but lacking real, demonstrable truth, myths do contain the truth of an idea or a principle. Their object is to get across the basic point by searing the image into the imagination—by embellishing it, distorting and reshaping it, repeating it and making it unforgettable by relating it to everyday occurrences. Oral myths lie at the foundation of our concept of time, having been fashioned in all their elaborate detail before anyone ever laid brush to paper. In creation myths, time could be represented as the history of succession of rulership, as the cyclic alternation of the opposing forces of good and evil, as an entity that had its origin in the separation of earth from sky, in the establishment of the first glimpse of order in an otherwise indescribable and meaningless universe.

THE BOOK OF GENESIS

The word *genesis* means "origination," and every genesis myth begins with a sense of time. Our modern scientific genesis began more than ten billion years ago in a colossal explosion from which all events and things have spun. Usually implicit in every genesis is a purpose, though for many of us that original explosion had as little purpose in a human sense as anything in history.

In the Judeo-Christian tradition of the Old Testament, the purpose of the creation stories in Genesis seems to be to demonstrate that all things were intended to be good. Perhaps all people need to believe in a world that can be conceived as orderly, intentional, purposeful, and as created specifically with them in mind—above all, a good world. Biblical Genesis satisfies that need by stating that God made it so:

> In the beginning, God created the heavens and the earth. The earth was without form and void, and darkness was upon the face of the deep; and the Spirit of God was moving over the face of the waters.
>
> And God said, "Let there be light"; and there was light. And God saw that the light was good; and God separated the light from the darkness. God called the light Day, and the darkness he called Night. And there was evening and there was morning, one day.
>
> And God said, "Let there be a firmament in the midst of the waters, and let it separate the waters from the waters." And God made the firmament and separated the waters which were under the firmament from the waters which were above the firmament. And it was so. And God called the firmament Heaven. And there was evening and there was morning, a second day.
>
> And God said, "Let the waters under the heavens be gathered together into one place, and let the dry land appear." And it was so.
>
> (1:1–9)[5]

As in the *Works and Days*, there is a rhythmic pulse that gives us a feeling of the way the cosmogony, or creation story, must have been told in

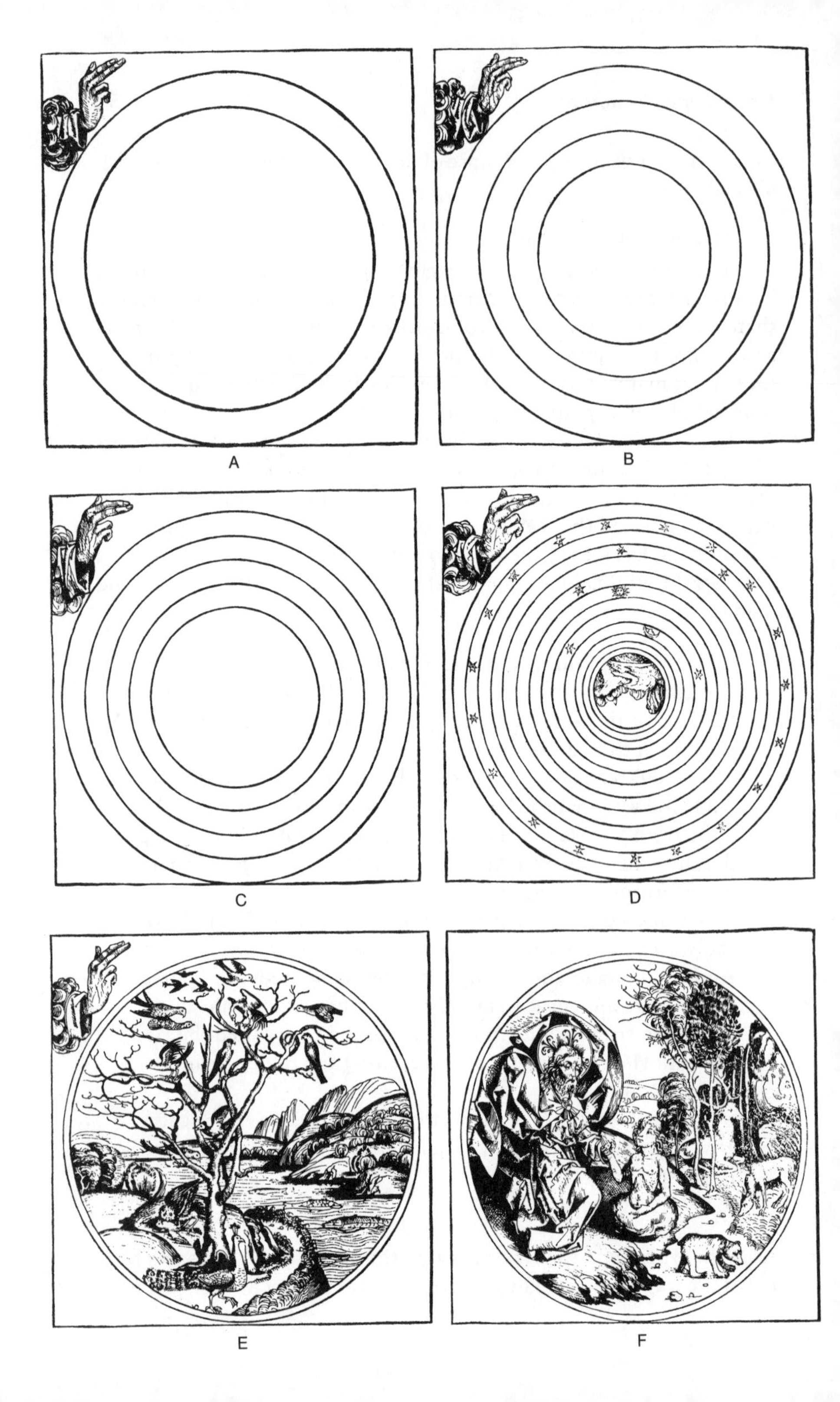
A
B
C
D
E
F

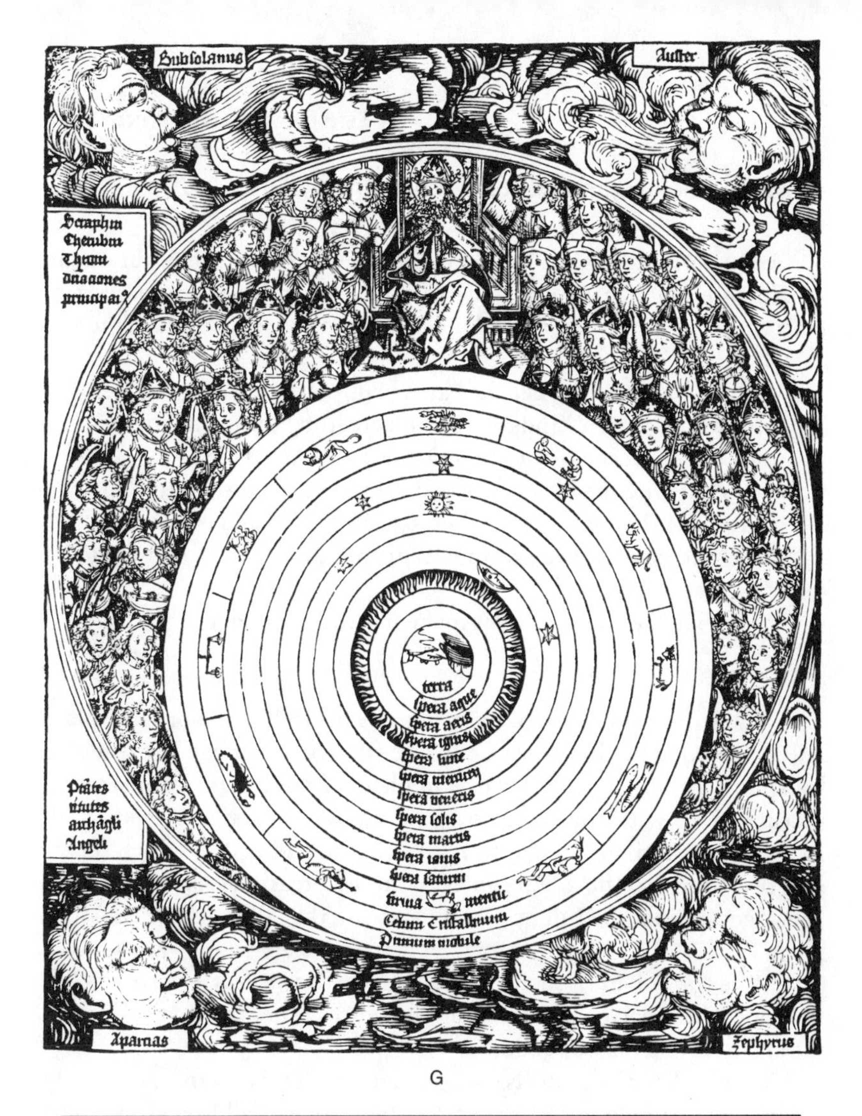

FIGURE 2.2 A pictorial interpretation of the Creation in Genesis from the late-fifteenth-century Nuremberg Chronicle reflects hierarchical order in time as well as in space. (*A*) The First Day: God's hand blesses His separation of light from dark (Genesis 1: 1–5). (*B*) The Second Day: He separates heaven from the waters (1: 6–8). (*C*) The Third Day: in a third act of separation, God puts the water in one place and leaves dry land in the other (1: 9–13). (*D*) The Fourth Day: time begins as objects in the firmament take up their motions; thus, we begin to see the world in its recognizable form (1: 14–19). (*E*) The Fifth Day: God makes the creatures who inhabit the world (1: 20–23). (*F*) The Sixth Day: God creates Adam out of earth, giving him a special place, while the animals roam passively in the background (1: 24–31). (*G*) The Seventh Day: God reposes above His finished product—a series of earth-centered spheres with everything in its proper place (2: 1–3). SOURCE: Hartmann Schedel, *Liber Chronicarum* (Nuremberg, 1493). The Huntington Library, San Marino, Calif.

words long before the advent of writing: "And God said, . . . And it was so. . . ."

In this orderly creation, God makes heaven first, then the earth; then day and night are separated out of chaos. Finally, God creates vegetation, celestial bodies (the calendar), animals, and, last of all, men.

> And God said, "Let the earth bring forth living creatures according to their kinds: cattle and creeping things and beasts of the earth according to their kinds." And it was so.
>
> And God made the beasts of the earth according to their kinds and the cattle according to their kinds, and everything that creeps upon the ground according to its kind. And God saw that it was good.
>
> Then God said, "Let us make man in our image. . . ."
>
> (1:24–26)

Man stands at the top of the hierarchy in God's time-ordered plan. All is intended for his dominion. The structure of time is paralleled by spatial order that is also hierarchical. If we try to reconstruct a picture to go with the words of Genesis, we end up with a stacked, multilayered universe something like that in figure 2.2, which serves as a testimony to the adaptability of an old creation story to a newer age. (See whether the full text convinces you that all the pieces fall into place.)

Note the power of the word in Genesis. It is the utterance of God that does the creating. And in every instance comes the rhythmic temporal downbeat: "it was good."

There is later in Genesis a second creation story. Man (specifically Adam) is fashioned out of the earth, the source of all life. God molds him from a ball of clay into a pre-existing environment. Later, God creates Eve, realizing it would be better if the human race consisted of paired opposites—male and female—rather than single individuals. The original pair confront evil and struggle to cope with the problems of living under a higher authority—the same problem encountered, as we shall see, by the newly created gods in the Greek *Theogony* and the Babylonian *Enuma Elish* creation stories. The fall of man surely begins with the first bite of the apple, but—like the tide to which all celestial bodies are subjected—man rises again. He returns from his chaotic origin via a series of acts oriented toward his ultimate salvation. Biblical scholars believe the second creation myth of Genesis probably existed alongside the first in the ancient Middle East; at least, most biblical scholars regard Genesis to be an amalgam of myths brought together from a number of different sources. The second creation story seems to be stressing the idea that, though past events alternated between harmony and discord, everything will lead ultimately to world order. First, God creates a world of righteousness, harmony, and well-being, replete with a covenant, a mutual consent by man to follow the law He has laid down. The injustice and sin in the world are the discord created by man. Each time he breaks the law by sinning, he is called in for judgment. Only an act of divine justice can restore the balance and return man—forgiven and transformed, repentant for his sins—back into orderly society. The stories of Cain and Abel, the temptation of Adam in the Garden of Eden, Noah and the Flood, and the Tower of Babel illustrate this rise and fall in a repetitive rhythmic pattern, a sense of tension followed by relaxation. Take the time sequence in the story of Cain and Abel:

1. Cain falls; he breaks the law by killing his brother out of jealousy.
2. He is called to judgment.
3. Punishment is administered (his land shall no longer be fruitful in spite of all his hard work).
4. He is restored. The Lord provides him protection by placing a mark upon him.

A time scale lies embedded within these biblical stories. As in most early myths, it is not the absolute time bar of modern geology and astronomy which is discernible in terms of millions or billions of years. What matters rather than length of time is the order of events and the rhythmic and harmonic overtones that characterize the cyclic flow of time. The head of time's arrow points to the tail. What happens once, happens again, though things are a little different the second time around. Cain falls; Cain is restored; but he is different the second time around.

HESIOD'S THEOGONY

This same concept of circular time occurs in other genesis myths. In Hesiod's *Theogony,* a fairy-tale-like ancient Greek myth of creation, there is so much concern about what time really means that it is personified by a god.

The story, which sounds like an X-rated TV soap opera, revolves around a battle of the generations. The history of the world is characterized as the history of the descent of the orderly government of Zeus by succession from his godly predecessors. Time begins after Gaia (the earth) and Ouranos (the sky) lie together. Evidently Gaia is (just like a woman in the eyes of Hesiod) sexually promiscuous, unpredictable, and incestuous, for Ouranos is her son. Ouranos had been born out of his mother's affair with Erebus, who, like the Night and Gaia herself, had arisen out of Chaos, the first power, a kind of indiscernible, dark abyss of many mixed qualities churning around in a disorderly state.

Mother Earth's labor is made difficult by Father Sky's jealousy and outright fear of each of his prospective offspring. He attempts to relieve his stress by shoving his children back into the womb as they are trying to be born. Mother Earth deals with this irritating situation by trying to trick her jealous husband and fashioning a secret weapon:

> . . . So she hid them away, each one as they came into being,
> and let them not rise to the light from down in the hollow
> of the earth; and this was an evil activity pleasing to
> Ouranos. But huge Gaia was groaning
> within and feeling constrained, and so she contrived an evil
> device. Swiftly producing a new kind of metal, gray adamant,
> she created of it a great sickle, and this she displayed to
> her children.
>
> (156–63)[6]

The next-born Kronos, Father Time himself, uses this famous sickle to castrate his father. Having separated his father's male member, he tosses it into the sea where its shining foam gives birth to the female deity Aphrodite, goddess of love.

Like the Old Testament's Genesis, the succession myth in *Theogony* then becomes a lengthy genealogical catalogue of alternating good and evil deities who represent different parts and powers of a highly animate, personified universe. Like the developing Greek state, this cosmos consists of both political and natural components. Zeus, king of the gods and bringer of all order to the world of mortals, is the end product of the genealogy of deities. He achieves his position by killing his father, Kronos, who apparently inherited from Ouranos the perverse attitude of doing away with his children "that no other lordly descendant of Ouranos should possess the honor of kingship among the immortals" (462). Rather than forcing them back into the womb, Kronos kills them by swallowing them as soon as they are born. But Zeus puts an end to the cycle of repetition by taking vengeance upon his father.

Unlike the others, Zeus is hidden away by his mother, who dupes her husband into swallowing a stone instead. Ouranos is forced to regurgitate the stone, which ultimately ends up on Mount Olympus, where it is used to establish an altar for the worship of Zeus. Reared and nurtured in secrecy, he is able to return to complete the conquest of his father. But opposing forces, seeking to avenge the death of his father, send forth the monster Typhoeus, an evil force powerful and hideous beyond all description:

> Wonderfully strong were the arms of Typhoeus
> to do all he wanted;
> he had the weariless feet of a mighty divinity and out of his
> shoulders a hundred heads of a serpent; a frightening dragon,
> rose, each of which shot forth a flickering black tongue;
> and out of his
> eyes flashed fire from under the brows of each of his heads,
> fire came blazing forth from each of his staring heads;
> and from each of his terrible heads he was able to speak and
> utter every imaginable sound. . . . Sometimes the sound was
> speech, the language the gods understand, sometimes the
> bellowing wail of a bull, a thunderous roll, an ominous rumble,
> sometimes the roar of a lion with heart unrelenting and
> mirthless, sometimes a hiss. The mountains around
> were ringing with echoes.
>
> (823–35)

Zeus, who must prove his legitimacy as ruler of all the cosmos by subduing Typhoeus (see figure 2.3), physically summons up his unsurpassable strength and cunning, wins the day, and is accorded the title of great establisher of order among both men and immortals.

A BABYLONIAN CREATION MYTH

This age-old tale of the rivalry among the generations and the punishment of men who suppress their children and cause their wives

FIGURE 2.3 Creation in the Greek *Theogony.* Zeus slays Typhoeus, here depicted with but a single head. Like Marduk who slew Tiamat in *Enuma Elish,* he separates her parts as one who slits open a mussel shell, thereby creating heaven and earth. SOURCE: From a painting on a sixth-century B.C. vase in the Antikensammlungen, Munich. From *The Poems of Hesiod,* trans. R. M. Frazer (Norman, Okla.: University of Oklahoma Press, 1983). Copyright © 1983 by the University of Oklahoma Press.

great pain has been told many times before. There are parallels between the seventh century B.C. *Theogony* and the earlier Babylonian creation myth known as *Enuma Elish,* or "when on high,"[7] after the first three words of an oral hand-me-down that ultimately was inscribed in cuneiform on clay tablets. The one-to-one parallel between the two casts of characters in these epic myths is no coincidence. Early Hellenic antecedents in philosophy, science, mathematics, and religion can be found in the Middle Eastern cultures.

In the *Enuma Elish* tale, Apsu (male) and Tiamat (female) represent the sweet and fresh waters that intermingle in the area where the Tigris and the Euphrates empty into the Persian Gulf. The couple creates five generations of descendants, each of whom personifies an aspect of nature. Ea, like Kronos, represents time, which does not begin to flow until earth and sky are separated. Also like Kronos, Ea is threatened by his father with whom he does away by performing an overpowering holy incantation which puts Apsu to sleep—but not before Tiamat asks that rhetorical question all of us wish we could phrase to the creator: "Why should we destroy that which we ourselves have brought forth?" (45).

The line of descent ultimately passes to Marduk, the youngest son. Like Zeus, he must overcome his last enemy, in this case his avenging mother. Only then can he be permitted to ascend to the position of chief tutelary god of the city of Babylon. The battle, ending with the slaying of Tiamat, is vividly described in the text:

> When Tiamat opened her mouth to devour him
> He drove in the evil wind, in order that she should
> not be able to close her lips.
>
> The raging winds filled her belly;
> Her belly became distended and she opened wide her mouth.
> He shot off an arrow, and it tore her interior.
> It cut through her inward parts, it split her heart.
> When he had subdued her, he destroyed her life.
>
> (97–104)
>
> He split her open like a mussel into two parts;
> Half of her he set in place and formed the sky therewith
> as a roof.
> He fixed the crossbar and posted guards;
> He commanded them not to let her waters escape.
>
> (137–41)

The idea behind this fantastic imagery is that at the root of the state lies divine kingship, and it must be established authoritatively once and for all through compulsive force. In like manner, a process of violent struggle also was necessary to create an orderly universe. As in Genesis, creation is conceived as an act of separation. Things do not simply materialize out of nothing; some essence called "chaos" is already present. Originally the sky and the earth were one. The force of the winds blowing down the valley of the Fertile Crescent still keep earth and sky apart in a universe completely surrounded by water—ground water; water from the sky, from the rivers and the sea. This dynamic creation can be contrasted with the watery beginning also described in the myth. Out of the union of Apsu and Tiamat, earth and sky are formed—slowly and gradually by the deposition of silt that builds up in the Tigris–Euphrates delta, accumulating all the way to the horizon as it accompanies the sweet river water on its downward course to the salty sea. What a reasonable way for the Babylonians to think about the passage of time—as steadily and slowly as the rich silt of the delta created a nurturing land in the place where the waters mixed. The violent version of the myth—the interruption of the established order—was yet to transpire. First, the heavy winter rains create a watery chaos by causing the rivers to overflow their banks. But in the spring, when strong winds blow the clouds away, the waters of earth and heaven are parted and the sun, Marduk's celestial parallel and his source of power, dries the land, restores order, and creates life. Two different stories of creation are interwoven in the same myth—one placid, the other violent. Each represents society and nature at the same time. Both scenarios are experienced in every season of the year. And both are true. Myth is reality.

The anthropologist Edmund Leach has drawn interesting conclusions about the social aspects of time expressed in the oscillatory nature of the themes in these succession myths.[8] Events in both *Theogony* and *Enuma Elish* alternate between activity and inactivity: order is established; there is a threat to that order; and finally, the order is re-established. This alternation between opposites governs the flow of the story. What is it about Kronos that makes him an appropriate symbol for time? What he represents is depicted by a series of actions that go back and forth between extremes: one action seems to be the reverse of the other, in contrast to our modern notion of time as an endless chain of events. Kronos castrates his father for pushing children back into his mother's womb, and later swallows his own children but, when Zeus overthrows him, is induced to vomit them up again.

One of the rituals associated with Kronos was a sort of new year's celebration in which masters and slaves reversed roles. The Greeks often spoke of an age that might follow that of Zeus: a time when men would be born from their graves and grow younger, when all the strife inflicted on the world would shrivel and disappear, when time would literally reverse itself and flow backward. Today we still associate Kronos as Old Father Time with the new year, a date when we propose to implement the most drastic changes in ourselves by making resolutions, those often-failed attempts to reverse our patterns of behavior.

In Hellenic times, ringing in the new year took on an agrarian symbolism. The act of castration, Leach says, can be thought to symbolize the annual cutting of the seed from the stalk which enables Mother Earth to become fruitful and bear a bounteous harvest. The reversal of roles experienced at year's end—for example, the replacement of Old Father Time by a newborn baby, of associating the beginning of life with the beginning of death in the form of the "grim reaper"—characterizes the way things really do appear to reverse themselves in nature at the end point of an oscillation. For just as a pendulum reverses its direction, the rising and setting sun glides on its annual course back and forth along the horizon, and the moon passes forward and backward through all of its aspects. In fact, when he creates the moon in *Enuma Elish*, Marduk commands it:

> Monthly without ceasing [to] go forth with a tiara
> At the beginning of the month, namely, of the rising over the land,
> Thou shalt shine with horns to make known six days;
> On the seventh day with half a tiara.
> At the full moon thou shalt stand in opposition to the sun,
> in the middle of each month.
> When the sun has overtaken thee on the foundation of heaven,
> Decrease the tiara of full light and form it backward.
>
> (14–21)

In the "Days" portion of the *Works and Days*, this alternation between contrary states is meted out daily, with one day of the moon connoting good, the next day bad.

Where do we get the idea that time is a succession of polar opposites? Biologically, the dramatic reversal of behavior of organisms at the peak time of their cycle is obvious to anyone who pays attention to nature. The Greek farmer, needless to say, sensed the reversals taking place in his field all the time. Just after the fruit is born, the leaves on the trees shrivel and drop off: they literally "ungrow"; they become "unborn" again. Animals burrow back into their holes whence they had come in the spring; and the warming sun, which had climbed ever higher into the sky to nourish the earth, redescends its celestial ladder.

For the Greeks, Kronos created the pattern when out of the homogeneous symmetry of chaos, he polarized the universe. He made time when he parted the earth and the sky; when he separated the male principle that fell into the sea to become its own opposite, the female principle in the form of Aphrodite. He also created what Leach calls the mobile element of time, the "becoming" that causes things to oscillate between the two extremes of existence.

Today we think of time as a linear chain of events, a sequence that began billions of years in the past and is likely to extend for an indefinitely long period into the future. But if we look a little more closely at the way we really use time in the everyday world, we may not be so remote from our ancestors. Think of how the average citizen spends a week. We begin with five hard days at the office, each essentially the same. We catch the subway to work at precisely the same time; lunch is a brief punctuation in a highly regulated and prescribed workday in which every minute counts. We return home in the evening to salvage a few precious minutes or maybe an hour for some specially pleasing activity. Then comes the weekend. Dramatically we reverse our patterned behavior: we sleep late; we don't set the alarm but, instead, devilishly permit our biological clocks to run wild. We do different things on the weekend. We travel, we feast, we celebrate. But then on Monday, it's back to work. Some of us speak of barely having survived the weekend. Literally, we "lose track of time"; we "re-create." Our weekends are special times—the sacred portion of our temporal two-phase cycle—while the weekdays are routine, common, or profane. Though lower orders of life also operate on a two-phase cycle, we know of no sacrality or profanity in oysters, leafy plants, or hamsters.

For us, time has meaning. Sacred and profane time, as the historian of religion Mircea Eliade called them,[9] seem to be strictly man-made—part of an order fashioned by moral beings in societies that participate in sacrificial and festal rites that emanate from their systems of belief (see figure 2.4). The radical transformation of our existence from weekday to weekend and back again—that distinctly human way of behaving—may be an evanescent social remnant of the days when the sacred rituals of our ancestors acted out the transition they believed humans all make from life to death and back again in the endless cycle of death and regeneration—two states characterized by two radically different kinds of time that impinge upon one another, perhaps more like what is represented in figure 2.4B than by the linear continuum depicted in figure 2.4A.

Time does not go on and on, as the metaphor of the geometrical straight line has it. As we live it, time is not the sheer succession of epochal durations and is not, above all, a concrete thing lying out there waiting to be parsed out. Rather, we create it by sensing the intervals between the events of our social lives and by imitating the pattern we feel and perceive in nature—the pattern we have come to believe characterizes all living things.

In *The Sacred and the Profane*, Mircea Eliade argues that, in the minds of most of our ancestors, there was no distinction between past and present. The heroic past of the gods remained alive in the present

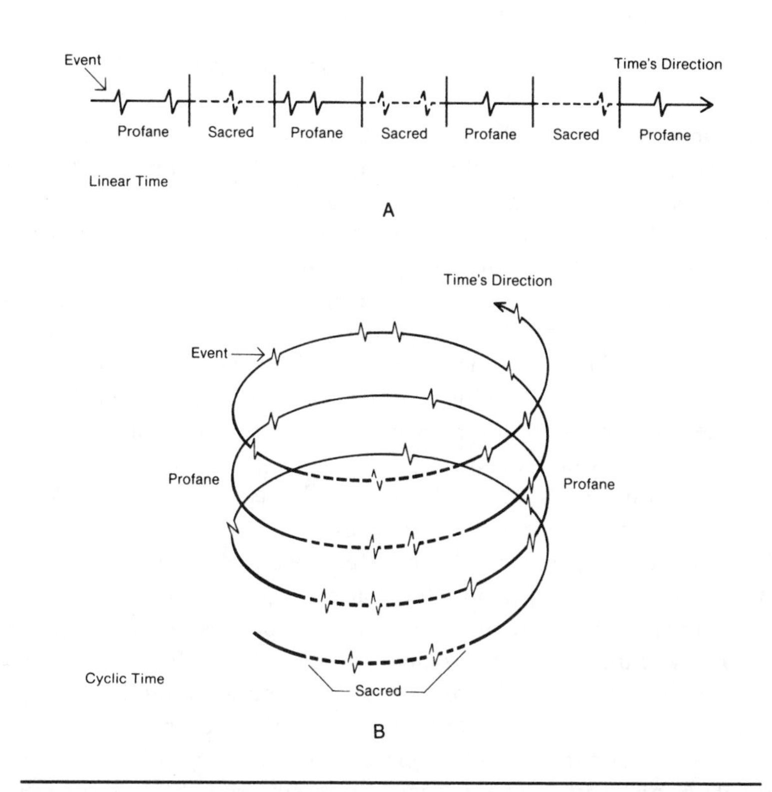

FIGURE 2.4 Models of segmented time. (*A*) division of linear time into sacred and profane; (*B*) cyclic time divided in the same manner. SOURCE: After E. Leach, *Rethinking Anthropology*, London School of Economics, Monographs in Social Anthropology, no. 22, fig. 17, 1971. By permission of The Athlone Press.

and was perpetuated by a ritual re-enactment of their deeds. The holder of political office did not simply *represent* the power of nature that governed all people; he *was* that power. In ancient Greece, the legitimate authority was ordained by the gods. A ruler usually continued in office for a full seasonal cycle or a multiple thereof, after which, just as in the *Theogony*, he was (usually) violently overthrown by a successor.

According to Eliade, archaic societies lived in a "paradise of archetypes" by creating a time structure based on renewal of the regenerative properties observable in all biorhythms. This periodic regeneration presupposes that every re-creation is a repetition of the first act of genesis or becoming. When a priest recited the *Enuma Elish* during ceremonies celebrating the creation, he was not just telling a story; he was acting out the final battle between Marduk and Tiamat at the end of the seasonal year. At the termination of the ritual, participants shouted, "May he subdue Tiamat, may he distress her life and shorten her days"; thus, they actualized their ancient cosmogony.[10] By re-enacting the creation myth, the believer relived over and over again the transition between chaos and order, never allowing past events to become a part of history—never, in effect, bearing the burden of time as we do. For Eliade, "religious man" lives in an atemporal world while we, "historical man," voluntarily create history by choosing to separate past events from ourselves.

We can only guess how far back oral calendar making goes. We are able to recognize humans like ourselves in the fossil record dating to 200,000 years ago, and we believe we have been sedentary only for the last 5 percent of that period. Not that pre-agricultural peoples were totally unspecialized in their division of labor. Modern studies of present-day hunter-gatherer societies like the !Kung bushmen of the Kalahari Desert reveal a surprising level of socio-economic strategy. These are not marginal people barely clinging to life who were denied the leisure, thought, and contemplation that comes with being sedentary. As hunter-gatherers, our ancestors likely would have learned the same celestial signs as Hesiod to judge the hunt, the blossoming, the running. And we have every reason to believe they communicated this information orally to one another as a means of coordinating their lives. We have, however, no tangible evidence about their calendar for their voices have become attenuated, their words lost forever, evaporated into thin air. If we want to know what they were doing with time we can only look at the sparse material record that has survived them.

The Written Mode

Lunar Calendars in the Ice Age

If we accept the traditional definition of writing as an ordered set of symbols that appear on a surface, a set that could be taken to represent a tally, we can trace the written calendar back more than twenty thousand years to the last ice age. In *The Roots of Civilization*, Alexander Marshack has interpreted engraved marks found on bits of bone from central Africa and from paleolithic caves in France to be rudimentary forms of an early lunar calendar.[11] The evidence lies in distinct clusterings of notches on the bones—marks that could not have been grouped together by chance. These marks take the shape of cuts that could have been made by drawing a sharp tool across a groove in a surface, the way one sharpens a knife in an electric knife sharpener (figure 2.5A), making the groove deeper with each pull; or by twisting a pointed object in a hole, gouging out a deeper hole as one twists (figure 2.5B).

To the untrained eye, these marks look random, almost accidental. But by handling the objects and examining the grooves under a microscope, Marshack was able to specify the direction in which several of the proposed notations were laid down as well as the manner in which they were grouped. For example, in the bone from the Dordogne Valley, the marks on one part of the sequence were made with a definite direction or turn of the tool, which is apparent from the comma-shaped gouges visible under a microscope. In another part of the chain, the commas take on a different direction. By holding the bone tablet in his hand and changing its direction, Marshack discovered that the marks must have been laid down in a "boustrophedon" or alternating left-right, right-left sequence (as shown in figure 2.5B), the way a farmer plows his field. In fact, Marshack deduced that the maker must have taken 24 turns to produce the 69 marks.

Most surprising of all, when he laid these gouge marks out in a linear sequence (figure 2.5C), he noticed that they could be divided into repeatable groupings. Could they have been counting days of a periodic cycle? Assuming each mark to symbolize a day, he fitted the pattern with its turns to a "lunar model": in other words, the groups of tallies could be so arranged that the major turns in the boustrophedon notation happened about every 14 or 15 days. This interval corresponds to the interval between the first sighting of the waxing crescent moon in the west after sunset and the full moon, or between full moon and the last sighting of the waning crescent in the west.

The motive for keeping a lunar record should be obvious. As I suggested earlier, a major portion of the lunar-phase cycle provides extended light for accomplishing many useful activities. Also, it helps to plan if one knows or can anticipate when light-time will come.

FIGURE 2.5 The first written calendar or a Neolithic tool sharpener? On this 30,000-year-old bone tablet (*A*) from the Dordogne Valley of western France, the engraved pits (viewed under a microscope) in *B* have been interpreted as a sequential serpentine pattern of notational marks. If we allow each mark to stand for a day, these (*C*) match a model of the phases of the moon in a pattern that can be extended over more than 2 months. In the model, the turning points in the pattern (*B*) are thought to coincide with the new (*solid circle*) and full (*open circle*) phases of the moon in *C*. (*D*) Carved bone with slash marks along the edge. SOURCE: Courtesy A. Marshack, *The Roots of Civilization* (New York: McGraw Hill, 1972).

A

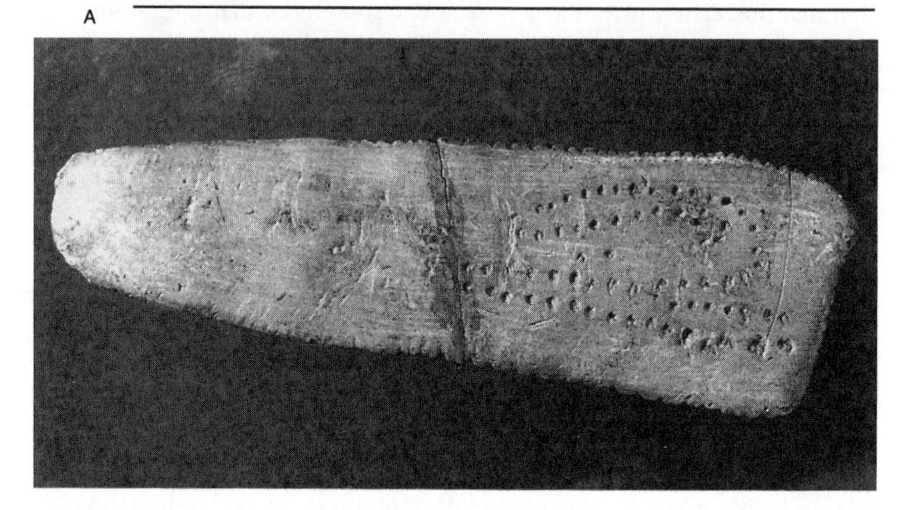

B

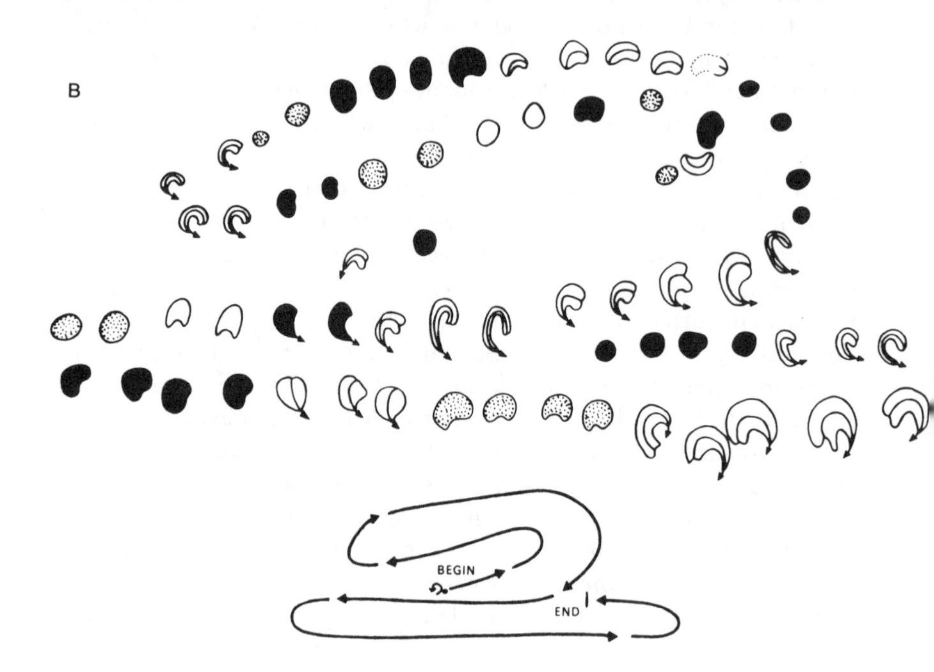

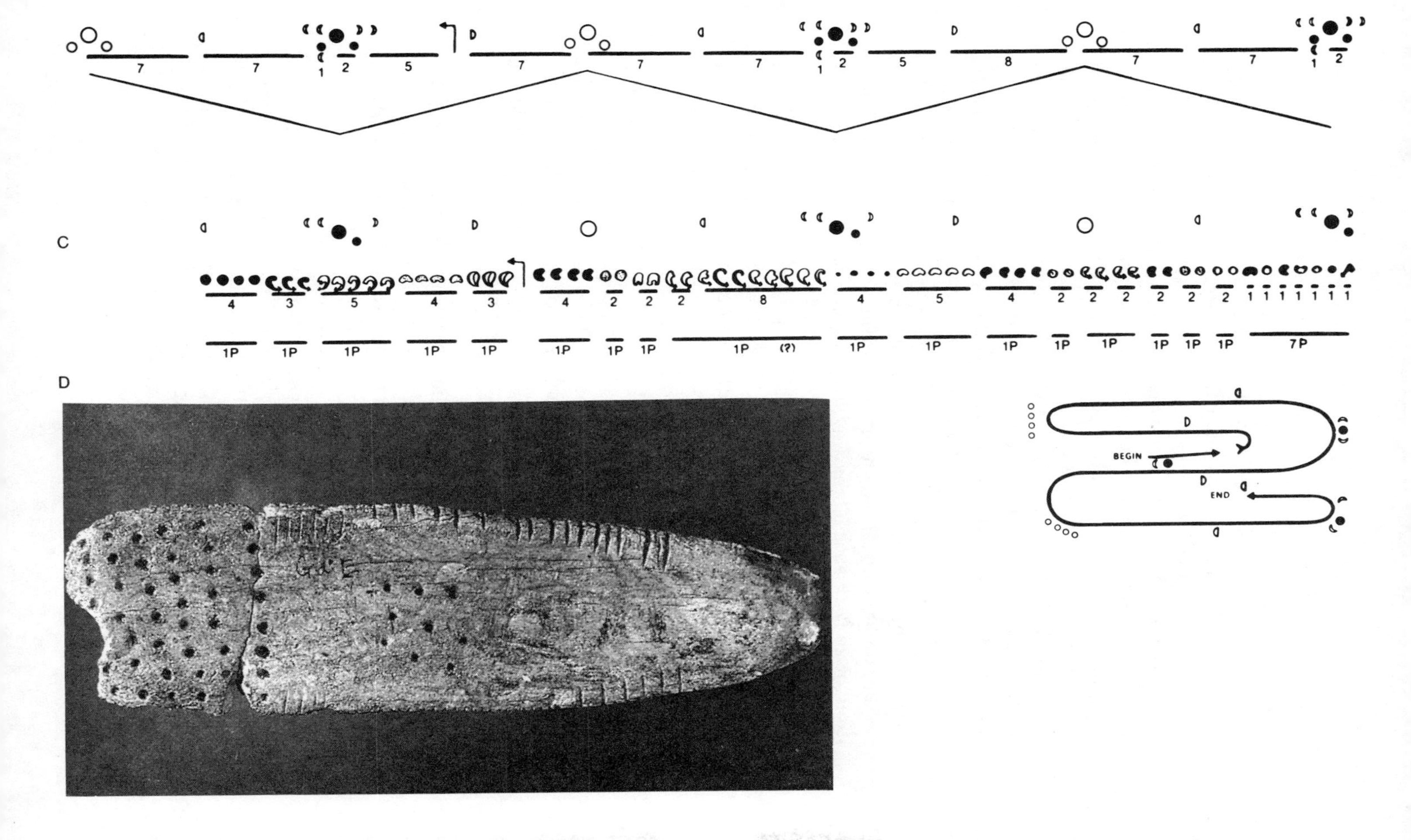
C
D
BEGIN
END

Keeping track of lunar events would offer the paleolithic inhabitants of Western Europe a means of abstractly correlating what Marshack calls "time-factored" events—those that occur sequentially in a scheduled manner—through a process that lends itself readily to measurement. These notations could represent the foundation of the associative process I referred to in Hesiod—the first step in the evolution of traditional writing, where a mark stands for a thing, in this case one day. Though the bone calendar of figure 2.5 stretches but two and one-half months, an extended series of such day tally markers could have led early hunter-gatherers to deduce that the period from human conception to birth was nine moons; that after two moons, a particular supply of berries would dry up; or that after every twelve or thirteen moons all the nearby streams would swell to capacity.

Psychologically, it is comforting to think that writing as we know it might go back such a long way. To imagine that our earliest ancesors were abstract thinkers like us—arithmeticians, perhaps even computists—offers a broader and higher historical pyramid to support our modern accomplishments. Though his basic ideas about the beginning of the arithmetic intellect in humans are accepted by a majority of anthropologists, Marshack's work, even after nearly twenty years, remains somewhat controversial. Some detractors say permanent calendar keeping is not consistent with what we know about the level of intelligence of these early people. Counting the days would have been a concept too narrow, too abstract for them to fathom. Besides, cave dwellers did not *need* to count days. They knew when to hunt, when to gather, and they certainly could tell when the extended light of the moon would come, simply by spotting the first lunar crescent in the west after sunset. Why bother to write it all down? The predictive capacity implied by Marshack's lunar hypothesis would constitute unnecessary baggage in their seminomadic way of life. (Marshack himself had never interpreted the marks to be arithmetical or predictive, only notational.) Still more conservative opponents have suggested that Marshack's bones contain no ordered pattern at all, that he has not provided enough examples, and that those he offers include a lot of imaginary interpretations at best. Are these marks only doodles and decorations, or were the bones only tool-sharpening devices, pure and simple? Slash marks along the edges of some of Marshack's bones (figure 2.5D) resemble the knife-sharpening grooves that can be seen all over the stone pillars and lintels up and down the Nile valley, where centuries of idle warriors whiled away spare time by keeping their basic equipment in shape. Likewise, the gouge marks on any of these bones could have been used to the same extent to fashion a point on an awl or a needle rather than to record a day in one's life. Recent experimentation with stone and bone artifacts suggests that the sort of markings that appear on Marshack's bones could have been made in a single sitting.[12] Therefore, if there is a pattern, it could have been intended to convey an overall result, rather than to constitute a continuous record—considerations that can be taken to argue against the calendrical hypothesis, though by no means do they disprove it.

Is Marshack's scheme only coincidental? True, you can make a pattern out of any series of marks if you try. Take the artifact in figure 2.5A. The full moon disks (the open circles in figure 2.5C) and new moon and surrounding visible last and first crescents (filled symbols) can be slid a bit to the left or right without much adverse effect on any perceived pattern. And, even if we admit these bones are not tools but arithmetic devices instead, why assume that only moons, days, or even time are being tallied? True calendrical tallies do have a repetitive nature, but wouldn't a cave dweller be more likely to count his kill or an inventory of weapons—items of more immediate concern to a Magdalenian tribe out fending for itself? Fitting an artifact to our impression of who made it is one of the most speculative areas in the discipline of archaeology.

In northwestern Mexico and at a much later period (2000–3000 B.C.), we find carved on stone similar patterns that can be interpreted as lunar calendrical tallies. These appear alongside "lists" of weapons and kills. The Presa de la Mula stone (figure 2.6) is a particularly interesting example. The record is carved on a 1-by-3-meter vertical slab located at the top of a promontory near Monterrey, Mexico. As Breen Murray's analysis has shown,[13] the horizontal series of linear arrays displays numerical properties of a characteristically lunar variety. Several dot-and-line grids appear to register not only the standard lunar month ranging between 27 and 30 days, but also multiples of that period. In several instances, the number 207 crops up; if each stroke mark represents a day, the total duration is just equal to 7 lunar months, a length that could represent the seasonal occupation of the area by the tribe; or, less likely, it could even be a lunar semester employed to predict eclipses.

FIGURE 2.6 A petroglyph or rock carving, from Presa de la Mula near Monterrey in northern Mexico, that may be an early calendar. Like the carved bone from Neolithic France, the numerical divisions of stroke marks, etched on the large stone slab over two thousand years ago by local hunter-gatherers, match the sequence of lunar phases. SOURCE: Courtesy W. B. Murray, Universidad de Monterrey, Mexico.

These old bones and stones may be time's earliest foundation documents—concrete evidence of the ingenious attempts of early humans to keep track of celestial rhythmicity through a written medium. Still, is the calendar really in the eye of the beholder? Is the 20,000-year-old bone from the Dordogne really our first history book—a tablet of time-factored information about the life of an ancient Ice-Age people, or is it a mere tool sharpener so full of slash and gouge marks that it was on the verge of being tossed away? How can we ever know? The further back in time we try to search out an example of a tangible calendrical artifact, the more controversial it becomes. Perhaps the most controversial time machine of all appears in figure 2.7.

FIGURE 2.7 Stonehenge. (*A*) View at ground level. Each upright that makes up the megalithic horseshoe weighs more than 30 tons. SOURCE: Photo by the author. (*B*) Plan showing some of the proposed time markers discussed in the text. Copyright © 1982 Hansen Planetarium, Salt Lake City, Utah. Reproduced with permission.

Stonehenge: Calendar and Observatory?

Stonehenge has been called the most extraordinary building in the world and for over five centuries has posed a riddle that has yet to be solved. Stonehenge is a circular structure made up of huge standing stones weighing 25 tons or more called *megaliths*, with "hanging stones," or lintels (*henges*), placed across the tops of certain pairs of them (figure 2.7A). It has stood starkly isolated on the barren Salisbury Plain of southern Great Britain for at least five thousand years. The archaeologists can tell us who built it and even when it was altered and modified several times over a period of two thousand years. The question that continues to intrigue us is why was it built? Was it a temple, a burial ground, a place of assembly for Bronze Age hunter-gatherers, a defensive structure? It is surrounded by a ditch outside of which rises a high bank. Or was it an observatory that the tribe employed as a kind of primitive timepiece to coordinate group activities—an unwritten calendar made out of stone?

Of all the ideas that have been put forth to explain Britain's famous prehistoric earthwork, the hypothesis that it was a calendar and an observatory is by far the most popular as well as the most disputed. The main axis of the structure is a dead giveaway. Built early on in the plan, the broad avenue that leads out of the horseshoe-shaped arrangement of the inner five henged stones, or *trilithons*, themselves laid down centuries later, is aligned toward the rising sun on the first day of summer. This may have been the beginning of an ancient tradition of marking the beginning of the seasonal cycle by registering the place of the sun at its greatest northerly extension on the horizon—a tradition preserved throughout history, for even today at Stonehenge the sun does not fail to keep its appointment (figure 2.7B, alignment A).

Suppose we take a series of rapid-fire snapshots of a swinging pendulum as it reaches the top of its swing. First, it begins to slow down gradually, then it comes to a complete standstill just before reversing its direction. This is exactly what the rising and setting sun does on its annual cycle. As viewed along the horizon, it executes a seasonal turnabout. In fact, the term *solstice*, which defines the first day of summer, means "sun-stand." It has been argued that the architectural clockwork of Stonehenge marks both standstill positions; the other occurs six months later in the year when the sun reaches its maximum southerly extreme on the horizon; the British call it midwinter's as opposed to midsummer's day. The astronomer Gerald Hawkins, who studied the ruins closely in the early 1960s, claims to have unearthed at Stonehenge other cyclic celestial phenomena that the builders deliberately encoded into the architecture.[14] For example, consider the alignments among the station stones (C), four large Sarsen boulders (named by archeologists after the region 20 miles north of the site from which the raw materials were hauled); these mark the corners of a perfect rectangle. Hawkins has argued that the edges, along with other directions (D, E, F, G) point to a continuation of the solar standstills and the standstill positions of the full moon on the horizon—locations it reaches every nineteen years rather than annually.

Stonehenge is both an observatory and a calendar for charting the movements of the sun and the moon: an *observatory* because someone standing in the correct place within the stone circle could witness the time-marking events directly; a *calendar* because—like notches on a bone or marks on parchment—standing stones could be thought, by an intelligent observer who knows the code, to constitute an unwritten, yet indelible record of the flow of events. With some experience, an attentive user could have employed this monumental timepiece to predict future events. Thus when the sun passed its southerly standstill position and was first spotted in its trilithon archway (alignment B), winter's cold blasts soon would sweep across Salisbury Plain; when the sun reverted to its position along the avenue at the opposite extreme, it would be the best time to plant. A more perspicacious observer could probe nature even more deeply: for example, when the full moon rose down the main avenue (A), then an eclipse of the moon would shortly follow. Some modern decoders of Stonehenge go even further: if you tally the number of *Aubrey holes*, a circle of evenly spaced chalk-covered pits located between the outermost circle and the ditch-and-bank structure, you get the number 56. This count of 56 has been used to model a computing device to predict eclipses that actually occur at Stonehenge.[15]

Critics of those who see one of Europe's oldest monuments as a scientific device—observatory and computer—argue that the calendrical approach lacks historical support. Given what we know of them, the Neolithic farmers who lived through Europe during the Bronze Age could hardly have been interested in accurate timekeeping. Archaeologists believe that Wessex, the area where

Stonehenge is located, may have developed by the fourth millennium B.C. into a set of chiefdoms with a definite social hierarchy, specialization of crafts, trade, and a religion incorporating ceremonials—all focusing on a single ruler of considerable prestige. In this social scenario, the monuments might have functioned more reasonably as territorial markers as well as centers in which rituals were enacted. If there had existed a federation of chiefdoms under one superchief, Stonehenge, being by far the largest of all the complexes, may well have been their supercenter.

The stone circle and giant causeway of Stonehenge may have been built as part of the ritual complex; maybe an approximate marker to the midsummer sun figured somewhere in its builders' religion. We know that early people throughout northern Europe erected colossal tumuli and stone rings; and there is sound evidence that early graves and tomb axes were preferentially aligned to the sun, likely as part of a sun-worshiping cult. But eclipses, precise time markers, 56-year cycles do not square with the historical and archaeological record. These people were surely less scientific, more ritualistic.

But if Stonehenge works as a computer, must it not have been designed as one? For the hard-scientist looking at ancient man, whether something "works" today often seems to constitute the ultimate test to determine the motive of the builder. One investigator stated that if he could see any alignment in the various parts of Stonehenge, then that fact would have been known to the builders.[16] But, in fact, the issues of original purpose and demonstrable function are separable. If I sail across the ocean from Asia to South America on a raft of reeds, I will have succeeded in proving that it was possible for our ancestors to have done so—an accomplishment that, taken by itself, does not prove that ancient Asiatic people actually did it. Likewise, or, if I use the tip of the knife I buttered my toast with this morning to tighten a loose hinge on the screen door of my study, it is not necessary for me to suppose that that knife was deliberately designed for household repair. The extension of the knife's use was purely my own doing—the result of laziness, of my disinclination to prowl about the cellar in search of my stray tool box.

And so Stonehenge, as well as many of the three hundred megalithic circles scattered throughout the British Isles and northern France, *could* have been used as solar and lunar calendars, perhaps even as rudimentary counting devices. Even if it were built at least during some stage of its long evolution with timekeeping in mind, the underlying motive for Stonehenge's construction is likely to have had much more to do with the worship of the gods and the legitimization of priestly power than with the collection, storage, and computation of precise data about the sky that characterizes modern science. Besides, three thousand years is a long time to uphold a tradition. Is there a building anywhere in the world today that still functions exactly as it did at the time of the birth of Christ?

"Every age has the Stonehenge it deserves—or desires."[17] The historian Jacquetta Hawkes's epigram has become almost a cliché to describe the way we shackle ourselves to the present, the way we deny the diversity of past peoples compared with ourselves by garbing our ancestors in our scientific clothing and pushing our ideas and motives into their now-empty heads.

I find it curious that we feel the need to choose between "ritual" and "science," between Stonehenge being a real observatory or just a place with symbolic astronomical orientations. Much of the disagreement across the disciplines of archaeology and astronomy on the uses of Stonehenge lies in the matter of degree. But isn't there a bit of science in a ritual conducted at a time of year signaled by the passage of sunlight down an avenue or through a stone archway? Not precise science like ours, but nonetheless predictive empirical science in the sense that one could formally anticipate the time to conduct the ceremony by tracking the position of sunrise. At the ritualistic level the architecture might have functioned to focus the power of the order of nature upon the re-enactment of some cosmogonic event that was chanted or sung in the local dialect. Just as the singer of songs uses nature's order as metaphor by comparing his love to the unfailing tide that ever swells and diminishes, so, too, the people of Stonehenge may have been striving to capture the rhythm of the sun's movement, fixing attention upon one of its critical turning points as a way of relating some quality of their myth to a phenomenon that would be visible to all true believers. But we cannot get into their heads. We can only speculate about that elusive mythic aspect, for—unlike its concrete representation in the passing of the sun through the archway—it is gone forever, vanished along with the minds of the people who conceived it.

We think of most early societies as adopting mathematical operations, graphic symbols for speech, along with a complex system of government, specialization in craft skills, and sedentary agrarianism

FIGURE 2.8 The first Indo-European writing system. *Left:* Clay tokens of various shapes were used to represent different commodities; often they were perforated so that they could be held together by a string. SOURCE: Courtesy of D. Schmandt-Besserat. *Right:* Later cuneiform writing shows part of a Babylonian list of the times of occurrence of new moons dating from the

fourth century B.C. SOURCE: A. Aaboe, "A Computed List of New Moons for 319 B.C. to 316 B.C. from Babylon: BM 40094," *Det Kongelige Danske Videnskaberres Selskab Matematisk-fysike Madeleleser* 37 (1969; Copenhagen): 3.

(all emphasizing an increasingly complex technology) in order ultimately to reach that condition we call civilized. But the truth is that we likely are not the end product of an unbroken sequence of scientific advancements that began on that plain in Wessex.

The Sumerian Token System

While we stand on shaky, though tantalizing, ground in the attempt to dredge up material proof of ancient timekeeping systems in the carved bones and standing stones of archaic Western Europe, there is no doubt that a decisive primary step in the development of our calendar was taken in Sumeria in the fourth millennium B.C.: that is, the token system which served as the precursor to the cuneiform writing upon which was founded the Babylonian calendar and numerical system. The transformations in this system, which took root in a sedentary, agricultural, and increasingly urbanized environment, are dramatic, well-documented, and thoroughly fascinating (see figure 2.8).

Economics and trade led to the invention of miniature clay forms fashioned into various geometrical shapes such as ovals, cones, pyramids, and triangles, which were used to keep track of different commodities, like oil and grain. Middle East archaeologists have discovered hordes of tokens of increasing complexity that seem to coincide with advancements in farming techniques accompanied by an increased variety of crops and enhanced productivity. According to Denise Schmandt-Besserat, the tokens seem to have functioned as mnemonic devices enabling the village bureaucracy to plan, store, and trade their seasonal yields.[18] They were found in stores of goods—foods, textiles, and other manufactured items in urban workshops—as well as in temples, where they are believed to have played a role in the earliest known system of taxation. In later times, accountants and merchants developed the habit of depositing these tokens in clay envelopes embossed with symbols. These symbols consisted of relatively few shapes but bore a clear relation to the shapes of the tokens inside. In fact, the same pointed stylus used to mark some of the more complex tokens was also used to fashion the roundish and wedge-shaped markings adorning the envelopes. Essentially the incised markings on each envelope told what was inside; the marked envelope literally gave away its contents. Eventually a more efficient and more abstract system of notation was developed, in which a flat clay tablet with wedge-shaped markings impressed upon it replaced the more cumbersome filled clay envelope. Special symbols were introduced for numbers in a hierarchical notation system so that, for example, twenty-two measures of grain could be specified by that number preceded by the symbol for a measure of grain.

This brilliant invention of an early system of writing and numeration did not occur in a flash; the archaeological record tells us that the cuneiform tablet was not conceived as a revolutionary device and adopted immediately by all people. Rather its ultimate success was achieved only gradually. Unmistakably, we owe a share of our tendency to keep efficient records of all quantities, including time, to this reliable and concise form of communication. Later the Greeks adopted their arithmetical and calendrical principles from the cultures of the Middle East, and the Romans, from whom we acquired today's calendar, borrowed in turn from the Greeks.

In this chapter, I have aimed to penetrate the fog of the distant past when our predecessors began overtly to celebrate the rhythms that beat within and about them—to remember those rhythms, to use them, to memorize them, to sing them, to record them, to carve them in stone. We feel we have a pretty fair grasp about what took place in time marking in ancient Sumeria and in the ancient Greek oral tradition. When we step back into the world of Stonehenge or grasp a mesolithic bone, we feel a little less certain about how to interpret what we see. Above all, our interpretations will always be colored by our present beliefs and motives—the most imminent danger when we try to reconstruct our temporal past.

Still, no matter how people choose to reckon time—whether *our* time or *theirs*—the method always reflects the basic periodicities induced upon us by the natural world. Rare is the element in any of the calendars I discuss that does not grow out of the repeatable phenomena of nature's cycles, both physical and biological.

PART II

Our Time: The Imposition of Order

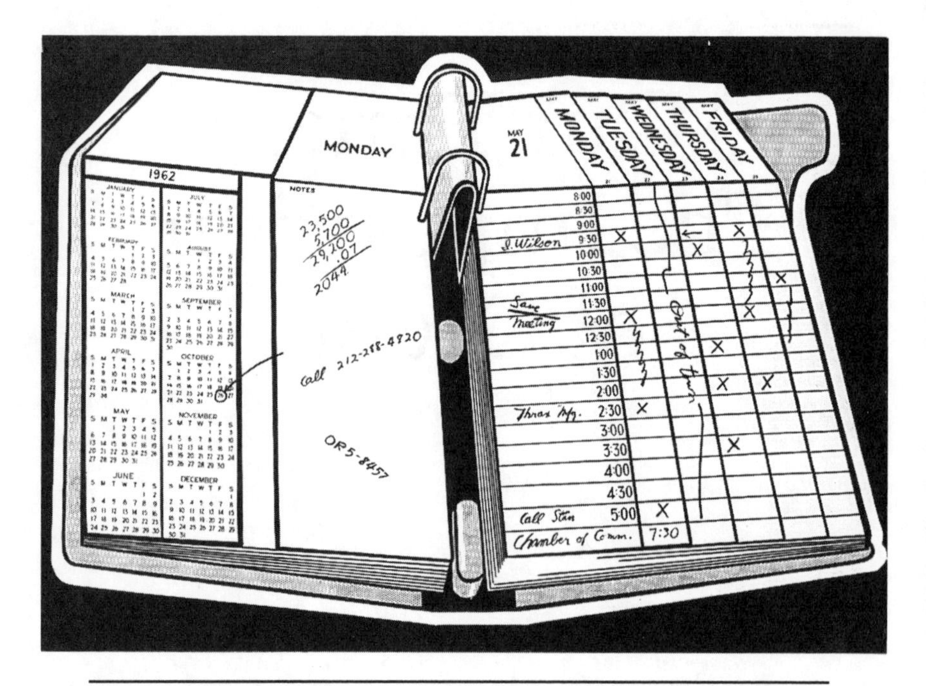

Desk Calendar. Roy Lichtenstein. The Museum of Contemporary Art, Los Angeles: The Panza Collection. Photo by Squidds & Nunns.

3

THE WESTERN CALENDAR

All the world's a stage,
And all the men and women merely players:
They have their exits and their entrances;
And one man in his time plays many parts,
His acts being seven ages. At first the infant,
Mewling and puking in the nurse's arms.
Then the whining school-boy, with his satchel
and shining morning face, creeping like snail
Unwillingly to school. And then the lover,
Sighing like furnace, with a woeful ballad
Made to his mistress' eyebrow. Then a soldier,
Full of strange oaths and bearded like the pard,
Jealous in honor, sudden and quick in quarrel,
Seeking the bubble reputation
Even in the cannon's mouth. And then the justice,
In fair round belly with good capon lined,
With eyes severe and beard of formal cut,
Full of wise saws and modern instances;
And so he plays his part. The sixth age shifts
Into the lean and slipper'd pantaloon,
With spectacles on nose and pouch on side,
His youthful hose, well sav'd, a world too wide
For his shrunk shank; and his big manly voice
Turning again toward childish treble, pipes
And whistles in his sound. Last scene of all,
That ends this strange eventful history,
Is second childishness and mere oblivion,
Sans teeth, sans eyes, sans taste, sans every thing.

SHAKESPEARE, *As You Like It* (II, 7)

TIME—we have socialized it, circularized it and linearized it, artificialized and corrupted it. In attempting to trace the origin of all time concepts, from pico-second to Great Year, that make up today's calendar, I have found it necessary to use these descriptive terms. Our days are numbered. Take down the nearest wall calendar and flip through it. Those 12 pages represent a packet of time, a full seasonal cycle—the period of revolution of the earth about the sun. That dozen pages also represents the time it takes the sun to pass all the way around the celestial background of constellations and to return to the same place among the stars; it is also a close approximation to the time it takes the daily sunrises and sunsets to make a complete oscillation from one extreme along the horizon to the other and back again—essentially what the builders of Stonehenge were supposed to have been marking. That packet of pages we call the calendar is, like the henges on Salisbury Plain, a device invented by our ancestors to subdivide the course of the sun.

Each of the 12 calendar pages, representing a month of the year, is compartmentalized into little square blocks that number 30 or 31, except for the deviant second page. While the year is the course of the sun, the month belongs to the moon, being related to the period of revolution of the moon around the earth relative to the position of the sun—another phenomenon we don't see.

There is nothing in the universe quite like the moon. If we were freed from the fetters of indoor lighting, telescopes, and abstract scientific explanations, we all might be a little more familiar with its apparent behavior, as was Hesiod or the people who fashioned Stonehenge or *Enuma Elish*. At the beginning of its cycle, the moon is a thin, sharp-horned crescent suspended above the western horizon, its cusps always pointing away from the sun, which has already set. On succeeding twilights, it waxes, or grows, appearing at sunset farther away toward the east from where it was at the beginning of the month. It passes its quarter phase and takes on that familiar "D shape" it always exhibits when it rides high on the *meridian* (the imaginary north-south line that passes overhead) in the south at sunset. Still later in the month, it enters the *gibbous* phase, when more than half its visible disk is lighted, and can be seen high in the east late in the afternoon. After half a cycle, when it reaches the full phase, the moon is most prominent of all, rising opposite the setting sun, illuminating the sky with a flood of pale yellow light all night long.

Two weeks have passed. Now the moon begins to backtrack in its cycle, its face slowly eroding away in reverse. It passes in reverse through gibbous phase, then back toward quarter as it wanes. When you see the moon in the morning daylight hours, this is usually what it looks like, for, having risen progressively later during the night, it remains in the sky well after the sun has come up. At month's end, the disk is reduced once again to a thin crescent, the vanishing sliver of which can be glimpsed low in the east before sunrise. Then it vanishes altogether for a couple of days as it becomes lost in the glow of the sun's light. When we see no moon, we speak of the "new" moon, but the moon is not visibly renewed until we see its crescent again in the evening twilight.

The whole process takes 29½ days—a *synodic month*. Although the cycle symbolized by any one of the calendar's 12 pages was intended originally to follow the moon's phases, some of the months we keep are now longer than others and not one matches the actual synodic lunar rhythm one can easily observe in nature. This discrepancy is a result of the convoluted, highly complex amalgam of motives that led to the formation of our calendar. Implicit in it are not only nature's rhythms but also religion, politics, and human intrigue. It is a safe bet that if travelers from another world tried to decipher our calendar, they would not stand the slightest chance of putting all the intricate pieces of the time puzzle together, however superior their intelligence.

The wall calendar contains both past and future. The days of a lunar cycle gone by, days beyond our reach except in memory stand conveniently blocked out so that one may recollect and reckon back to where one was or what one was doing at a particular point in that last lunar cycle. Next to the month past lies the month of the future—that which has yet to happen, time unreckoned, time to anticipate, time that has never existed. The unmarked calendar on my wall implies that the two are the same, that past and future do not really differ at all. But I can determine the truth or falsity of that statement when the moon recycles and I turn the page.

Any page of the wall calendar will reveal another subdivision to the time units Westerners have created: seven vertical divisions to the sequence of numbered blocks, each column with its own name. The four or five horizontal bands, called *weeks,* into which we group the moon's days constitute a peculiar time division. There is no single celestial body such as sun or moon, no obvious natural cycle to which we can directly attribute this little packet of time. Moreover, many other cultures and traditions also tally an interval of about the same general duration in their calendars.

Finally, there is the day—the simplest and most obvious unit of time. There is a number for each one. Today is the 21st; tomorrow, the 22nd. Change numbers at midnight, and you have measured out one complete rotation of the earth on its axis—but once again that is not what our senses detect. We witness, rather, a two-part cycle, the first of which is marked by the passage of the sun across the sky from horizon to horizon, followed by an extended period of darkness, after which the sun returns to approximately the same place. (Recall that the sunrise position varies from day to day.) The day is thus really an abbreviated solar rhythmic cycle that provides a counterpoint to the longer sun cycle of the year, with the moon's beat lying in between.

Circadian, circ-annual, circa-lunar—to use the biologist's terms—are the principal cycles that make up human time. We in the West have re-envisioned on sheets of paper the very intervals to which our environment has already entrained us. No one willfully set out to create these intervals. The desire had always been to reach for the future, to set up a date-reaching scheme—a plan or model for describing patterning in the passage of events which might help people to know when in a series of measured intervals a particular event would take place. The fabrication of such a model depended upon knowledge of when a similar event had taken place in the past. Man-made calendars became both past and future time-marking devices. Still, Westerners have fashioned artificial divisions of time out of rhythms derived from living within nature's environment. We have not been content to allow ourselves to float about aimlessly on the waves of nature's ocean of time. We want to count the waves, to classify them, to package them neatly into recognizable patterns. Now, 20,000 years after having picked up a fragment of bone in one hand and a piece of flint in the other, perhaps to assist our memories, we have begun to control those waves—even to make waves of our own. We have done far more than count the moons and the suns.

On your wrist, you may see yet another mechanism we have made to control both the passage of events and the rhythm of our behavior. This device's very shape conveys the notion of the daily round of time and the natural form of the sky—unless you happen to be wearing a stylized rectangular model with distorted numerals, or a digital without hands.

Wristwatches partition time still further—the day into hours, 12 of them read twice over. The little hand is the last remnant of any direct contact between the day as we know it and the place of the sun in the sky, a pointer that shows where the sun is situated—east in the

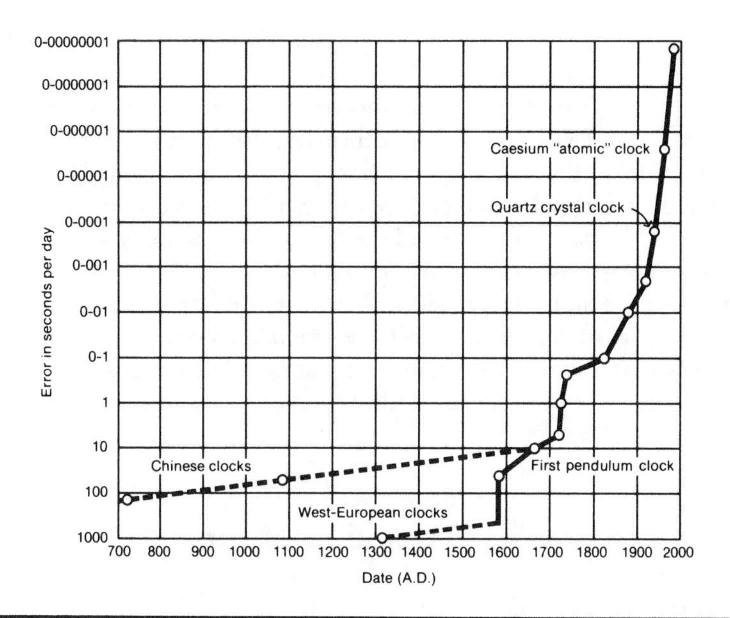

FIGURE 3.1 The progress of precision in time measurement. This plot shows the increase in the accuracy of mechanical clocks from the time of their invention up to the present. SOURCE: An updated redrawing of a graph in H. A. Lloyd, "Time-Keepers—An Historical Sketch," in *The Voices of Time*, J. T. Fraser, ed. (Amherst, Mass.: University of Massachusetts Press, 1981), fig. 3, p. 400. Copyright © 1981 by J. T. Fraser.

morning, when it points left of the 12; west in the afternoon, when it points to the right; and straight up at noontime. We also divide the hours into sixtieths called *minutes* (not tenths or hundredths the way we subdivide most things) and the minutes into sixtieths named *seconds.* Recent preoccupation with time-setting records in Olympic competition has led to watches that read to hundredths, even thousandths (not sixtieths or thirty-six hundredths), of a second. In scientific laboratories the divisions proceed to *micro* (millionths) and *pico* (billionths), even *femto* (quadrillionths) seconds. Paradoxically, we seem to have expended considerable effort trying to measure an entity we have difficulty defining (see figure 3.1). Recently Tiffany's in New York displayed a Patek Philippe Swiss watch said to be the most complicated in the world. Weighing 2.4 pounds, held together by 332 screws and exhibiting 24 hands, it performs more than three dozen different chores—among them the calculation of Easter Sunday, the times of sunrise and sunset, and the orientation of the Milky Way in the night sky. Value: $3,000,000 to $6,000,000.[1]

At the *largo* end of the time spectrum, our imagination is equally boundless. Our years grow into centuries, the centuries into millennia. The longer the time cycle, the less precisely is it defined. We have invented ages, epochs, and eras. Many non-Western calendars such as that of the Maya of Yucatan and the East Indians employ similar concepts. Scientists own the record at both ends of time's spectrum. Modern geologists and astrophysicists have conjured up the longest intervals: the lines of demarcation between them are unevenly spaced time markers that seem to be significant along the course of evolution of life, the solar system, the Milky Way galaxy, and even the entire universe.

The Day and Its Hours

Let us begin in the middle with the day and its parts, and then work up the hierarchy of cycles from short to long, concentrating on how time's parts were both conceived in our ancestors' heads and brought together. There is no need to speculate about where the day cycle came from. It is the most basic of all natural periods—the one responsible for the circadian cycles found in practically all living things. Behaviorally, little influences our lives more deeply than the variation of day and night, light and dark, sleep and wakefulness. What seems problematic about the diurnal period is where to begin and end it and how to split it up. In most cases our ancient predecessors indicated the time of day by concrete phenomena rather than by mechanical abstract devices, like marks or numbers, whose operation has nothing to do with the defining event. To express a particular "hour" of the day, one would simply point to the place in the sky where the sun could be found at that time. This is a common technique. The Konso of central Africa, for example, reckon the parts of their daylight hours approximately as follows—all of them by pointing:

9 A.M.–11 A.M.	gudada
11 A.M.–2 P.M.	guuda' guta
2 P.M.–4 P.M.	kalagalla
4 P.M.–5 P.M.	harsheda akalgalla
5 P.M.–6 P.M.	kakalseema
6 P.M.–7 P.M.	dumateta
7 P.M.–8 P.M.	shisheeba

Each interval is not numbered but, rather, is named after the activity that takes place at that time of day. Later intervals are compressed because, the people say, this is the most important time of the day for performing activities. Thus, the name of the 5-to-6 P.M. interval means "when the cattle return home." Those longer intervals during the heat of the day reflect periods of relative inactivity. While such vaguely defined boundaries and durations are in stark contrast with our regimented modern way of precisely dividing up the day, the human hour-clock is more accurate than we technology-dependent moderns might think. Though simple morning, noon, and afternoon can easily be sorted out simply by observing which side of the sky the sun is in, even 8- or 10-, perhaps even 12-, fold divisions of the day can be reckoned unambiguously simply by gesturing with the arms. Were an hour's accuracy all we ever needed, the entire world might be completely devoid of clocks today. More than accuracy, the desire for uniformity affected our Babylonian ancestors' rational way of marking time.

The invention of the simple technology of shoving a stick vertically into the ground signaled the first break away from nature which culminated in our uniform timekeeping system. The times of sunrise and sunset usually designated as beginning and ending the day are variable, and so are the lengths of daylight and nighttime hours. The partitioning of day and night into 24 hours, another human imposition upon nature, came perhaps from the division of the zodiac into 12 equal parts, each one marked by a constellation through which the sun passed over the course of one moon cycle. Because it takes the sun approximately 360 days to make a complete annual circuit through the stars, an easy system for dividing up seasonal time seems to have suggested itself: that is, intervals divisible by 6 and 12. Thus, the sexagesimal notation came to be a part of time reckoning, with 60 minutes to the hour and 60 seconds per minute, 12 hours per day and night, and so on. Spatially, the circle representing the round of the sun on its celestial course was also divided into 360 degrees. By Roman times, day and night were joined as a unit that began and ended at midnight, a more convenient time to make the transition, at least in the business world; even though midnight was a more abstract point in time than the distinctly visible sunset or sunrise. Still, the hours reckoned by the sun's shadow casting were unequal. To circumvent these nonuniformities, the ancient Babylonians invented artificial hours, which they defined as one twenty-fourth part of the day. Like the beginning and ending of the Roman day, such hours had nothing to do with the position and rate of movement of the sun across the sky, but at least they were equal and rational; they were fair. This artificial system of time units was not practical, however, and did not achieve widespread use until the end of the Middle Ages with the glorious invention of the mechanical clock (see figure 3.2C).

In the developed state societies of Europe and America, public or shared time was the general rule. In the Roman forum, a timekeeper shouted out the noon hour as the sun passed between the Rostrum and the Grecostasis, two of its most prominent buildings. His counterpart in the ancient Aztec capital of Tenochtitlan (now Mexico City) stood in a round temple in the plaza fronting the great Templo Mayor, and announced to all citizens the opening and closing of the market, the departure of the warriors, the beginning of the games, the time to worship. Organized society demands a highly regulated daytime period in order to coordinate its principal activities. In ancient Greece and Rome, this perceived pattern of solar angles called hours was transformed into a series of crude public sundials (figure 3.2A)—objective and impersonal entities at which people could steal a glance to help keep themselves on track. (Compare the sundial from Indonesia in figure 3.2B.)

The science historian Derek Price characterizes the earliest Greek water clocks and dials as "simulacra," or simulations of how things work.[2] The original intention of the builders seems to have been the aesthetic satisfaction derived from making a device that imitated the heavens. Our first real clocks, such as the Tower of the Winds in Athens, were models of the cosmos composed and constructed to glorify the beauty and simplicity of heavenly motion, not necessarily to satisfy an appetite for precision.

Religious demands, more than the desire for quantitative science, played a major role in the increased precision that threads through much of the history of timekeeping in our culture. The Christian way demanded a controlled, disciplined life; and in the Christian Middle Ages of Western Europe, it was the strict rule that the call to prayer, particularly in the monasteries, be made at the correct time of day. In fact, the mechanical clock was the *result* of an interest in measuring time, not the *origin* of that interest. Like many artifacts emanating from Christian worship in that age of religious preoccupation, the timepiece of the Middle Ages developed into a work of beauty and complexity. Clocks became showpieces. The one-hand version used to mark only the hour gave rise to two hands, as the hours and their parts chimed, gonged, and clattered their way into our cycle of daily activities, which now began to be submerged in the noise of still other machinery.

Our medieval ancestors devised other clever ways to mark time: clocks that run by gravity as water drips out of a filled tank through a tiny hole and into the base of a calibrated tank (figure 3.2C); slow-burning candles with the hours ticked off in bright colors on the banded wax; and the familiar ship's clock of sand falling from one globe to another through a tiny hole, also gravity-operated like the water clock.

"Time is money." As the secular dictates of production and trade in an expanding, interlocking market economy began to rival and surpass the sacred demands religion had placed upon timekeeping, each of these types of clock was found to have its problems. Gravity clocks start out faster, but then run more slowly; sundials do not keep equal hours, because the shadow lengthens more rapidly in late afternoon. And even if one adjusts the markers to account for the non-uniformity of "shadow time," the hours differ in different parts of the year and in different locations within the state. Should people who work shifts of different lengths be given the same pay? Why should one person be required to start work earlier than another? Or stay later? Is it possible to adopt a reliable system of uniform hours? These concerns propelled pre-Renaissance Europe farther and faster along the road of artificial mechanical timekeeping which today has come to span all units of our calendar, large and small.

By the late Middle Ages, a new kind of public time system had been created. Born out of religion, commerce, the rise of towns, and the bureaucracy that accompanied them, the "work clock" slowly began to dominate life in the thickly settled European urban communities of the twelfth and thirteenth centuries. Bells of different pitch and duration pealed from the turret that marked the center of town. They tolled when to begin and end work, when to open and close the market, when to call the people to assembly or to begin the curfew. Only in remote rural areas did folks still keep time by the natural course of the heavens, which served well enough to organize a set of daily farm activities such as milking time, meal time, market time—the way it was done in Hesiod's day or by the Konso in Africa today. This new way of marking time was automated and unnatural; it was

FIGURE 3.2 Time is motion. Some early mechanical ways of keeping track of time: (*A*) This Roman sundial from the ruins of Pompeii is an inverse map of the heavenly sphere. The shadow of the pointer moves over the marked or unnumbered bowl-shaped surface to give the hour of the day and the day of the year. SOURCE: Courtesy S. L. Gibbs. (*B*) A sundial from Java that divides years into months. The length of the month is measured by equal spaces through which passes the shadow of the gnomon cast by the noontime sun; this scheme results in unequal time intervals for the months. SOURCE: Drawing by J. Meyerson. (*C*) Depiction of a waterclock (clepsydra) in a medieval manuscript. Most waterclocks were powered by a falling float that sat in a container out of which water dripped; the float was attached to a pointer that moved over a calibrated surface. In this painting, the prophet Isaiah is shown intervening with the timekeeping process. He sets back the clock to give King Hezekiah more time to recover from his illness so that he can put his house in order: "And Isaiah the prophet cried to the Lord; and he brought the shadow back ten steps, by which the sun had declined on the dial of Ahaz" (2 Kings 20: 11). SOURCE: J. Needham, *Clerks and Craftsmen in China and the West* (New York: Cambridge University Press, 1970), fig. 78. Reprinted with the permission of Cambridge University Press. (*D*) Engraving of an early-eighteenth-century time-ball apparatus from Portsmouth, England. The dropping of the ball, usually at high noon, served as a visual time signal by which the chronometers of ships in port could be set. SOURCE: *New Philosophical Journal* (Edinburgh) 8 (1830): 290. Courtesy, editor, *Journal for the History of Astronomy.*

A

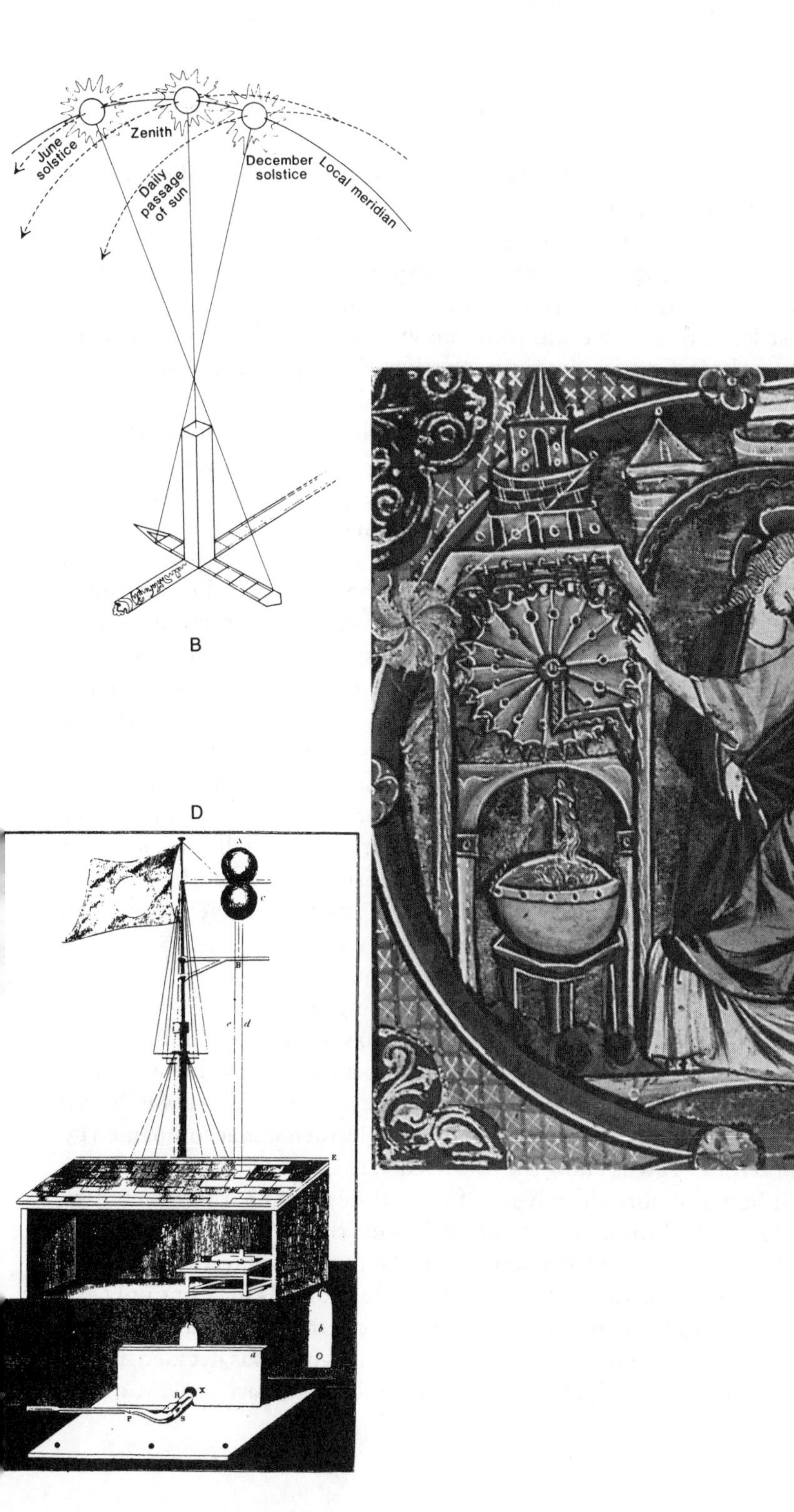
June solstice
Zenith
December solstice
Local meridian
Daily passage of sun
B
D
C

based on reason. Time had been extracted from nature and molded to fit religious and bureaucratic uses.

After the Middle Ages, Europeans moved faster toward harnessing the day. An age of discovery and exploration necessitated traveling vast distances, especially in a westerly direction. To navigate, one must keep fairly accurate time: moving across different longitudes implies keeping different times, for there is a slippage of suntime with longitude. When the rising sun is a handspan above the horizon in New York, it lies on the horizon in Chicago. At the same time in Denver, twilight has barely begun, while it is still pitch-dark in San Francisco. To chart their courses and keep track of the time, early explorers carried a ship's *chronometer*—essentially a well-balanced, accurate clock. The same requirement for a reliable, independent, and accurate means by which to track time held true for the tradesmen who journeyed with their wares between Venice, Munich, and Basel, each city having kept its own separate system of hours. Because businessmen were constantly changing their clocks, a set of conversion tables became an absolute necessity so that one could tell which city kept what system of hours, who began the day at noon, who at midnight, who at sunrise or sunset. (Noon started the day in Basel, but they called it one o'clock!) Up until the twentieth century, every respectable harbor city had its own time ball that fell once a day at noon to send ocean navigators a precise visual time signal by which to adjust their chronometers (figure 3.2D). (The famous New Year's ball in Manhattan's Times Square is a distinct survivor.)

The advent of railroads in Europe and the United States further quickened the pace of precision timekeeping and placed a heavier burden on the clockmaker who now needed to serve the demands of timetables accurate to the minute. Goods, especially perishable ones, had to arrive and depart on time so as not to be left to mold in a warehouse or to run up a huge bill. The technological burden was accompanied by a social one. Towns along the line needed to agree on a system of standard hours. Before the advent of zone time in 1883, there were two kinds of travel time: "railroad time" inside the train, and "local time" in the towns outside. The latter differed from town to town, for at a given time the angle of the sun from the meridian is the same only at a given longitude. Step one pace east or west of that line, and the natural hour changes. Even noon and midnight change. It was decided, therefore that everyone lying within a fixed distance east or west of the nearest whole multiple of 15 degrees of longitude would keep time by that parallel. Thus, the eastern standard time zone is defined (except for a small correction) by the sun's angle with respect to the meridian 75 degrees west of Greenwich, England; while central standard time is reckoned by the 90 degree meridian, mountain standard time by the 105th, and Pacific standard time by the 120th parallel of longitude. Everyone 7½ degrees either side of these standard meridians keeps the time of that meridian, unless the line of demarcation bisects a heavily populated area, in which case it is shifted somewhat to avoid confusion. During the portion of the year when we keep daylight saving time, we simply make an instantaneous leap to the time zone directly east of us. Then, everything, including the setting of the sun, happens an hour later by clock time, giving us the illusion of having more daylight. At the other end of the cycle, the sun also rises an hour later—no real problem for most of us but an annoyance to early-rising farmers, who often object to legislation to extend daylight savings time over a wider portion of the year.

Today we live by legalized time. The continuous time differential experienced in nature has, for the sake of convenience, become discontinuous and partitioned as we move from coast to coast. On an afternoon-long intercontinental flight, it is easier to change my watch three times, in each instance either tacking on or knocking off a huge chunk of time, rather than speeding the watch up or slowing it down in a continuous manner (changing it at the rate of 1 minute for every 14 miles traveled in an east-west direction), but that is what would be required if we were to remain subject to natural time.

There is no choice: if you want uniform hours in the day, then you must cast your eyes away from the sun in the sky, and so in the eighteenth century, astronomers banished the *apparent sun* and put in its place the *mean sun*. This rational yet fictive sun moves at precisely the correct rate to compensate for the nonuniform rate of movement of that visible yellow disk our ancestors once worshiped. This fictive sun and its time, which we keep on our machines, gives us the equal hours and days we cannot seem to do without.

The Smallest Units of Time

Our smaller subdivisions of time are even more rationalistic than nationalistic when compared with the larger ones. While the clock's dial still tell us the day—the very word *dial* is named for the day—the minute and the second are total figments of our imagination. If you tune your all-band radio to WWV, the radio station of the National Bureau of Standards (2.5, 5.0, 10, 15, 20 megahertz), you can hear the seconds being ticked off officially, 86,400 of them to every earth rotation—just as the medieval clock tower in Augsburg or London chimed out bigger time bytes to our predecessors. The really persistent listener would note days, like 31 December 1987, when an 86,401st second was added, for now even the once-supposed constancy of the rate of the earth's rotation, the very day itself, has become anachronistic. Its variable length is too irregular for our needs (it changes by up to 1/1,000 of a second in some years); worse still, we cannot anticipate with reasonable certainty the rate of change of day length, though we do have some idea why the earth slows down and speeds up by almost imperceptible amounts. One cause is surely the drag effect on rotation produced by the earth's atmosphere. The same kind of friction occurs at the contact between the ocean and the surface of the solid earth—literally, the rubbing of the ocean against the solid earth. This effect is particularly pronounced in shallow bays where the tide breaking against the shore dissipates vast quantities of energy—millions of horsepower in the Bay of Fundy in Canada alone, much of it derived from the energy of rotation of the earth. Similar energy dissipation occurs at the contact points between different layers in the earth's interior. It is as if a huge brake were applied to a turning wheel, in this case the wheel of the days. Ultimately they will become longer and longer as the rotating earth clock runs down.

The modern quest for precision in timing the parts of the day has achieved revolutionary proportions late in the twentieth century. Once the clocks we were capable of building began to surpass the accuracy with which we could measure the earth's rotation, we were compelled to seek a more reliable standard. The vibration rate of a cesium atomic resonator was first installed as an alternative time standard in the 1950s. This new "pendulum" measures out how long it takes an electron of an atom to pass from one energy state to another. But because no one can see individual atoms, such oscillations are measured only collectively and statistically. Furthermore, energy levels, even electrons and atoms themselves, are the creations of the human mind: they are microscopic metaphors we devise as a way of explaining the microcosmic behavior of nature. Thus, when we look at a line in the emission spectrum of cesium atoms, we say that a piece of matter locked away in a box behaves *as if* it consists of atoms, surrounded by electrons, certain ones of which are rapidly making the transition from one atomic orbit to another, exchanging energy with the environment as they go. When we hook up other black boxes to the one containing the cesium oscillator, we are able to tap that energy—to count the transitions or oscillations made by these moving electrons. And when we do so, we find that 9,192,631,770 of them can be used to define the international second. We do it with the obsessive accuracy of better than one part in a hundred trillion. (Expressed in spatial terms, this would be like knowing the distance to the moon to less than the width of a human hair!) Heralded as a triumph of watchmaking over astronomy, the "atomic second" became the official world time standard in 1967. It is this time unit that we usefully divide into milli-, micro-, pico-, and femto-seconds. Now I can replace my long-winded statement with a more meaningful one: the cesium atom oscillates between the two hyperfine levels of its ground state in 9.192,631,770 picoseconds.

Recently the concept of the natural oscillation of microscopic matter as a time standard has directly entered the world of everyday public time in the form of the digital watch. Here the oscillators are not individual atoms but quartz crystals energized by batteries. Typically these crystals vibrate many thousands of times a second with a remarkably steady rhythm because of their stable molecular structure. The vibrations are monitored, stored, counted, and packaged into different units of time through microchip circuitry, and then fed into liquid crystal display units that give time, day, date, and month, as well as a lot of other information of dubious digital necessity. I recently saw a watch on sale that offered times around the world, stored up to a dozen addresses and phone numbers, and intoned a diverse array of musical signals—all for only $19.95! It seems astonishing that all these feats can be accomplished without moving mechanical parts, thanks to human technological ingenuity. Our quest for the precise time of day may go down in history as the greatest obsession of the twentieth century. Originally motivated by religion and business, rather than pure scientific concerns, our quest for the ideal model, the standard upon which to construct a rational way of marking time, has carried us from the vast cosmos down to the abstract invisible world of the atom. For the genius of our precision, we have paid a dear price. An objective outside observer might say that in the modern world time controls us, rather than the other way around. Consider what the anthropologist E. E. Evans-Pritchard once said of the Nuer tribe of central Africa:

> I do not think that they ever experience the same feeling of fighting against time or having to coordinate activities with an abstract passage of time, because their points of reference are mainly the activities themselves, which are generally of a leisurely character. Events follow a logical order, but they are not controlled by an abstract system, there being no autonomous points of reference to which activities have to conform with precision.[3]

The Week and Its Days

After the day, the week—represented by the horizontal bars on most calendars—is among the most recognizable and commonly used parcels of time and also one of the most convenient. The week seems to have developed because it was simply a sensible way of dividing up the days of the month; and of all longer time periods, it is among the few that does not vary in length. Some biologists believe the week is self-determined. The 7-day biorhythm in the human body is one of the recent discoveries of modern chronobiology. It manifests itself in the form of small variations in blood pressure and heartbeat as well as response to infection and even organ transplant: for example, the probability of rejection of certain organs is now known to peak at weekly intervals following an implant.

We are not unique in broadcasting this beat: even simple organisms, down to bacteria and one-celled animals, seem to share it with us. There is, for example, a 7-day rhythm in the *mermaid's wineglass*, a species of algae whose configuration resembles a champagne glass with a long stem and a large flowery globe at the end. This organism can be entrained to reduce its rate of growth only when exposed to an alternating light-dark period of 7 days—no more, no less.

Does social time *entrain* biological time? We might be able to connect the faint *circa-septan*, or 7-day, periods in our biological makeup to the week cycle upon which economically motivated human beings thrive. However, we know that the Romans, to whom we owe most of our temporal habits, worked on an 8-day cycle, the last day of which was a market day. Similarly, our word *sabbath* comes from the Jewish concept of the periodic recurrence of a day of abstention from work in the cycle—the seventh day—which the Jews gave over to the worship of the deity. So important was this round of the calendar that the most famous creation story in Genesis was built around the everyday structure of social time. The creation lasts 7 days—not 3 days or 6 months. Parallels between human origin and the cycle of social life experienced by ordinary people served to emphasize the hierarchical order depicted in the creation story. The week may not be a memorial of the creation of the world; instead, the creation is a mnemonic of the work week.

There are other reasons to think that this work-rest cycle of several days' duration is natural to the human condition. Most cultures with a developed system of trade and commerce employ a cycle something like it. Five-day market weeks are kept in parts of Africa and Central America. The Inca employed 8-day weeks, at the end of each of which the king changed wives. Villagers and field workers, having spent 8 days in their fields, would come to market on the ninth day. Short market weeks of 3 or 4 days were known among the Bantu of Angola in the last century; then there are more lengthy artificial periods of time, like the 16-day market week of the Yoruba people of the coast of West Africa and the well-known 20-day week cycle employed throughout ancient Mesoamerica (which I will deal with in detail in chapter 6). The Kedangese of eastern Indonesia are the creative record setters when it comes to fabricating week cycles. They have ten kinds ranging in length from 1 to 10 days, each with its own set of names, all running at the same time. Furthermore, they seem to be able to calculate in their heads when every conceivable combination will recur in the maze of cycles that eternally preoccupies them.

Verses in the *Enuma Elish* seem to tie the week to the moon. After Marduk creates the "ornament of the night," he commands her "to make known the days" by going forth with a tiara:

> At the beginning of the month, namely, of the rising the land,
> Thou shalt shine with hours to make known 6 days.
>
> (15–16)

On the seventh day "with half a tiara"; or, according to another translation, "On the 7th day halve thy disk."[4] The implication seems to be that the 7-day cycle is a fair equivalent to a readily visible celestial phenomenon—when the moon passes from first visibility with its tiara-like crescent, to the familiar and well-defined "D-shape" quarter; then, though the *Enuma Elish* does not specifically state it, the next 7-day period might be associated with the time between quarter and full; and so on. The whole month of the phases is conveniently divisible into four quarterly periods in which the number 7 predominates. The British term *fortnight,* which stands for 2 weeks (14 nights), may have been a spinoff of this early relationship between the week and the month, an interval splitting the month into its pair of familiar half-cycles from new to full and from full to new moon. We cannot, however, be sure that the recognition of the period of the moon's phases came before the creation of the septenary cycle. (The moon's phase cycle is 29.5 days and not simply 4 times 7, so the fit is not as precise as we might like.)

If the moon does not lie at the foundation of the week, then perhaps the planets do. There is more than a hint of a planetary origin in the mere listing of the names of the days. The following table shows the Anglo-Saxon and English equivalents for each of the Latin named divinities in the day sequence:

LATIN	*ANGLO-SAXON*	*ENGLISH*
Dies Solis	Sun's day	Sunday
Dies Lunae	Moon's day	Monday
Dies Martis	Tiw's day	Tuesday
Dies Mercurii	Woden's day	Wednesday
Dies Jovis	Thor's day	Thursday
Dies Veneris	Frigg's day	Friday
Dies Saturni	Seterne's day	Saturday

Obvious in this list are the sun and the moon; less so, the match-up between Jove, also known as Jupiter, and his Nordic counterpart, Thor. The planet Saturn appears in the list as well. In fact all celestial bodies that move regularly through the zodiac and were known before the invention of the telescope appear in the 7-day list, though arranged in an order that makes little sense to us: Sun, Moon, Mars, Mercury, Jupiter, Venus, and Saturn. Today we line up the planets in the order of their distance from the sun—Mercury, Venus, Earth, Mars, and so on—but that was not what mattered in Babylonian astronomy. And the Babylonians—indeed, everybody in the world before the Renaissance, as far as we know—did not conceive of the sun-centered system that gives rise to the modern way we in the West learn to list the planets. For them a more sensible order consisted of listing the celestial wanderers according to *how fast they moved* from one constellation of the zodiac to the next, from the slowest to the fastest: Saturn, Jupiter, Mars, Sun, Venus, Mercury, and Moon. In the more astrologically motivated astronomy of the past, each hour of the 24-hour day was thought to be presided over or ruled by a planet, the order of designation being given by consecutive runs through the 7-day list. A day received its name from the planet that presided over its first hour. Thus, the first hour of the first day was given over to Saturn, and that day was named Saturn's day. On that day Saturn also governed the 8th, the 15th, and the 22nd hours. Counting down the list, the 23rd hour of the first day would belong to Jupiter; the 24th and last to Mars. Therefore, the first hour of the next day would be enjoined to the sun, and the name of the second day of the week would be Sunday. Because 24 divided by 7 leaves a remainder of 3, to finish up the order of the day names in the Babylonian week cycle, all we need do is count off by 3s. Three positions forward from the sun lies the moon; thus we have "moon-day" or Monday as the third in the week cycle. A count of 3 more takes us to Mars, whose day is the equivalent of Tuesday (Mardi Gras means "fat Tuesday"). Next in the list come Mercury's day, Woden's day or Wednesday; then Jupiter (Thursday); and finally Friday, Venus being the equivalent of Frigg or Fria. The cycle is completed when we return to the first hour of the 8th day which, like the first day, turns out to be represented by Saturn. While Saturday was the first day of the week in the world of Exodus, upon their flight from Egypt the Jews made it the last day of the week—out of pure hatred, so it is said, of their oppressors.

While the weekday scheme seems to be of pre-biblical origin and was used extensively throughout the Middle East, it was not formally introduced into our calendar until late in Roman times, not until the reign of the Christian emperor Theodosius late in the fourth century, though it had been initiated a few generations earlier by Constantine. Earlier the pagan Romans had been using an intricate backward counting scheme with a threefold unequal division of time that was unmistakably tied to the month.

The calends, from which we derive our word *calendar*, were the first days of the month, traditionally the time when religious leaders called people together to outline the festal and sacred days to be kept during that month. The ides fell at midmonth, usually the 13th or 15th day; and the nones, on the 9th day before the ides. The 4 to 6 days between calends and nones were named the "days before the nones." The 8 days between the nones and the ides were the "days before the ides." Finally, all the rest were the "days before the calends" and could be as many as 9. This backward way of counting time is similar to our reckoning time before the hour ("a quarter to six") or to the countdown before the launching of a space vehicle.

The scheme for naming the 7 days of the week, as outlined, grew no doubt out of the ancient practice of scientific astrology. The term *scientific* is appropriate here, for astronomers were required to predict precisely where the planets would be in order to prognosticate the course of human history. The influence of the planets on society was thought to be determined by where they lay among the stars, and the predictive power of a planet was tied directly to the "magnitude of its sphere." In other words, a planet's distance from the earth determined the size of its orbit *around the earth*, and that in turn was reflected in its speed. The slowest (Saturn) was thought to be farthest, and the farthest was considered the most powerful—a notion diametrically opposed to present knowledge, the power of gravity weakening dramatically with distance. So strong was this ancient astrological influence that 7 became a sort of magic number. The number of the planets served as the core of an associative principle that had significance for designating other entities: the metals (mercury is quicksilver); trees, plants, and animals (onions and donkeys belong to Saturn!); and, getting back to time, the supposed seven ages of man. In the scheme of the Greek astronomer Ptolemy of Alexandria (*c.* A.D. 150) the human time order is heaven bound, going from the lowest to the highest orbits. The period of infancy up to four years old was ruled by the moon; childhood, or ages four to fourteen, by Mercury; adolescence (fourteen to twenty-two) by Venus; youth (twenty-two to forty-one) belonged to the sun; full manhood (forty-one to fifty-six) to Mars; early old age, which lasted to sixty-eight years, was dominated by Jupiter; and whatever time was left belonged to Saturn. (A millennium and a half after Ptolemy laid down these seven ages, Shakespeare celebrated them in the well-known passage that heads this chapter.)

This associative way of thinking is, as I have said, an easy way to recall perceived patterns in the universe, a way of fitting patterns and events into a scheme or system covering all the mutual influences that might occur among them. If I assign an entity in a particular position in a list to one of a series of rotating states or qualities, then that entity automatically acquires a relationship to the quality; there is no need to look for a causal connection. I may say, for example, that just as I name the parts of my body from my head to my foot, so too may I name the stages of my life from birth to old age by passing along a continuum from top to bottom and by identifying each stage in one list with a segment or joint in the other. I can say that my head represents childhood and my foot old age. By arranging—in a single super list, table, or diagram—all the things and events that make up my universe of existence—elements, seasons, even the constellations of the zodiac—I can fashion a hierarchically ordered system whose core is the associative principle.* Often the properties will be found to oscillate between extremes in like manner as I pass down the continuum; that they are extremes may be a motive for their fitting together with each other in the first place. Thus, the seasons and the winds can be said to alternate like one's temperament or disposition, just as the good and evil properties of the planets balance one another out as we pass downward through their shell-like orbits from heaven to earth, Saturn being evil, Jupiter good, Mars evil, and so on.

*This is the principle embodied in the old nursery rhyme:

> This is the way we wash our clothes,
> Wash our clothes, wash our clothes,
> This is the way we wash our clothes,
> So early Monday morning.
>
> This is the way we iron our clothes, etc.
> So early Tuesday morning.

And so on through Wednesday (mending), Thursday (scrubbing the floor), Friday (sweeping the house), Saturday (baking bread), and Sunday (going to church).

In some cultures of the Amazon, the tapping of a woodpecker's beak upon an infected tooth is said to alleviate the pain. Of this association the French anthropologist Claude Lévi-Strauss has remarked:

> The real question is not whether the touch of a woodpecker's beak does in fact cure toothache. It is rather whether there is a point of view from which a woodpecker's beak and a man's tooth can be seen as "going together" (the use of this congruity for therapeutic purposes being only one of its possible uses), and whether some initial order can be introduced into the universe by means of these groupings.[5]

Such order was achieved by most of our forebears, including the early Christians in their association of Christ with Sunday. They met on the sun's day because it was believed to be the first day in Genesis in which the creation of the world was concluded. It was also the day Christ, their savior, rose from the dead. Thus, Christ, like the sun, became the light of the world.[6]

The Month and Its Moon

While we are unsure of the origin of the weekly cycle, there is no doubt about where the month comes from. The moon intrudes itself upon us, displaying its variability in shape, light, and position as well as a reversibility in its cycle. The moon is the epitome of continuous time reckoning, for its behavior suggests both continuity and duration; yet its changing aspects make it nature's ideal event marker as it exhibits noticeable differences from night to night.

Not only does the moon seem to beg for our attention—especially if we live in the country and can see it unimpeded by tall buildings—the menstrual period tied the human body to its light early in the distant past. Our word *menstruation* signifies "moon change," the Latin *mens* meaning "moon." German peasants call the menstrual period simply *die mond*; in France, it is *le moment de la lune.* For every woman the position of the moon in her cycle is different; but for a given woman, it is generally the same. At some point early on, a mental connection was made between a woman's failure to menstruate at the appropriate lunar phase and the birth of a child nine moons later. The Maya *tzolkin*, or 260-day cycle—the most important time unit in their calendar—was likely related to the nine-moon (265.7 day) interval between human conception and birth (see chapter 6).

At first, there was no attempt to count the full complement of days in a month, for people needed to know not *how long* one actually was, but only *when* an event would take place. Like the sunsets that move from day to day along the horizon, the moon is its own time indicator. To those early observers, when it came the month began, and when it was gone, the month ended. Meantime, the thickness of its crescent or the bulge of its gibbous phase was enough to mark the time between. When people first applied arithmetic to the creation of a time-reckoning system—which, as we have seen, may have happened over 20,000 years ago—it must not have been difficult to discern that the full cycle of lunar phases transpired in about 29 or 30 days. Our ancestors probably initiated the tally with the appearance of the thin crescent of the "new moon" in the west after sunset, which many tribal cultures greeted with great joy, and still inspires moon watchers in our own culture. At least, this seems the most logical and decisive point to begin the count cycle: either you see it or you don't. In respect to the full moon, it is not always easy to decide on which of two nights its disk most closely matches a perfect circle. Our modern versions of where the cycle commences makes even less sense: that is, the point of conjunction or the time when the moon in its orbit passes directly between sun and earth, which we call "new moon" though the moon is not actually visible then. The first thin crescent is the young moon. We know for sure that in the ancient Middle East the month began with the actual observation of the first crescent. Astronomers were delegated to stand on a high place and peer low into the west at dusk to spy the visible signal that would indicate that they did not need to add a 30th or a 31st day to the present month but instead could wipe the slate clean and begin the cycle over again with "day 1." Imagine the difficulties we would have today—paying rent, collecting debts, meeting deadlines—had such a system persisted!

For some the first crescent is a symbolic representation of the resurrected old moon—the one that returns after having become obscured in the light of dawn a day or two before. For our ancestors, the phases became celestial parallels in micro-time of the waxing and waning cycle of the hero's life in the old succession myth. In one version, the visible young moon conquers the devil of darkness as he waxes to maturity: you can even see the face of this man-in-the-moon who blazes forth triumphantly at the peak of his career—the full moon. But then the devil begins to eat away at him, and he starts to wane, lose power and wither away in old age. His feeble remnant falls from the eastern pre-dawn sky, and he disappears. During the time of new moon, some tribes used to say, "The moon is dead." But soon his son returns anew to avenge the death of his father.

That we number the days of this cycle suggests we prefer to deal with time abstractly. In most tribal societies, the phases are broken down into units that, rather than being numbered, are listed and named after concrete descriptions of the phase and position of the moon or of the way other parts of the living world behave at that time of the cycle. The very name of the day of the month gave away information about where the moon was in its cycle. This kind of naming system is laden with natural qualities. In the East Indies, for example, "little pig moon" and "big pig moon" correspond approximately to the 11th and the 12th days of our cycle, when the moon is well into the gibbous phase: it is at this time that the pigs, excited by the lunar light, are likely to break out of their pens and scramble about the fields. The 14th day is "lying," an apt description of the way the full moon rests on the horizon when it rises in the east at sunset. The 16th is "burner" because the pre-dawn moonlight shines through the door of the house. They call the 26th and the 27th days of the phase cycle "long tree trunk" and "short stump," respectively—words that may simply represent the declining shape of the crescent. Perhaps woodcutting activity might once have been associated with that time of the month. The anthropologist who gathered the data before the turn of the twentieth century seems not to have asked. "Going inside" and "inside" refer to the last visible crescent (day 28) and the next night's vanished moon because, the natives say, "they are going back to their house to rest." In Polynesia, the 3rd day after full moon, when it rises well after dark but still remains luminous, has a beautiful and descriptively apt name: "the sea sparkles at the rising." Also in parts of Polynesia, the first crescent meant "to twist" because it looked like a thread; the second day of the month was "crescent"; the third and fourth, "the moon has cast a light," because you have seen your shadow by it for the first time; the eleventh, "conceal," because the sharp points of the crescent are lost; the thirteenth, "egg," which describes the roundness at this stage of the gibbous phase; and so on.

Our division of the month into roughly four weeks seems unimaginative, even casual, by comparison with these indigenous calendars that reveal a people who operate close to their natural world. Indeed, it seems to be a general rule that the more complex a bureaucracy, the more abstract and contrived the month calendar. Where the Romans preferred a non-uniform three-fold division into seemingly inconvenient calends, nones, and ides, the earlier Greeks were more exacting, dividing the month into 3 decades, or periods of 10 days*—perhaps a derivative of counting on the fingers. The names of these intervals, however, retained the old descriptive elements: the Greeks called them "waxing," "middle," and "waning"; and the middle period may have consisted of time borrowed from the halves of an even more natural 2-part lunar-phase cycle. Whatever the case, this uniform method of dividing lunar time possesses certain advantages for the conduct of state business: each decade becomes an exact measure of the 30-day month; also, the number of the day of a decade is connected to the number of the days of the month: for example, the 3rd of a decade is the 3rd, the 13th, or the 23rd of the month. Compare this scheme with ours in which the name of a day of the week has no obvious connection with the number of that day in the month. When we reckon by weeks and then try to switch over to the time of the month, we need to consult a calendar in order to determine on what day of the week the month begins.

The Months and Their Politics

The names given to the individual months that make up the cycle of the year can, in some societies, be as informative as those given to the days that comprise the month. The sequences of actual month names on page 110 say a lot about the people who created them and the environment they lived in:

*Had the French Revolution succeeded, this custom might well have been reintroduced into the European calendar in 1789.

A.
Spawning month
Pine-sapwood month
Birch-sapwood month
Salmon-weir month
Month of hay harvest
Ducks-and-geese-go-away month
Naked-tree month
Pedestrian month (when man goes home on foot while ice still remains)
Month in which men go on horseback
Great month
Little-winter-ridge month
Wind month; month of crows

B.
Deer month
Strawberries month
Little-corn month
Watermelon month
Peach month
Mulberry month
Nut month (when they are crushed and mixed with flour to make bread)
Maize or great-corn month
Turkey month
Bison month
Bear month
Cold-meal month
Chestnut month

C.
Millet-is-cut month
Winter month
Beans-flower month
Month in which tamarinds of the north are ripe
Tamarinds-and-beans-are-ripe month
Month in which cythere tree flowers
Bulls-seek-shade-of-Sakoa-tree month
The-guinea-fowls-sleep month
Rains-rot-the-ropes (with which the calves are tied) month
Gourds-flower month
Month in which the grains of Fano are ripe

Clearly, the people of group A must be sedentary and live in a cold climate. Some of the life forms in their month list are familiar to those who reside in the northwestern part of the United States and the southwestern Canadian provinces: the running of sap and the spawning of salmon are time markers as distinctive as any celestial event. Actually, the people responsible for that lunar calendar live in the tundra of northern Siberia (the Ugric Ostiak). On the other hand, those who devised calendar B must be from a much warmer climate and appear to be hunter-gatherers, at least during part of the year. The names of their warm months indicate the variety of vegetables and fruits they collect, while colder moons are assigned to the animals they hunt. We can sense the turn of the seasons around harvest time when their concerns pass from corn to turkeys, and the change from the grinding of nuts in the dead of winter to the picking of strawberries in early spring. This is the lunar calendar of the Natchez tribes of the lower Mississippi River valley. Finally, group C surely must reside in a tropical zone that has flowering trees and a moist climate where monsoon rains "rot the ropes." These are the Malagasy, the people of the coast of the island of Madagascar, off the east coast of central Africa.

Our calendar has vestiges of a similar descriptive lunar-naming system in the harvest moon (the first full moon after the autumn equinox) and the hunter's moon (the next one following), which appear to be among the few temporal constructs passed down to us by the native Americans who lived in the continental United States before contact with Europeans. Otherwise, our months were named according to an abstract set of concerns. Rather than wanting to keep track of when to pick berries, we in the West have long tried to make the months fit exactly into the seasonal cycle of a year. This seems never to have been much of an issue for other people; indeed, while the Siberian Ostiak had 12 months, the Natchez had 13, and the Malagasy, 11.

The story of the attempts to make the months fit into the years constitutes a major chapter of our Western calendar. We are the beneficiaries of the old Roman calendar, which, though first written down about 300 B.C., probably dates back to before the founding of the republic in 510 B.C. This calendar, which began simply as a convenience for farmers, consisted of a listing of festivals and *fas*, or "right days," for conducting business. Not until society became urbanized did the calendar become formal, complicated, and subject to the many irrational vagaries still present in it today—vagaries that have led to much calendar reform over the centuries.

As we have seen, the Roman calendar utilized the two shortest and most obvious astronomical time units: the day, measured by the successive appearances of the sun; and the synodic month, starting with the successive appearances of the crescent following the new moon; these time units easily lent themselves to measurement via counting. In addition, an 8-day market week of uncertain origin was meshed with nature's two cycles to produce a 120-day year, or *annus*, that consisted of four months bearing the names of deities—Martius, Aprilis, Maius, and Junius; each consisted of 30 days, an approximation of the lunar-phase cycle. Later, six months were added—Quintilis, Sextilis, September, October, November, December. Their names, unlike those of the months preceding, signified their numerical order—fifth, sixth, and so on. This form of the year, made up of ten lunations or 300 days, was probably devised to correspond roughly to the gestation period of both humans and cattle. Because it seemed more orderly to keep the weekday names in step with the month—that is, to begin each new year on the same day name—four additional days were added later (traditionally by Romulus, with his brother Remus, the mythical founder of Rome), one each to the months of March, May, July and October. That brought the number of days in the year to 304, a number divisible by 8 without remainder. The rest of the seasonal cycle simply was not tallied: it fell during that portion of the year when the fields lay fallow, and therefore was not significant. While this notion of unaccounted time might bother us, it made sense to the Romans, who also related time to activity and apparently found it logical to count more for the time when things happened and to count not at all when nothing happened. Though the early Romans had the concept of a span of time equivalent to our year and related to the seasonal or solar cycle, their calendar was primarily a lunar cycle.

Perhaps the desire to incorporate natural periods of longer duration into a date-reckoning scheme is universal among all societies, the lengthier sequences of events serving usefully as checks against one another which might be employed in planning the future. In the Roman world, the first attempt to coordinate the lunar month with the solar year can be traced to Numa Pompilius, the second king of Rome. He simply added 50 days to the 304-day calendar of Romulus. In the revised calendar, 8 days were subtracted from among the other months, the sum being added to 50 to make a 58-day period; this interval then was divided in half to form two new 29-day months—January and February. They were tacked on to the end of the month list to convert the 10-month year into one of 12 months, the closest approximation to a solar-seasonal year that would also accommodate a whole number of lunar synodic months. By 150 B.C., an additional day was added, to the month of January, perhaps to keep the total days in the year an odd number (which was considered lucky). The Roman year had thus grown from a humble agrarian time count to a cycle of 355 days, a number that would not accommodate a whole number of traditional market weeks without the occasional insertion of extra days into the calendar, a process known as *intercalation*. It had taken nearly six hundred years from the time of the inception of the calendar before

Roman chronologists took on a challenge that was to occupy them, if not with equal intensity at all times, through the zenith of their empire: how to devise a manageable, lasting method for inserting days into the calendar in such a way that it would keep in step precisely with both the moon's and the sun's cycles?

The goal of the calendrical game was to write the perfect canon, the heavenly musical score that most ideally captures the eternal beat of a pair of natural celestial rhythm makers. The closer one listens to the rhythm, the more nearly perfect the possibilities that can be devised.

To understand the problem of intercalation confronting Roman chronologists, let us look at the periodicities of the moon's and the sun's cycles, which today can be measured with extreme precision: (a) the year of the seasons, or the tropical year (365.2422 days), is defined as the interval between successive passages of the sun across the vernal equinox, an imaginary starting point on the zodiac; (b) the lunar synodic month (29.5306 days) is defined as the interval between two successive passages of the moon by the sun; this is the familiar month of whose phases I have been speaking. Now, if the tropical year is divided by the lunar month, the quotient is: $A/B = 12.3683$. Thus, while the closest approximation to a tropical year, the 12-month count falls short by 0.3683 of a month, about 11 days. Intercalation by, say, adding a 13th month into the list about every three years can make the lunar and solar event sequences more harmonious.

There are other ways to solve this problem, each of which was attempted at one time or another in the history of the Western calendar:

1. Insert 3 months over the course of 8 years, thus approximating the remainder of the quotient A/B by 3/8 or .3750.
2. Intercalate 4 months over the course of 11 years, which would approximate the remainder by 4/11 or .3636.
3. Intercalate 7 months over the course of 19 years, making the remainder 7/19 or .3684. This scheme is known as the Metonic cycle after the fourth-century B.C. Greek astronomer Meton, who first recognized it.
4. Intercalate 31 months over the course of 84 years, thus approximating the remainder by 31/84 or .3690.

The Romans had, however, a solution of their own. In 150 B.C., they created a 22- or 23-day month called Mercedonius, which was inserted after 23 February in every other year or as needed, thus advancing the average year count from 355 to about 366. In the Old Roman calendar, 23 February marked the Festival of Terminalia, a day of sacrifice to Terminus, the god of boundaries. His festal date had been the last day of the year in an older form of the calendar.

Now the politics of intercalation became an issue in the development of the calendar, for the pontiffs, those Romans responsible for the administration of the cults of the state, ordered the calendric observances. Periodically they allowed the full years to be those during which their friends were in office.

The intercalation problem was not confronted with any rigor until a century later, when Sosigenes, chief adviser on calendric matters to Julius Caesar, recognized the shift of the sun's position in the heavens relative to the dates being tallied in the seasonal year. By then, civil time was already out of joint with nature's astronomical time by three months. In 45 B.C., Sosigenes suggested the following set of remedial measures, which have since collectively come to be known as the Julian calendar reform:

1. Make in the year count an adjustment that results in the restoration of the date of the passage of the sun by the vernal equinox to the correct date, then 25 Martius. This necessitated the intercalation of three whole months in 46 B.C., a year that came to be known as the "Year of Confusion"—with 445 days!
2. Abandon totally the true lunar synodic month as a calendar base by introducing 12 alternating 30- and 31-day months that added up to a year of 365 days. This signal event in our history ended once and for all the habit of month intercalation; for the first time, the annual cycle of the sun became the only natural period to serve as the fundamental time unit in the calendar.
3. To plan for the future, in every fourth year let one day be added to the year, at the end of Februarius. Averaged over a 4-year period then, the length of the man-made year would become 365.25 days, a close fit to the natural one.

With this step the difference between the canonic and the seasonal years became:

Man-made year	365^d	6^h	0^m	0^s
Natural year	365^d	5^h	48^m	46^s
Difference per year			11^m	14^s

Thus, solar events, even with the intercalation program of the Julian reform in place, would fall behind the tally kept by the chronologists by just one day in a little over a century, hardly enough to concern the Romans given the temporal quagmire out of which they had just extricated themselves. The Julian reform would suffice for sixteen hundred years.

Anti-Semitism and the need to determine the date of Easter Sunday were jointly responsible for the next action in the reform of the Western calendar—an action decidedly ecclesiastical in nature. With the emphasis of unity in the Roman empire being placed upon the Christian church in the fourth century A.D., pagan cults and rituals along with the old Roman calendar were abolished and in their place was instituted a calendar based upon ecclesiastical reckonings and the commemoration of saints. One goal of the Council of Nicaea (Iznik in Asia Minor) held in A.D. 325 was to set a single date for the celebration of Easter by Christians in both the Eastern and the Western Holy Roman empires. Traditionally, the date had been fixed in the Hebrew calendar as the 14th day of the month of Nisan, which began with the appearance of the crescent of the spring moon, the one nearest the spring equinox. The Hebrew calendar, with its strictly lunar base, did not, however, observe the equinox with any precision. In fact, we have no details about how they actually determined the month of Nisan. The Christians were compelled to celebrate the anniversary of the Paschal date close to the time of year when the event was documented as having occurred, and to fix it on a Sunday. Above all, church law dictated that Easter must not fall on the Hebraic equivalent. The *computists*, specialists in charge of Christian calendric computations, believed a conflict would be avoided if they selected the first Sunday after the full moon that followed the vernal equinox. Now, since it already had been recognized that the equinox was falling earlier and earlier in the calendar year (it had shifted backward about three days in the four centuries since the Julian reform), the religious problem of determining Easter Sunday became tied to the astronomical problem of fixing the place of the vernal equinox in the zodiac.

Given the set of rules for determining the movable feast, the computists were called upon to reconcile three periods that do not readily mesh: the week, the lunar synodic month, and the solar year. It turned out that Easter could occur any time between 22 March and 25 April; and it would take over five million years before the dates of the Paschal observance would recur in the same order. Having discovered the impossibility of devising a mathematical formula to set dates in the future, the computists created complicated tables based on averaged full-moon intervals. These long-winded tables of *epacts* listed the age or phase of the moon on 1 January, from which the Easter date easily could be computed for that year. But these artificial tables gave only approximate information about the phase of the moon on New Year's Day (the error could amount to 2 or 3 days); the true motion of the moon was far too complex to be formulated precisely in tabular form. One of the calendrical debates that followed after the Middle Ages centered around the extent to which religious festivals really needed to be calculated with astronomical accuracy. After all, as the great German astronomer Johannes Kepler is reputed to have said, "Easter is a feast, not a planet."[7]

By the sixteenth century, the recession of the real year with respect to our artificial account of it had grown to 11 days (figure 3.3). Easter Sunday began on the average to fall later and later in the season and thus to appear warmer and warmer. In 1582, Pope Gregory XIII appointed a commission to deal anew with the reform issue. As was the case a millennium and a half before, two actions were needed to assure that the future festival date would arrive at the proper location in the year of the seasons: the equinox needed to be restored to its proper place in the year cycle; and the commission needed to devise a mechanism to hold it fixed.

After much debate about whether the lost time might be made up in small parcels over a long interval, the first problem was solved, as in Caesar's time, in a single bold stroke simply by dropping 10 days out of the calendar. Thus, the equinox was moved from the 11 March date on which it had occurred in 1582, to 21 March. Pope Gregory XIII wrote a decree declaring that the day after 4 October of that year would be 15 October.

The second step of the Gregorian reform consisted of changing the leap-year rule by decreeing that among century years, only those divisible by 400 shall be leap years. Thus, while A.D. 1600 and 2000 would be leap years according to the new system, A.D. 1700, 1800, and 1900 would not. This recipe had far-reaching consequences for it dras-

FIGURE 3.3 The Gregorian calendar reform. A member of Pope Gregory's commission on the calendar alarmingly points out the backslide between the Julian calendar and the year of the seasons, each represented by a set of day marks framing a zodiac. After several years of bickering, the Pope's commission took decisive action and dropped ten days out of the calendar, thus putting artificial time and nature's time back in step. SOURCE: Archivio di Stato, Siena, Italy.

tically reduced the shortfall inherent in the Julian leap-year system by cutting the length of the calendar year, averaged over long periods of time, below 365.2500 days to 365.2425 days, or only .0003 longer than the true year of the seasons. So near perfect was the new rule that the man-made year cycle would now roll ahead of the seasons by just 1 day in 3,300 years!

As might be expected, the Gregorian reform was immediately adopted by all Catholic countries, but was resisted rigidly by the Protestant-dominated nations like Germany and at first given little attention by non-Western cultures. Great Britain and its colonies did not adopt the new calendar until 1752, by which time they needed to drop 11 days from the count. Indeed, in early life, George Washington celebrated his own birthday on 11 February. The calendar was not accepted by Russia until the Bolshevik Revolution of 1917; by then, even more days needed to be eliminated.

Minor reforms have been suggested since the time of Pope Gregory, such as agreeing to convert A.D. 4000, 8000, and 12,000 to common years, which would reduce the difference to 1 day in 20,000 years. Finally, at an Eastern Orthodox congress held in Constantinople in 1923, yet another rule was adopted in the parade of legislation engendered by the drive for accuracy. It stated that century years divisible by 900 will be leap years only if the remainder is 200 or 600. The resulting calendar is accurate to 1 day in 44,000 years.

All reforms seem to aim for pristine completion of the year cycle as well as precise arrival at a solar date. The more fingers in the bureaucratic pie, the greater the concern to build up and tightly interlock larger and larger cycles to gain control of the future. In the last analysis, any new rules we might devise in the future to improve accuracy would be futile because we know now that the length of the tropical year is itself changing because of the gravitational disturbances the earth encounters from other bodies in the solar system, and the rate of change is comparable to the difference between the true and the man-made years.

It has taken Western culture more than twenty-five centuries to devise an eternal and unchanging artificial framework that can assure the exact setting of a date far in the future with maximal certainty. Our goal, begun modestly, had always been to foretell the actions of nature's heaven as unerringly and precisely as possible. In pursuit of that goal, people traveled a sinuous, often irrational course, one involving accommodation among religious and political as well as among astronomical concerns. Along the way, Westerners modified that goal by electing to sidestep a part of the problem that could not be mastered—the synchronization of the natural, lunar, and solar cycles. The development of our calendar has also been accompanied by increased complexity and abstraction, as ever more detailed correction programs carry us farther away from nature. Yet it is ironic that our attempts to devise modes of time reckoning arose out of the eternal wish to keep our lives in tune with nature. With the development of technology—from sun watching at the horizon to weight-driven clocks to atomic timekeeping devices—we have removed our marking of time from nature's realm and in the process, become ever less in touch with the events that occur there.

4

THE YEAR AND ITS ACCUMULATION IN HISTORY

> The more the universe seems comprehensible, the more it also seems pointless.
>
> —STEVEN WEINBERG, *The First Three Minutes*

THOUGH THE YEAR is the longest time-reckoning cycle in the formal calendar, the one on a wall or desk that you consult on a regular basis, we have nevertheless fashioned even bigger bundles of time, to tie events of the relatively remote past to those of more immediate recollection. Large time units like century, era, and epoch lead one to wonder how events separated by such physically insensible intervals can be thought to have any meaningful relation to one another.

As I shall discuss in part III, the Western tendency to reckon time in huge parcels is hardly unique. Some of the philosophical repercussions of the West's approach are, however, unusual. The idea that time comprises a linear succession of events that are related through cause and effect has replaced the earlier associative way of conceiving time, where rituals are acts performed or festivals celebrated in response to cyclic phenomena in nature, such as certain phases of the moon or the solstices.

In discussing the macroscopic makeup of time, I shall confront the problem of how we in the West have chosen to conceive of the idea of history, and the post-Renaissance discoveries of the vast time scales of the universe made by the sciences of geology and astronomy and of the species that populate it by the naturalists and biologists. Such discoveries have demythologized the old concept of creation, rendered it obedient to a set of natural laws, and transformed it into a mechanistic and evolutionary happening quite apart from the theater of human action. We arrive on the threshold of the twenty-first century propelled by machine-made time. While our old dependence once lay in the gods, our modern deity has become the clock.

The Western Year

A Seasonal Cycle

The year is like a circle. Seasonal events and holidays succeed one another and become united in a complete cycle when they begin to repeat themselves in a recognizable pattern. For the Greeks, the deified symbol of this great circle was the most famous time god himself. According to the mythologist Jane Harrison: "Kronos is the Accomplisher of the full circle of the year. He is not the Sun or the Moon, but the circle of the Heavens, of Ouranos, husband of Ge [Gaia]; of Ouranos, in whose great dancing-place the planets move."[1] Although we tend to think of the year as purely a solar cycle, actually it is outlined and delineated by the movement of most celestial bodies. Kronos is not a sun god, or even the god of the heavens; he is the god of the circle of the year, the personified expression of the calendar of Hesiod's *Works and Days* in which the passage of time is marked by the celestial dance steps performed by sun, moon, and planets.

While the change of events during a year is obvious, not so is its beginning and its end—its duration. We humans generally reckon the years and our passage through them in reference to significant, landmark events. Every year has signal events one memorizes subconsciously and in sequence to order the happenings in everyday life. Though some people may recall "that was in '87" or "at the end of '85," most of us say, "It was the year we moved to the West Coast," or, "It happened during the holiday season." Sometimes we can pin down the event in question by locating it between definable temporal landmarks, with only the vaguest hint of seasonality: "I'm sure of it because it was shortly after you returned from school but before we bought the new car in the spring"—hardly a mode of time reckoning on which everyone can agree.

On the other hand, because in the West we equate time with money, in the business and professional world we are compelled to use precise, legally defined station points in the calendar: "Let's see, today is the 13th, so our bill of lading ought to be sent out to you by the 20th." But the time written in our record is not usually the one that gets impressed in our memory file. We know we need to be back home to receive the goods after a stop in Chicago but before we see a client in Pittsburgh. These key events, like marriage or the move to a new apartment, become the signposts on time's road in our and other cultures; to these signposts we attach peripheral happenings and form a chain of events to hang on to as we try to grope our way back into the labyrinth of memory or to link ourselves up with the uncertainties of the future.*

As I have said, we all sense time as an event sequence in our culture as well. If you were to write down in chronological order the ten most significant events, good or bad, that have influenced the course of your life, the odds are high that you would make little or no reference to a numerical-year calendar when you try to order them. A good part of your early chronology would undoubtedly be made up of the school years, which serve as the basic chronological landmarks. After high school and maybe college graduation, the events in job advancement would probably take over as your personal historical mileposts. The most obvious year references would be those that have been

*Native American tribes also developed abstract symbols to stand for the years. Perhaps the idea of equating years and circles is universal, for the Dakota Sioux used little round circles to denote them. Other tribes marked the years by notches on a stick. But these symbols were not counted; the Sioux and other tribes did not perform arithmetic with their years as we do: "The postwar period has lasted 44 years," or, "It has been 8 years since the start of the AIDS epidemic." Instead, each circle or notch stood for an event. The year-narrator used the stick as a mnemonic, not as an arithmetical device—less like a wall or desk calendar and more like the way one ties a string around a finger or memorizes a series of first letters of items to remember to purchase at the supermarket: "*B* means bread, *O* means orange juice, *Y* means yogurt." Sequential arrangement of these items into a list in that particular order makes the task even simpler for now one need remember only the single world *BOY.*

FIGURE 4.1 The years of the Dakota Sioux winter count. One can read the events off the calendar and count years at the same time: each year in a sequence is, instead of being numbered, identified by the principal tribal or intertribal event that took place in it. In the complete sequence (*not shown*), the distance between any two pictures represents a time span; scanning the series of pictures, the mind moves from the *event*, sensed directly, to the *duration* between—a much more difficult entity to conceive, and that numbers help us keep track of. (*A*) The winter people were burned (their camp was destroyed by a prairie fire; 1762–63); (*B*) Oglalas engaged in a drunken brawl which resulted in a tribal division (1841–42); (*C*) Nine white men came to trade with the Dakotas (1800–1801); (*D*) A trader brought the first guns (1801–2); (*E*) Dakotas saw wagons for the first time (1830–31); (*F*) Crows came to the Dakota camp and scalped a boy (1862–63). SOURCE: After G. Mallery, *Picture Writing of the American Indians*, 10th Annual Report, Bureau of American Ethnology, 1888–89 (Washington, D.C.: Government Printing Office, 1893).

forced into your memory by convention—say, your class of 1968 high school graduation or the college class of 1972.*

If the duration of a year is felt as the churn of events that make it up, then the year is a whole, equal to the sum of its parts. It is not so

*In the Dakota Sioux year-count system, absolute time was marked not by the date or the number in a cycle, but rather by the most significant event occurring between consecutive winters (see figure 4.1). They painted each event in the form of a sequence of picture writings on buffalo hide: the inundation, the war, the great snowfall, the disastrous hunt, the construction of a new town by the whites, and so on. These major events constituted the Sioux's winter count—a year calendar not unlike our month calendar. In the Dakota Sioux's calendar, time and history appear to be the same.

much *divided* into months or seasons as *comprised* of all of the bits and pieces whose repetition makes their union recognizable. Life's social events are acted out against nature's seasonal background—agricultural operations, the coming and going of the birds, the rain, the leaves on and off the trees. We think of these events as taking place over twelve months or, more broadly, four seasons—though, in many areas of the southern United States and in most tropical regions of the world, it makes better sense to think of only two, the wet and the dry alternating endlessly.

Today's relatively useless notion of a four-season calendar probably began as a two- or a three-season sequence. At least we know that in ancient Greece fall went unrecognized. There was only winter, spring, and summer; and these were associated with the qualities of cold, wet, and dry, respectively. The fruit harvest time did not become regarded as a season in its own right until about the time of Homer. Then, like spring, it came to represent a transition period between the two other seasonal extremes of the year. It was only when the four seasons were rigidly affixed to the two equinoxes and two solstices on the sun's course that we developed the four periods of almost precisely equal length still used today—although they have little practical meaning. Does summer's heat really begin when the calendar implies, late in June? Not when most parts of our country suffer heat waves in April and May. At the other end of the real meteorological scale, there is the winter of November which occurs, according to the calendar, in most northern climes in the middle of autumn. In reality, there is an overlapping of the seasons—a sense of instability to the event sequences that make up the cycle of nature's behavior. Sometimes the drought is prolonged, winter comes early, spring rains and the rising of the river are devastating. Thus, allocating nature's behavior among the four seasons gives us only the illusion of control.

In actual practice, our seasonal calendar pays less attention to nature's sequential events and more to those of society. Of course, we still have a hunting season, and the harvest season is on when the vegetables in the supermarket are least expensive—if not the best tasting. But many seasons that we choose to recognize are only marginally connected to the four climatological ones. We celebrate the holiday season and shop during the fall fashion season. We usually try to suppress all thought of the income tax season, which arrives toward the end of the fiscal year. A host of other seasons, even more vaguely connected to the solar cycle, nevertheless still manage to occur annually: the flu season (obviously weather related); the back-to-school season, which once commenced after harvest time when children were needed at home on the farm; and, in some locations, the "oysters-R-in" season. Finally, there are seasons one would be hard pressed to connect causally to the cycles of nature, except to suggest, "That's when we always do it." I have in mind the bridal season, the concert theater and ballet season, and television's many "new seasons"; the new car season; the football, baseball, and other sports seasons (with the advent of indoor stadiums, the latter seem to merge into a single continuous yearlong season)—not to mention the January white-sale season and other after-the-holiday bargain seasons.

Where for centuries public festal days were tied in the Christian West to sacred days, when all commerce stopped, today few such days are observed, the exceptions being Christmas and, in some parts of the country, the Jewish high holy days. Today most stores remain open on holidays, and the holidays when some businesses and schools close are no longer sacred. In the United States, more civic than religious are the Fourth of July, Memorial Day, Veterans Day, Labor Day, Thanksgiving Day, and all of the President's days, along with Martin Luther King Jr. Day. Likewise, the French have Bastille Day on 14 July; and the Mexicans, 20 November commemorating one of their great revolutions. It is ironic that while we want every year cycle to be precisely the same length, we are flexible on the observation of patriotic holidays within the round, many of which are now observed on the Monday nearest to their original date.

Fixing on New Year's Day

Had one sufficient patience, one could reckon a year's interval by simply watching a full cycle of passages of the daily sunrises or sunsets along the horizon, as did the ancient builders of Stonehenge or the Hopi–Navajo, who followed the course of the year by the convenient natural markers of distant peaks and valleys along the horizon. Or one could mark out the year as the Greeks and Romans did, by noting the height of the noontime sun about the south horizon. In wintertime a stick placed on the ground casts its longest shadow at noon and it is shortest in the summer—a cycle that, like the rhythmic oscillation of the sun along the horizon, repeats itself after a full year.

Today we base the duration of our year on the interval between successive passages of the sun by the same point among the stars. We once began and ended the year when the sun passed the spring equinox late in March, a point when the periods of day and night were equal; but since Roman times the starting point has been the dead cold period in January. For most cultures, the best time to begin seems to lie somewhere within an interval of rest or inactivity—say, just before the awakening point in the annual cycle of life; often it is marked by some easily recognized sky event. Thus, in the ancient Inca capital of Cuzco, the new year began with the heliacal rising of the Pleiades (see page 300), the same universally conspicuous star group that also shows up at the appropriate time in the *Works and Days* (see page 42). In Cuzco these stars made their appearance toward the end of the period when the fields were dormant, between harvest time and the time for sowing.

In most cultures, New Year's Day is celebrated as a time of renewal and is always marked by a kind of festival of rebirth, of the cycling of death into life. Think of the New Year's resolutions in our culture, of the promise to turn over a new leaf—a fresh page in your book of life; of casting off the old and ringing in the new. Physically, this act of turning means raising your eyes from the bottom of the old and up to the top of a new page. This uplifting activity often takes place close to the time when the sun, having plunged to its lowest depths in the sky around the time of winter solstice, now offers concrete evidence that it will return to its ascendant position and thus reawaken the land and restore our lives with its light and warmth. The principal Christian holiday falls close to the winter solstice. Symbolically, the celebration of the birth of the Christian saviour was set, in the old pagan calendar, to fall at that time in the cycle when life was in the greatest need of being renurtured. In the pre-Christian world, year's end and beginning was the time of the resurrection of the dead, of funerary feasts, of the extinction and rekindling of the fires. In the Christian vein, spiritual nourishment could be provided only by the new light the coming of Christ cast upon the world. The 12 days of Christmas, over which the festival of Christ's birth is celebrated, may represent those uncounted days left over at the end of a 12-moon cycle that one needed to tack on to the end of the lunar year in order to complete a full cycle of the sun. For some peasants of central Europe, these exceptional 12 days between Christmas and Epiphany lay outside the year, and each came to be regarded as a prediction of things to come during one of the 12 successive months of the New Year. Mircea Eliade regards this interval as a suspension of *profane time*—a deliberate halt to life's chain of events, a time for recollection and reconsideration of the past as well as prognostication of the future.[2] Thus, the weather on each of the 12 days offered a forecast for the succeeding 12 months. Many farmers' almanacs still employ this regenerative concept of time in which the future is contained within the past, the New Year repeating in miniature the whole cycle of creation.

Most calendars associated with organized societies seem to be fixed by a count in which the year's end and beginning take on the same special connotations. The Egyptians, for example, kept 12 months, each of 30 days. At the end of the 360-day period, they tallied 5 special days outside of normal time, each of the 5 special days represented the birthday of one of their principal gods. The evolution of the annual calendar from 360 to 365 days may have taken place when, in Egyptian eyes, the 365-day period came to be viewed as a more relevant indicator of the inundation cycle of the Nile upon which all life depended. Ancient Mesoamerican calendars also reckoned a 360-day year cycle, except that they utilized an 18-month scheme, each month having 20 days (see pages 195–96); and as for the Egyptians, the 5 days remaining at the end of the Mesoamerican year constituted a special "month" of unlucky days—a period during which any activity, especially birthing, was to be avoided at all cost. We know too little of their history to be able to tell whether, as in the Christian calendar, the Mesoamerican year end was used for prognostication. Still, the similarity of the year-ending periods in these far-flung calendars does seem to show that all civilizations need to regenerate themselves by temporarily suspending or annulling the passage of time. At least for the Maya, the end of the year was apparently a time for flushing out all bad ideas and thoughts. In Central Mexico, this period is associated with the kindling of "new fire" in the hearths of all houses and temples, an act accompanied by the trashing of many household items, such as dishes to eat from and mats to sit upon, and by their replacement with new items.

Just as events could be added to history, so, too, they could be subtracted. The disinformation campaign conducted by the Soviet Union to "de-Stalinize" their history provides a contemporary example of a phenomenon that must have been rampant in ancient times, to judge by the number of effaced statues of kings and emperors archaeologists have unearthed. Bureaucratic societies require a calendrical ideology that can both become known, and at the same time make sense, to every participant.

There is a sense of order in reckoning history by the years. One gains a firmer hold on both past and future. As we shall see in part III, Maya kings were able to impress their subjects by numerically linking their ancestral origins to the birth of the gods thousands of years in the past. And by giving numbers to their years they were able by mathematical calculation to predict eclipses and other natural events many cycles in advance. In the West the numerically based liturgical calendar permitted the setting of the celebration of Christian holidays at precisely the proper moment through the establishment of complex formulae devised by the high priests who controlled and monitored time's passage.

Formally, the serial numbering of the years as we know them did not actually begin until the sixth century, in the reign of Dionysius Exiguus in the Eastern (Christian) Holy Roman Empire. The sequence of years B.C., or before Christ, was not initiated until more than a thousand years later, in A.D. 1627, by the astronomer Petavius.* This abstract, rational arithmetic scheme still serves as the framework with which we order all events. It is history's time carriage, not history itself.

Long-Time Reckoning and Great Years

As we attempt to mark time in units that transcend years, we think of the ten years of a *decade* and the hundred of a *century* which, though both numerical indicators of duration, nevertheless seem to acquire a character of their own. It has become fashionable in this century to employ the decade as a way of characterizing patterns of social behav-

*The Christian Era is defined as the interval of time between the birth of Christ and the present: a number of years "A.D." (*anno domini*) refers to the lapsed interval from Christ's birth; "B.C." refers to a given number of years before the Christian era. In the running count there is no "year zero," the sequence of years proceeding B.C. 2, B.C. 1, A.D. 1, A.D. 2, and so on.

ior, although such behavior either occurred in one corner of a particular decade or spilled over into the next. Thus, the liberalism of the 1960s really became overt only toward the end of that decade, while its seeds were sown in the 1950s. Not all the twenties were "roaring," and didn't the Gay Nineties really overlap with the twentieth century?

For intervals in excess of centuries, we bundle the years into periods of unspecified duration called *ages.* These less quantifiable intervals are basically hindsight reckonings, determined largely by the density of social, political, or religious events that are later regarded as having occurred within them, and especially by major turning points in history—the Age of Reason or Enlightenment; the Middle Ages, to which historians have assigned practically an entire millennium; the age of the Renaissance or of the discovery and exploration of the New World.

A chronological *era* is even longer than an age. Characterized as a succession of years proceeding from one fixed point in time to another, historically an era can be distinguished by a spectacular event, such as the birth of Christ, which gave rise to the Christian era. For the staunch believer, the end of this era will arrive with the second coming of Christ. This longest of time lines was bolstered in the early Christian mind by the belief that all things are made expressly for the end they fulfill. For the Christian, the end lies in the arrival of the kingdom of God which will initiate a new era—a timeless eternal existence to be experienced only by the true believer. Western history has been guided at least since the beginning of the Middle Ages by this teleological concept of a time line. Long time in our calendar is linear, not circular as nature would have it. Our existence is like that of a bead gliding along on a string. One extremity of the string represents the beginning and the other the end of the era. Tilt the string slightly downward, and the bead is pulled along by gravity from the beginning to the end of its journey. So, too, faith in the Second Coming pulls the true believer forward through the good life toward the ultimate judgment that gives reason to it all. In modern times, this notion of advancement along a time line still prevails, except that technology has replaced religion as the force that propels events to succeed one another; nonetheless, the doctrine remains teleological. As Augustine expressed it: "Would to God that people could avoid these roundabout aberrations, discovered by false and deceiving philosophers, and stick to the straight path of sound doctrine!"[3]

Prior to Augustine, Western man's sense of long duration had been rhythmic and repetitive, events being real only if they re-enacted the original archetypal ones. The Babylonians and the Greeks conceived of a "Great Year" as the period of the world from creation to destruction to rebirth. Formally they considered it to be mirrored by the time required for sun, moon and planets to attain the same positions with respect to one another that they had occupied at a previous time. These celestial positions marked the time to re-create the events of the past—to literally witness them happening all over again. The lives of the great philosophers would be relived. Everyone would re-experience what had already happened. Even towns and cities would be re-created just as they once were. As we shall see, this outlook is echoed in modern theories of cosmology that propose a continual destruction and rebirth of the entire universe ad infinitum, a scenario that produces enough rolls of the dice to re-create identical chemical, atomic, and subatomic states many times over.

Though the old cyclic way of reckoning eras seems outdated, it nevertheless still lies at the root of precise astronomical timekeeping. Consider the difficulty of subtracting one date from another in the present calendar, say the date of your birth from today's date. Feeling the need to devise a system for reckoning long time intervals in a less cumbersome way than by tables of variable months and years, astronomers devised a system of days numbered in a continuous sequence from a fixed point in time. The starting point, the Julian era or period, 4713 B.C., was fabricated in the sixteenth century by the French philologist Joseph Scaliger, who rolled back three major time cycles—two purely astronomical and one socio-political—to a fixed point of coincidence: the period that includes all possible combinations of the days of the week with the first of the year (28 years); the period over which a given phase of the moon is returned to the same date of the seasonal year (19 years, the so-called Metonic cycle [see page 113]; and the nonnatural 15-year cycle of *indiction*, a period during which certain governmental acts took place.* The product of these three cycles—or 28 times 19 times 15—is 7,980 years. In the course of this period, no two years can be written down with the identical set of numbers in all

*First instituted by Constantine the Great in the fourth century A.D., indiction was a period relating to the imposition of taxes, specifically those to be paid troops upon being discharged from the army after 15 years, the maximum length of service he permitted; this led to the general practice of keeping monetary accounts on a 15-year cycle.

three subcycles. Time calculations in this system are relatively simple. For example, Pearl Harbor was bombed on Julian day 2,430,336 and the invasion of Normandy took place on JD 2,431,247. By simple subtraction, we can readily find that these major events of the Second World War were 911 days apart.

Great Year cycles like the Julian era exist in calendars all over the world. The Indian calendar adopted from the earlier Sanskrit tallies a 2,850-year Great Year composed of 150 Metonic cycles (the second cycle in the Julian era). As we shall see later, theirs was a calendar composed of mathematically precise relationships as well as socially significant time intervals. And not unlike the Julian system, it has a celestial zero point—a conjunction of all of the visible planets in the constellation of Aries, an event calculated to have occurred at midnight on the night of 17–18 February 3100 B.C.

The rational basis of the Greek astrological theory of the flood utilizes the same conjunction date and at the same time offers an excellent example of the associative principle. The zodiac—which charts the course of the sun, moon, and planets among the stars—is divided into 4 triplicities, each tied to one of the 4 elemental qualities:

Aries, Leo, and Sagittarius are Fiery.
Taurus, Virgo, and Capricornus are Earthy.
Gemini, Libra, and Aquarius are Airy.
Cancer, Scorpio, and Pisces are Watery.

Close encounters of the planets (conjunctions) seen in the sky in any of the first three signs portend destruction by conflagration (the multiple conjunction of 3100 B.C. happened in this triplicity), while a conjunction in any of the last group indicates a flood. A few interesting zodiacs from different times are depicted in figure 4.2.

It is possible to make as many planetary conjunction periods as there are planetary pairings. The Jupiter-Saturn conjunction has traditionally been one of the most influential in long-term cyclic timekeeping, for these two planets possess the largest planetary orbs. Faster Jupiter passes slower Saturn every 20 years as both journey on the zodiacal roadway. Therefore, Jupiter-Saturn conjunctions would occur in the signs of a given triplicity for approximately 240 years before sliding over into the next set. A full cycle through all 4 triplicities would take about 4 times 240, or 960 years. For many of our predecessors, these intervals took on great significance. The shift between triplicities portended major dynastic changes; a complete full cycle, the arrival of a major prophet. Such cosmic re-creation cycles operate in frameworks of different lengths among different cultures of the world, but the celestial associations that guide their underlying structure is pretty much the same. For the Chaldeans the universe would be deluged when the 7 planets were assembled in Cancer, and destruction would proceed by conflagration when they arrived in Capricorn. The ancient Mayas of the New World dated the last creation from 11 August 3113 B.C. for reasons we cannot yet grasp with certainty, but which likely possessed celestial guidance. Politics depended on horoscopes that paid particular attention to conjunction cycles. Political life, like the celestial lives it reflected, was discontinuous. A planet resided either in one constellation of the zodiac or another, never in between. So, too, life is made up of transitions or passages through a series of stages. One can't be—at one and the same time—in and out of office, alive and dead, in and out of love.

Another similarity about Great Years in the calendars of the world is that their starting points all seem to occur a few thousand years B.C. This fact, together with the strong dependence that many diverse Great Year calculations have upon movements in the zodiac, has led to speculation that there may be a single celestial base period recognizable by all humans and underlying all calendars, and that somehow everyone is subconsciously driven to pattern eras and Great Years after it. The *precession of the equinoxes* is the time it takes a predefined starting point on the zodiac (the intersection point between the *ecliptic* [the plane of the earth's orbit about the sun] and the *celestial equator* [the extension of the earth's geographic equator onto the plane of the sky]) to make a complete circuit through all 12 member constellations—a period of approximately 26,000 years. Modern science explains precession by referring to the wobbling of the earth's axis of rotation about the axis of its plane of revolution about the sun, as manifested in the classic motion of an agitated spinning top; but once again, this is not what people on earth experience directly. For our predecessors, precession revealed itself through a slow variation in the date of the year on which a star made its heliacal rising. Thus, a festival once marked by the time of the heliacal rising of a bright star would begin to shift its position within the seasonal year. This gradual phase shift of sun time relative to star time could be

FIGURE 4.2 Time bands the world over. (*A*) A wheel of time for common folk. The round of the unnamed months surrounds a man and a woman. As in the *Works and Days,* each month is represented by a human activity that fits with what is going on in nature's background. For example, June (in the eleven o'clock position) is the beginning of summer—when the sun's heat drives out the moisture in the moors and all things reopen; thus, the farmer is depicted mowing his hay. The zodiacal constellations in heaven enframe the rest of the world. Beginning in the eleven o'clock position and moving counterclockwise: Cancer, Leo, Virgo, Libra, Scorpio, Sagittarius, Capricorn, Aquarius, Pisces, Aries, Taurus, Gemini. SOURCE: Bartholomaeus Angelicus, *De Proprietatibus Rerum* (Lyons, 1485). The Huntington Library, San Marino, Calif. (*B*) Detail of signs of the zodiac and labors of months in the royal portal, central bay tympanum of Chartres Cathedral. *Lower left,* July, a harvester on his knees reaps wheat with a sickle; *upper left,* Cancer; *lower right,* April, man grasping a tree branch covered with flowers and leaves; *upper right,* Aries. SOURCE: Photo by L. Aveni. (*C*) The burden of the spheres of time. Less earthy and spiritual, more mechanical and scientific, in this earth-centered planetary model from the early Renaissance, Atlas holds the world upon his shoulders (compare figure 6.7A in which Maya deities carry the burden of time). SOURCE: W. Cuningham, *Cosmographical Glasse,* (London, 1559). The Huntington Library, San Marino, Calif. (*D*) The band of the zodiac is but one of a multitude of circles on the heavens by which to reckon time and the existence of zodiacs in both the Old and the New worlds raises the possibility that the concept may be universal. In this purported Maya zodiac from the Paris Codex, a parade of animals, their jaws tightly clenching Maya day signs, hang from the body of a banded serpent. Serpent, bird, scorpion, tortoise, and rattlesnake are easily recognizable among the thirteen constellations. SOURCE: Codex Paris, Bibliothèque Nationale, Paris: Codices Selecti, vol. IX (Graz: Akademische Druck-u. Verlagsanstalt, 1968).

A

B
C
PRIMVM MOBILE
CRISTALLINE
FIRMAMENT
CŒLIFER
ATLAS
Hic canet erranteꝭ Lunam, Solisq; labores
Arcturūq;, pluuiasq; hyad. gēinosq; triões
D

detected within a single lifetime. If observed with great accuracy, the amount of shift determined could be used to calculate the time it would take the heliacal event to migrate all the way through the calendar year; thus the 26,000-year cycle could be computed, at least approximately.

Originally the parts of the great cycle of precession were taken to symbolize the different ages of man in the Western calendar, each age (1/12 of 26,000, or approximately 2,200 years) being signaled by the passage of the vernal equinox into a new house of the zodiac and the consciousness of that age being passed on through the telling and retelling of myth. Thus the Christian era opened when the sun rose in the constellation Pisces at the spring equinox. The previous age, that of Aries, is the one in which Moses arrived on Mt. Sinai; and before that, in the age of Taurus, people worshiped the golden calf or Taurus the Bull, which still appears in the headdress symbolism of certain Egyptian deities as well as in the Minoan architecture of the palace at Knossos. And a popular song from America's 1960s revolution extolled the peace and love that would characterize the forthcoming Age of Aquarius (slated to begin when the equinox sun enters that constellation in A.D. 2137).

Though the grand cycle of precession is one of nature's longest celestially based periods, there is little evidence that it, along with its division into star ages, really occupied the pivotal role in the calendars of all the cultures attributed to it by some scholars. For example, no one has satisfactorily demonstrated that civilizations remote from early European and Middle Eastern contact, such as those of central Africa, Mesoamerica, and South America, paid significant attention to the zodiac and the planetary conjunctions and other events that transpire among the celestial bodies moving through it, even though they all saw the world as beginning a handful of millennia back: 4004 B.C. for Western Christians before Darwin; 4713 B.C. in the Julian calendar; 3113 B.C. in the Mayan calendar; and 2850 B.C. in the Indian. While we may derive a sense of cross-cultural cosmic unity out of the belief that somehow we can all trace our origins back to a single point through a universal scheme, the chroniclers have, in fact, conceived many combinations of celestial and/or historical events. In the seventeenth century, Archbishop James Ussher pegged the origin of the world at 4004 B.C. by tabulating Old Testament genealogy—a period of a few *thousands*, as opposed to hundreds or millions, of years that may be more representative of the perceived dividing line between human and divine history than of reality. One hundred generations amounts to only 2,500 years, an interval not beyond imagination in contrast to the millions and billions of years offered by paleontology and astrophysics. If time is actually beyond history and religion, then what is it—and when did it begin?

Our Creation Stories

Time is many different concepts and ideas: it is change, it is movement, it is sequentiality. Of two ways of looking at the universe, we in the West seem to be attempting either to seek a singular order that we already have faith exists in the universe (we need only to uncover it), or to create an order out of all that we perceive and all that happens to us. Whether it exists or whether we make it up, order is something we cannot live without: at least it is natural for us to want to live in such a world than in one that is purely chaotic. If there is a single motive to characterize why we keep time, that motive must surely be the search for order.

The first view—that the order is already there—is widely held among modern scientists, while the second is more in line with the way humanists tend to look at the world. So far as our time reckoning originates, as I have demonstrated, in the simple inborn rhythms of nature the scientist is correct: the calendar is already in us. But the humanist is correct, too, for in the past 3,000 years, we have taken time outside of ourselves. We have made our own expressions of it—in poetry and song, with hammer and chisel. We have framed it in tiny blocks and hung it on a wall. We have linearized and circularized it, endowed it with a quality of irreversibility, even artificialized it by wrapping it around our wrists, and exalted it in the turrets of our religious buildings.

Linear Time and God's Process

Why have we insisted on expressing time as a linear entity? How have we come to think of this intangible entity in the same way that we conceive of length-time as the fourth dimension, or of time lines?

According to Marshack's explanation for the notches on 20,000-year-old bones (see pages 67–71), our paleolithic ancestors already were representing change linearly and sequentially. The wiggly boustrophedon notation in figure 2.5B is nothing but a time line with event marks indicated one per day, the days then being grouped into months—is this not time represented as distance? The spatialization of time in the paleolithic record finds its most abstract representation in Einstein's theory of relativity.

Before the advent of clock time, people thought of time as distance traveled. We still speak of that two-hour drive into the city or of going back *farther* (rather than further) in time. The long ago and faraway are somehow linked in our minds: they share a common foreignness and strangeness. We automatically think of the Bushmen of remotest Africa and the aborigines of central Australia as people not unlike our primitive ancestors. Looking back at the celestial derivatives discussed earlier, we can think of time having been created out of the journey of the sun and the moon among the stars. The so-called lunar *mansions* of the zodiac, through which the planets or "wanderers" pass, are derived from the Latin word *manere* ("to sojourn"). Time travel is anything but a space-age concept!

The Greeks had a lot to do with cultivating our linear time perspective: after all, they were the inventors of geometry. They also idealized time and associated it with movement in a rational way. In Platonic philosophy time was thought of as a moving image of eternity. It is the real mark of rational order in a space-bound universe of change. For Plato, space takes priority. We witness the extent of objects in space and describe them by the three dimensions—length, width, and height. We view all changes in these material things in space. To describe change, we employ the concept of time. Now comes the idealization: Plato believed that what we fail to see with our own eyes, amidst all of these changing forms is the eternal or fixed form, the one that lies beyond time and beyond sensibility. For example, all the triangles that might be constructed on a blackboard—or a piece of paper, or out of wooden forms nailed together, or out of pieces of cardboard pasted together—are mere representations and imitations of the idealized form of the triangle that we all must share in our minds—but that is intangible in the experiential world. How else could we all come to know what a triangle is? Just think about it, says Plato. Use your memory and you will realize that there is a perfect triangleness, a form fixed and unchanging in the mind, a quality that transcends time. For Plato, memory does far more than grasp a past that no longer exists; it seizes truth eternal. Only manifestations of forms like this—△—exist in the sensible world of space. When we create such forms (as I did in my written draft and the compositor has done in the previous line), we come as close as we can to reaching the real, timeless truth that lies beyond us all.

Aristotle was the Greek responsible for putting change into quantitative form. He conceived the notion that to be aware of time you need to be aware not only of the objects of experience that reside in three-dimensional space, but also of yourself (in his regard of self-awareness, he seems to have anticipated modern psychologists by a couple of millennia). For Aristotle, time was the "number of change," the "measure of motion."[4] It comes across to us as activity-related. And that is still the way we measure motion, except we have expanded the notion beyond Aristotle's purely celestial concerns. The root of his many arguments about time seems to be that one perceives neither motion without time nor time without motion. But time is not the counting process used to describe change, and it is not the calendar hanging on the wall; it is instead the stuff that gets counted. Likewise, length is not the ruler we use to measure it; thus, Aristotle's time is "the number of movements in respect to before and after."[5] Therein lies the origin of the idea that time is a yardstick with numbers on it. Today, 2,500 years later, time takes on the material form of a numbered dial.

Where did change in the material world originate? For Plato, the eternal principles of mathematics and geometry were first given material form by a sort of divine craftsman. First, as in Genesis, there was chaos; then the creator brought ordered change into existence. In Plato's cosmogony, time was the prime characteristic of ordered change in the universe. It began when the ideals, the forms, the true realities achieved material embodiment. The Greeks of the fifth century B.C. were concerned not so much about what time is as about what is real. Is it, on the one hand, the flux, the change, and the becoming; or, on the other hand, the being, the permanence? At least for Aristotle and Plato, the real truth was rooted in the permanent and the eternal, which could be experienced at best only in vague and illusory form in the changing three-dimensional world of the senses—and time charted the course of change.

I have suggested that the Christian teleological view of the world, which did not begin to be taken seriously by a substantial segment of European society until a few centuries after the crucifixion of Christ, greatly enhanced this Greek-derived linear view of time. Christianity gave the time line a more subjective, human perspective than the Greek rational objective view of the earlier philosophers. Augustine, perhaps noted more for the questions he raised out of his capacity to doubt than for the statements he made as an ardent believer in the God of Christianity, was one of the first to pose certain down-to-earth questions: If in the beginning God created time, what was He doing before that? In other words, if a new will came into existence within Him to do something He hadn't done before, then Genesis can hardly be considered as *the* beginning—at least not for Him. Of the question of time, Augustine remarked, "If one asks me, I know what it is. If I wish to explain it to him who asks me, I do not know."[6]

But for Augustine, who wrote in the fifth century A.D., time comes out of mind, not out of space. Past and future arise from the present. Memory is present-past, and expectation is present-future. We bring out of our memory words that come from images of events that passed through a sequence and left an imprint on our minds. *We* are the subject. Time tells about what happens to *us*—not to the external things Aristotle and the Greek philosophers dwelt upon. For Augustine, space does not need to take precedence in the discussion. His early Christian world of existence is really a stage set up for the events that will lead to the salvation of a fallen world—a salvation that will be experienced only with the coming again of Christ. This event was awaited with great anticipation in the time of Augustine, nearly 500 years having passed since His resurrection.

The Christians and the Roman Catholic Church adopted the Roman system of timekeeping and, like the Romans, used it to perpetuate and maintain the church's increasing power. By the end of the Middle Ages, roughly the fourteenth century, Christianity pervaded Europe and had a tenacious grip on the European way of thinking about time, creation, and history. Although the medieval view of time was allegorical, attaching religious symbolism to all of nature's happenings, Christian thinkers were incorporating Aristotle's views about physical time, the number of motion, and so on, into the teleological, onward- and upward-pulling orientation implicit in the Bible's stories of salvation and redemption. In the natural world, time's scale began to take on the metaphysical form of a ladder, whose rungs were represented by each of God's plant and animal creations. The most rudimentary lower-ordered ones lay at the bottom of the ladder, while God's best attempt at a perfect being stood at the top rung. Only man was created in the image of God, and only he was set up to have dominion over all of God's other creations, just as Genesis told it directly. This ladder-of-time of the Middle Ages was one of the models employed in the molding of our modern concept of evolutionary time. True, the time scale of creation was thought to be vastly shorter then, and the relationship between the various forms God created and the processes that led from one form to another, were nothing like what Darwin would propose centuries later. Nature's order was the order of God, and was ordained by revelation—the only form of acquiring truth in this prescientific age. The perfect order produced a one-to-one correspondence between all things that comprise the universe: the parts of the body and the metals were thought, for example, to be related to the planets. However different the thinking process, the roots of the time structure we would employ centuries later were already firmly implanted.

Harnessing Time to Science

I have spoken several times about the difference between the way we measure time and how we live with it in practice, as if we are of two minds—the scientific one that attempts to exactify and quantify a set of temporal rules, and the societal one that tries to break them. This dichotomy developed during the scientific Renaissance and the Enlightenment in sixteenth- and seventeenth-century Europe, as people came to believe that they might be able to inquire for themselves about the future. As some intellectuals developed a skepticism about the old preordained fixed order of things, the focus began to shift away from the issue of what happens to humans toward what is going on in nature; and by the late sixteenth century, the new inquiry into nature spread. In the middle of that century, the Polish astronomer Copernicus, dissatisfied with the complexity of motion required in any earth-centered model of the universe, proposed that the sun and not the earth was the center of the solar system. His followers considered the implications of such an idea. If the earth really were in motion, then all the stars in heaven need not be confined any longer to a sphere with all the luminaries pinned on it and forced to rotate about the earth in 24 hours. Instead, they could be imagined to lie at any distance from the earth. If space were permitted greater extent, why not time? Perhaps its scale, too, had been underestimated.

The thinker Copernicus was followed by the doer Galileo, who made use of the recently invented telescope (then called "optic reed") to collect evidence to demonstrate Copernicus's hypothesis. Examined up close by the empirical and inquiring Renaissance mind, nature began to take a far more intricate and complicated form. Jupiter was attended by moons like the earth's moon: you could count four of them; and if you looked at them from night to night, you could see them change position. Saturn possessed rings; there were mountains on the moon, spots on the sun; and the number of stars that could be viewed through the nascent spyglass was more than one could ever have perceived with the unaided eye. New instrumentation had revealed all of these things directly to the senses. The view was intoxicating, almost hypnotic.

Heaven was not the only room of nature's vast house to attract the interest of the new empiricists: Galileo also experimented with earthbound things. He became interested in the behavior of falling bodies—little round balls made of different substances dropped from towers of various heights, rolled down planes inclined at different angles, and fashioned into swinging pendulums by hanging them on the ends of different lengths of string. He timed their motion by pulse beat and water clock. The data from his experiments became the raw material for molding a new objective kind of truth—a truth about nature that was concerned with the "thing," and no longer with self.

This careful, precise, and quantitative probing and examination of the material parts of the world redirected long-term intellectual goals and motives. What was the first cause of things? Could an underlying cause be rooted in the old Greek idea that certain basic principles or laws—mathematically expressible statements—govern the behavior of all things? Mathematics, the gift of the Greeks and the Babylonians recovered during the Renaissance, seemed to work elegantly as a way of expressing the motion of the moon or a falling stone. Was God a mathematician? Ought the beginning of time really be conceived when He devised that ideal fixed body of laws that set nature to work? Perhaps we need to be concerned with the process that makes things behave as they do, with the cause that underlies the thing rather than with the thing itself.

In the mid-seventeenth century the French philosopher and mathematician René Descartes conceived of God's process as mechanical, an ideal metaphor for its time since machinery and mechanical processes had been at the center of European life, with the invention not only of the mechanical clock (see pages 88–89) and the telescope, but also of the magnetic compass and the printing press. To explain the unknown, we can use only the concepts and the language we already know. It is clear from Descartes's language, as he explains how the earth formed, that he was thinking of the universe as a piece of machinery, like a clock made up of gears and parts joined together, a mechanism that, once set in motion, operates automatically, fueled by its own self-evident principles:

> this Earth on which we are was formerly a star composed of purest matter of the first self luminous element, occupying the centre of one of these 14 planetary virtues . . . as the less subtile parts of its matter joined up bit by bit, they collected together on its surface, and there made up clouds or other thicker and darker bodies, similar to the spots that can be seen on the sun's surface, continually forming and later dissipating. . . . Several layers of such bodies were perhaps piled up one above the other, so reducing the vortex containing the Earth that it was completely destroyed and . . . the Earth together with the air and the dark bodies surrounding it descended towards the Sun as far as the place where it is at present.[7]

Descartes's illustration of the coglike vortices out of which humans were created graphically resembles the wheels and gears of a gigantic machine (see figure 4.3). There was great force and power in Descartes's idea. If we could decipher not *why*, but *how* the machine worked, then the mathematical set of laws that tells how the universe works offered the potential to predict the condition of the physical world at all times in the future. We could test specific predictions regarding such conditions by directly examining the phenomena that had been forecast. Any differences between the prediction given by the laws of the machine and the results of experiment could be attributed to misunderstanding of the law—differences we could correct by changing or adjusting one of the cogs in the blueprint of the machine.

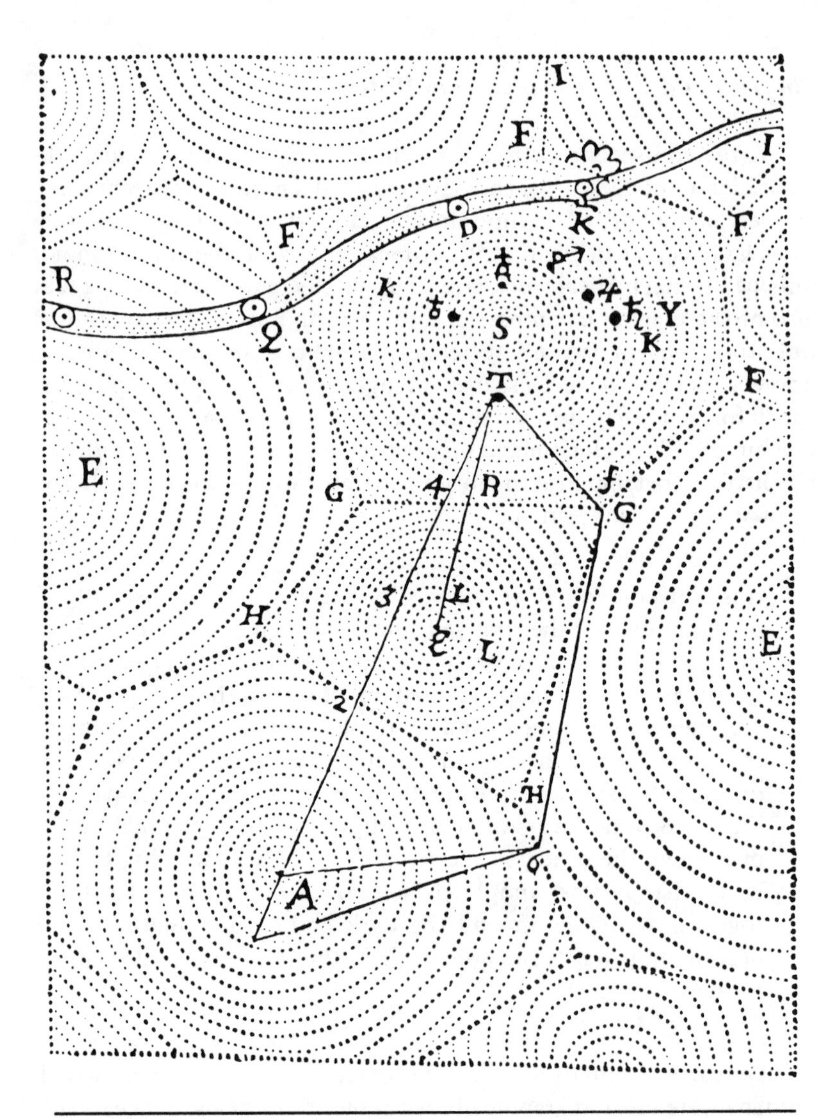

FIGURE 4.3 The "world machine" in the seventeenth-century vortex-cosmology of Descartes conceives of the formation of planetary systems as a result of the gearlike mechanical flow of matter—all subject to the laws of nature. Each vortex churns about its center, point S being at the center of one of the vortices out of which the sun was created. Symbols adjacent to the sun represent the planets; more remote symbols are other stars. Dotted circles represent lines of flow on which the planets are said to have formed; the strip across the top is the orbit of a passing comet. SOURCE: R. Descartes, "The World or a Treatise on Light," in C. Adam and P. Tannery, *Oeuvres de Descartes,* XI (Paris, 1909).

There was another revolution going on in the setting of nature's long-term clock. Once quite comfortable together, science and religion were on a collision course, and time was the issue. Today we are still experiencing the fallout. Scientific naturalists who probed the physical processes governing the behavior of the surface of the earth began to realize that the biblical time line, revealed in Genesis to be no more than six thousand years long from creation to the present, could not possibly square with the evidence from the geological record, which suggested a far greater antiquity for the earth. What were marine fossils, for example, doing on the tops of mountains or buried beneath layer after layer of soil deposits of varying composition? And why were identical species found in the same layers on one coast that were also found on another? How long must it have taken for two continents to float apart or a river to carve its way through solid rock down to the level of the valley where it now flows? How long for the sides of a mountain to be eroded away into the rocky deposit at its base? The processes of deposition and erosion that led to these conditions could not possibly have taken place in years—or even generations—not if they ran at the same slow, uniform rate in the past as can be seen operating today. Thousands, hundreds of thousands, conceivably millions of years would have been required to produce such changes. Time could still be a line, but to account for all these processes it would have to stretch almost to infinity.

The biblical answer to all these questions about "How long?" lay in catastrophe. Time's events are not continuous and evenly spaced. There are breaks in time. The changes wrought by the flood epoch were sudden. And even today large sections of earth can be turned upside down in an instant by an earthquake or covered over in a flash by the eruption of a volcano. Catastrophe has no need of "process."

Biological turf became the great battleground in the debate about whether the time scale is continuous or catastrophic. Specifically, the debate focused on whether the species of plants and animals that could be found in both the fossilized past and live present worlds had remained fixed or had changed one into the other. The resemblance between apes and humans was all too obvious, but one fact became clear to those who studied the animals with the same care and rigor that Galileo had applied to the stars with his telescope and to falling bodies in his laboratory: for every kind of animal one could identify, another could be discovered closely allied to it in appearance and behavior—as if they lay next to one another like the links of a chain. For example, amphibians are aquatic—like fishes—but also terrestrial—like warm-blooded animals. There are fishes that fly like birds and birds that dive into the water like fishes. God's creatures could be laid out on a line, from the simplest to the most complex along a scale of interrelations, in a sort of continuum.

At first it was thought that all living forms in this great Chain of Being (from which derives the once-popular notion of the "missing link") were created each unto themselves, with man residing at the top of the hierarchical pile of complexity, just as Genesis implies. But the progressive, process-oriented, developmental way of thinking about nature had taken hold and, by the mid-nineteenth century, had become encased in a feeling of social optimism about what the new machine technology could do for the human race.

The calendar reform of the republic created by the French Revolution, one of the late eighteenth century artifacts to emanate from the worship of nature's time machine, constitutes one of the most thorough attempts to wipe out the traditional calendar system. In a single stroke, all connections to the past were erased, and all units of time abolished and replaced with new, more uniform ones. Months were made the same (12 each of 30 days with a 5-day period tacked on at the end) and, instead of retaining the names of emperors and deities, were named after seasonal aspects characteristic of each: mist, frost, snow, germination, harvest. Days were divided decimally into 10 hours each of 100 minutes, every minute containing 100 seconds. There were 10 days in a week instead of 7, which meant 9 consecutive days of toil instead of 6 before a day of rest!

Now the Republican era replaced the Christian era; the count of the years would begin with 1792; and the new New Year's Day would be 22 September—the equinox. But French Revolutionary time ended as abruptly as it began. In 1806, only 14 years after it had been introduced, Napoleon brought the French Revolutionary era to an end and restored the Gregorian calendar.

All that remains of the failed calendar today are a handful of odd-looking timepieces gathering museum dust (see figure 4.4). Perhaps the attempts of the government to gain control of the people by forcing upon them a new secular rhythm in the name of progress ran too much against the grain of religious tradition. Maybe the break with the past and with all other time-reckoning systems outside France was too sudden, too radical, too confusing, to live with. Whatever the reasons, there is some irony in this little-known secular attempt to put natural phenomena at the center of the time system of a new-age

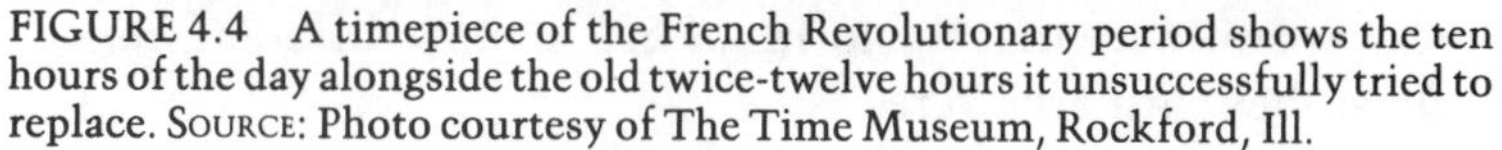

FIGURE 4.4 A timepiece of the French Revolutionary period shows the ten hours of the day alongside the old twice-twelve hours it unsuccessfully tried to replace. SOURCE: Photo courtesy of The Time Museum, Rockford, Ill.

French society. The Enlightenment philosophy emphasized that science, reason, and the natural order were the principles man was designed to live by, and yet the method of implementation was discontinuous and forced. The new calendar was a social creation rather than a natural one.

Evolutionary Time

When the progressivist attitude was applied in the 1700s to the question of seeking the laws and processes—those first principles by which the biological species operate—the idea of evolution was born. The key lay not in thinking about biological laws that operate on individuals but rather on those that apply to species—collections of individuals all having the same detailed organizational characteristics. Attention was turned to the search for laws that govern descent and inheritance. Darwin gave us one kind of law we were looking for. It embodied both process and progress, but no one had any idea how drastically it would deflate the human position in the great scheme of things:

> This preservation of favorable conditions and the rejection of injurious variations, I call Natural Selection. It leads to the improvement of each creature in relation to its organic and inorganic conditions of life; and consequently, in most cases, to what must be regarded as an advance in organization.[8]

> It may be said that Natural Selection is daily and hourly scrutinizing, throughout the world, every variation, even the slightest; rejecting that which is bad; preserving and adding up all that is good; silently and insensibly working, wherever and whenever opportunity offers, at the improvement of each organic being in relation to its organic and inorganic conditions of life. We see nothing of these slow changes in progress, until the hand of time has marked the long lapses of ages, and then so imperfect is our view into long past geological ages that we only see that the forms of life are now different from what they formerly were.[9]

With a stroke of the pen, human beings became a mere accident on nature's randomized ladder of progress.

In the twentieth century, three major developments have taken place in the way we think about time, all extensions of trends well under way in the last century. First, time and space have become tightly welded together; second, the time-space scale has been vastly extended; and third, we have honed the idea of evolution acquired from Darwin and his predecessors into the model par excellence for understanding creation and cosmology. Let me examine each of these premises.

When we look far out into space with our most powerful telescopes we also look backward in time. This is because the speed with which information comes to us from those distant places—the speed of light—is not instantaneous. The space-time map in figure 4.5 illus-

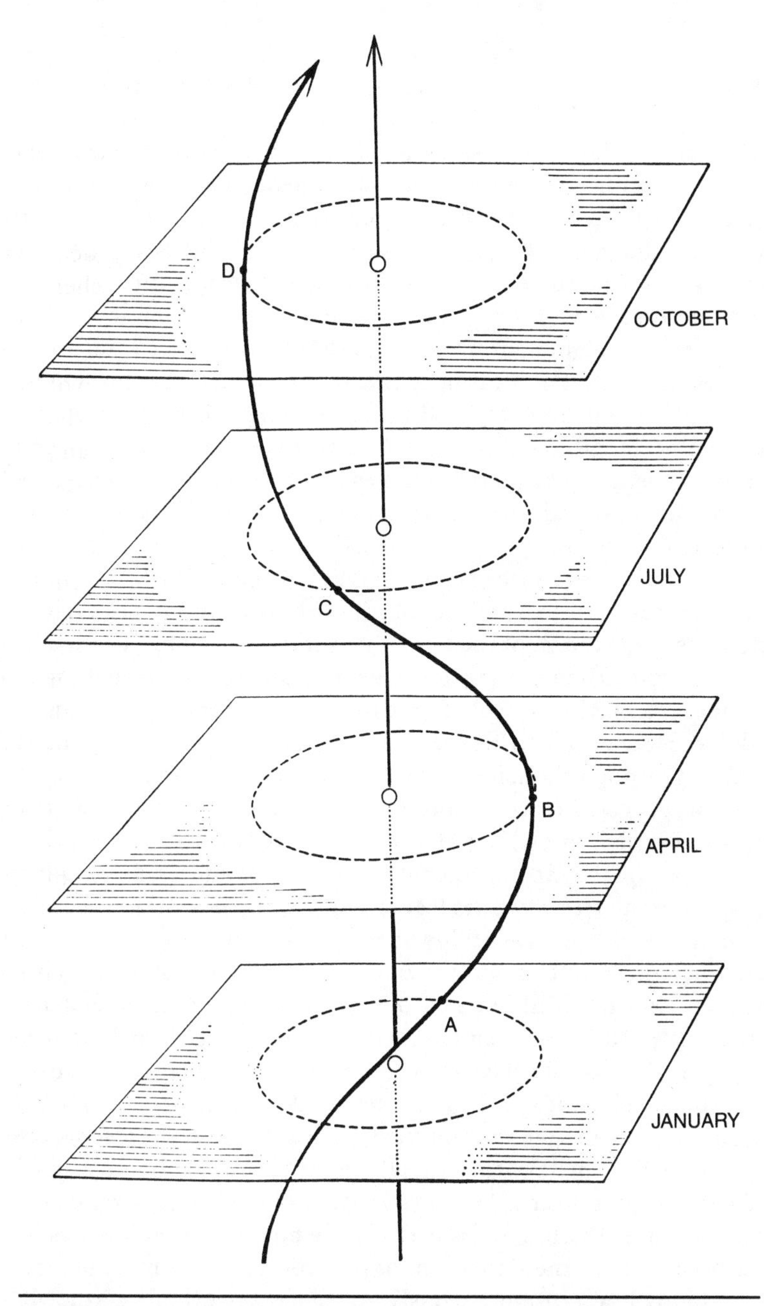

FIGURE 4.5 How we make time into space. In the drawing, the earth's position is represented in four different planes of existence stacked up along a vertical time axis. A "world line" (*A, B, C, D*) connects these positions. After an illustration by James Egleson in G. Gamow, "The Evolutionary Universe," *Scientific American* 190 (3 [1956]):55–63.

trates the sensible universe as a series of vertically stacked time-planes, each one a hypothetical two-dimensional cross-section of space. Conceived by the physicists who developed the theory of relativity, this diagram represents *time* as *space.* It is the best picture we can make to pictoralize what is hypothesized as happening at a given location in space over the course of time. In the direction we ordinarily use for the height coordinate we have substituted the time dimension so that instead of creating a three-dimensional diagram showing the length, width, and height dimensions of a static spatial universe, we employ length, width, and *time* to depict a dynamic *spatio-temporal* cosmos. The picture implies that *now* and *then*, words used to describe temporal arrangements, really are tied directly to other words we employ for spatial concepts, like *here* and *there.*

Specifically, the space-time diagram tells me that if I look out to any distance—say, to a tree about 40 yards away in my backyard—I actually see the image of the tree, not as it looks *now* but as it looked *then*—at some time in the past. This must be true because it takes a definite, calculable interval of time for the light signal that ultimately makes the tree visible to me to bounce off the tree, make its way across the lawn, through the screen door, and into my eyes and brain. Not a lot of time—just a mere ten millionth of a second for a tree 30 yards away. Certainly not enough of a time delay for me to conclude that I am not getting the most up-to-date, on-the-spot impression of what's going on in the tree's life. Its leaves and bark are essentially the same age *now*, when I see it with my eye, compared with *then*, the recent past when the light first left the tree's surface on its way to my visual channel. And the bird perched on one of its branches cannot have shifted its position or even have experienced a heartbeat in that ten millionth of a second delay between transmission and reception of the light signal. The ant crawling up the trunk of that tree cannot have advanced a step in that brief interval. Still, as I gaze at the tree so close to where I sit, I believe that I really do not see it in the present but rather in the past. Even if I get up out of my chair, walk out to that tree, and press an eyeball up to its bark, I have no hope of seeing it as it is *now* because the time signal cannot pass through my optic apparatus and into my brain instantaneously. Indeed, every event I witness lies in the past, the more remote from me in distance, the further back in time. Twentieth-century time offers me no *now*, no *present*, except for the flash of an instant I conceive in my mind.

This revelation may not matter much in the everyday world, but when you turn your eyes upward and outward and begin to deal with distances far greater than those separating me from the tree or New York from Tokyo, the lapse in time between here and there proceeds to grow. Moonlight is already 1½ seconds old when it gets here; the light of the sun, 8 minutes. Light from the nearest star has aged like a fine wine for more than 4 years, the nearest sizable galaxy over 2 million, the most distant perhaps 15 billion years. The realization of such measured distances has stretched our time line from the biblical few thousand years, past the geological dimensions of millions, tens and hundreds of millions of years, to the new astronomical and cosmological time frame of billions of years.

Looking along the time axis of figure 4.5 to those earlier planes of existence, we ask, If species and societies have evolved, how has the universe evolved? This is the logical question given the post-Darwinian emphasis on development and process. If the world is changing, is the universe changing, too? For the astronomer, this question translates to one that is observationally testable: Does anything look different in the early stages of the universe which we can see by looking at things very far away compared with what we see in nearby things that are closer to the present?

The answer is decidedly yes. First of all, the farther the galaxies are, the faster they seem to be moving away from us. The universe is expanding, and the galaxies that comprise it are all flying away from one another at colossal speeds. Run the tape backward, and you can imagine all of them having been united together at the time of creation. Yes, too, because astronomers have detected *quasars*, superluminous celestial objects that appear to resemble galaxies in their earliest stages of formation. Since we find quasars only at great distances from us, they must have existed only in the far-distant past, either having become extinct or transformed into something else in the nearby present. It is as if the great birthing period in the universe has long since passed, and our only view into that cradle lies in the remote images we acquire with the world's largest telescopes.

Our modern genesis story is the reigning evolutionary theory of cosmology, which suggests that the whole universe started with a bang—a *big bang* about 15 billion years ago as near as we can calculate it given the measured rate of universal expansion. The earliest form of the evolutionary universe theory was devised around the turn of the century, barely 50 years after Darwin announced his theory of biological evolution. At first it was referred to as the "primeval atom theory" because it advocated that the entire universe resulted from the radioactive decay of a single super-atom. The work on *radioactivity* of the French physicists Pierre and Marie Curie had just become known to the world, and this exotic notion of atoms erupting like popcorn had changed Western ways of thinking about the material world. Whether we see the universe as one giant atom that runs down from a more complex to a less complex state, as the early form of the evolutionary theory seemed to indicate, or as a build-up process from a simpler to a more complex entity, as today's Big Bang conceives it, the truth is that we think of the universe we live in as a developing, dynamic, evolving aggregate of matter and energy. Modern "Genesis" offers us a universe in transformation, and today's questions about time and creation emphasize process, not stasis. Our interest lies in change. What we really want to know is how the universe progresses.

The act of separation is one of the major themes that has emerged out of our modern understanding of how the early universe underwent change. This is a striking parallel with Babylonian, Greek, and even the Old Testament creation myths. Originally the universe consisted of a kind of formless aggregate of matter and energy linked together. (The cosmologist George Gamow called it "Ylem" after the biblical Hebraic term meaning that stuff out of which all was created;[10] modern astrophysicists call it *plasma.*) In the first few moments following the flash of creation, the initial expansion was accompanied by a drop in temperature and the unified substance decoupled into matter and energy, the two formal entities that make up the universe we recognize today. It is fascinating to hear these developments described by one of the early proponents of the evolutionary theory:

> The transition from the reign of thermal radiation to the reign of matter must have been characterized by a very important event: formation of giant gaseous clouds. From these "protogalaxies" the galaxies of today must have developed, somewhat later, by the condensation of gas into individual stars. During the period when matter had played only a secondary role in the infinite ocean of thermal radiation, it had had, so to speak, no will of its own; the particles of matter were

> "dissolved" in the thermal radiation, much as molecules of salt are dissolved in water. As soon as matter took the upper hand, the forces of gravity acting between the particles must have caused a growing inhomogeneity of the matter in space.[11]

Though modern physics may have lost the nerve to address ultimate cause, this passage describes the act of separation in the same vivid way as is described the battle between Apsu and Tiamat in the Babylonian version of creation. Here the terminology remains rife with animation and personification, except that more abstract powers seem to be engaged in the struggle that will ultimately lead to human existence. This evolutionary theory is our creation myth. And just as the Babylonians inquired about what was happening before Apsu and Tiamat, it is reasonable for us to ask what might have preceded the universal expansion. In their mind's eye, some evolutionary cosmologists have extrapolated creation back to an earlier state during which they imagine it to have been contracting down from a previous expansion.[12] Having reached a maximum state of compression, the contents rebounded so that today we find all things in an expansive mode. There may come a day, they propose, when the universe will reach its limit of expansion; the mutual forces of gravity tugging all the while on the component parts will pull it all back together again into a dense ball, but then it may re-expand again. This scenario, the so-called oscillating universe, while not as popular as it used to be, nevertheless offers the satisfying attractions of a time scale that is both continuous and eternal. But while things go on and on, we as *Homo sapiens* won't be around to participate, for the next big crunch will squeeze us out of existence. Furthermore, when it happens, "we" won't be *Homo sapiens* anymore, because the principle of natural selection, which runs on a much faster clock than cosmic evolution, will have carried us much farther upward and outward along one of the branches of Darwin's budding and blossoming evolutionary tree. If and when the big crunch finally comes, you and I will be only fossilized remnants of an indeterminate species that could scarcely be recognized as our progeny. Of course, this is all hypothetical. Whether there will be a collapse, much less a future expansion, remains a serious issue for modern cosmologists. Today evidence weighs against it.[13]

Another innately satisfying quality of an oscillating universe is its rhythmicity. The theme of alternating opposites and polarities is already familiar in the many schemes that have contributed to our cosmological heritage and occurs in non-Western creation stories (see part III). The attribution of paired states to things and essences—from plus to minus, yin to yang, good to evil, proton to anti-proton—emerges in all creation stories, one of those universally desirable, symmetric ways of portraying the human condition.

As far as can be determined empirically, the present condition of the universe seems to be entirely independent of what took place before the Big Bang. Prior to the creation event, all matter, radiation—indeed, all information about existence—was trapped in a domain of time-space that remains forever inaccessible to us. Our cameras can be run back to only about 15 billion years—before that, the film self-destructs. For us, time began then. And as long as we are unable to acquire any *experiential* evidence from the pre-expansion period, we can never know, through scientific inquiry, what happened before the Big Bang. Any answers we might seek to this question must be metaphysical rather than physical. But metaphysics for most of us has a negative connotation. By its very definition, it jumbles things together from outside physics—even from religion. We place our faith in today's Genesis upon truth as revealed through fact—fact based upon the testimony of the material world, not upon the spoken word. Physical inquiry has come to be regarded as the only viable means of getting the facts, through what we experience and what we can measure. A millennium and a half ago, Augustine put the task of going back beyond time, for those who have the fortitude to inquire into it, into terms that still hold true for us: they would enter into a chaotic pool of fire and brimstone called Gehanna, created by God when he made heaven and earth, "where there are dreams or those visions of immaterial reality which are perceived in ecstasies."[14]

Cosmologists today are expending considerable mental resources on the first few millionths of a second after creation because these instants represent our primal experiential contact with the world outside. One of the most significant questions being raised about those first moments concerns whether the basic mathematical laws we have been using to describe and predict the behavior of the physical universe are, in fact, the first and only laws. Are the laws and the very constants of nature changing? Darwin had shown unequivocally that life has a history. So does the earth. We even have given names to geological periods to describe how the earth progressed. But, unlike biology, geology, and astronomy, physics has until recently been the one branch of science thought not to possess an evolutionary time scale; its laws have been nature's untouchables, immutable and invariant, assumed always to be the same. Today some physicists seriously argue that, in the earliest stages of the universe, the constants of nature might have been a bit different. So physics, too, may evolve.

As the twentieth century concludes, we begin to realize that every aspect of existence, both human and physical, has succumbed to the paradigm of development. Nothing remains fixed in nature; all things progress continually along a time stream that flows in one direction. The evolving universe is infected with the same germ of indifferent, even pessimistic, fatalism detectable in Darwin's biology. We have sought to harness and control nature, but the payoff seems to be that we find it to be all out of control. While we can stave off the effects of the deterministic laws of biological evolution by playing the dangerous game of altering the structure of our genes, cosmic determinism is sure to defeat us. Today we face the issue of whether we can ever significantly change the course of life on earth by our own direct intervention.

We find ourselves in the role of bystanders, not participants, who have been written out of the script. Our determinism has led us to the attitude evoked by Weinberg's statement quoted at the beginning of this chapter. Is it time for a new course of action—a change of philosophy like that experienced in the Renaissance, an ideological change that gets us back into the picture as active participants who *can* make a difference? There have been recent attempts to climb back into natural history. Though not given much support by traditional cosmologists, one novel idea suggests that the universe has all the properties observed in it today because, if those properties had been significantly different, we would not be around to witness them.[15] The *anthropic cosmological principle* may sound like double talk, but the argument makes sense, its advocates say, if you look at the relationships between some of nature's constants and other measurable physical quantities that are time-dependent, like the rate of expansion of the universe. The principle's advocates contend that different combinations of these numbers produce the very same equalities—a result that, in their judgment, cannot be pure coincidence. For example, the age of the universe expressed in atomic units—that is, the time it takes light to travel a distance equal to the radius of a proton—turns out to be the same as the gravitational coupling constant that measures the strength of the universe's gravitational force (the number is 10^{-40}). Now, if these constants were much different, galaxies, stars, and planets could not have formed as we know them, and consequently we could not be here. Because the observer requires a world in which to live, the existence of the spectator becomes linked to every constant of the universe. Imagine alternative worlds in which there were no Judas to betray Christ, no Mozart to compose a symphony, or no Lee Harvey Oswald to assassinate President Kennedy—each an unreal world because we were not there to participate in these events. So goes the logic. But to say the universe is here *because* of us seems like backward logic. This way of reasoning runs counter to the deductive logic of science, which depends upon a scale of time that places cause before effect. Clearly, we cannot have been here before the universe! Traditional cosmologists have expressed their indignation. They see it as too teleological, as lacking in scientific principles, as a gratuitous abandonment of the "successful program of conventional physical science"[16] that tries to explain natural phenomena by appealing to universal laws.

Biologists also have been making efforts to extricate humanity from the abyss of determinism, and some have raised the question whether life's processes exhibit directional properties. Would the living world take on a different appearance if we ran the tape backward? Stephen Jay Gould and other biologists have employed "clade" diversity diagrams to decipher whether time is like an arrow that points only in one direction.[17] *Clades* are segments of the branches of the evolutionary tree whose thickness denotes the diversity of species within those segments (see figure 4.6).

Viewed through time, all biological taxonomic groups develop out of a single species, grow to maximum diversity as measured by the number of different types within species, then decline, and become extinct. A typical diagram showing clade development plots time vertically against degree of diversity horizontally. Ideally, a clade diagram would be a battleship-shaped configuration where the linear tip at the bottom indicates the creation of the species, which then waxes up to the widest point, typifying the largest number of members that occur at the peak of success of the species. Finally, the configuration wanes and narrows down to a point of extinction at the top. Given the ran-

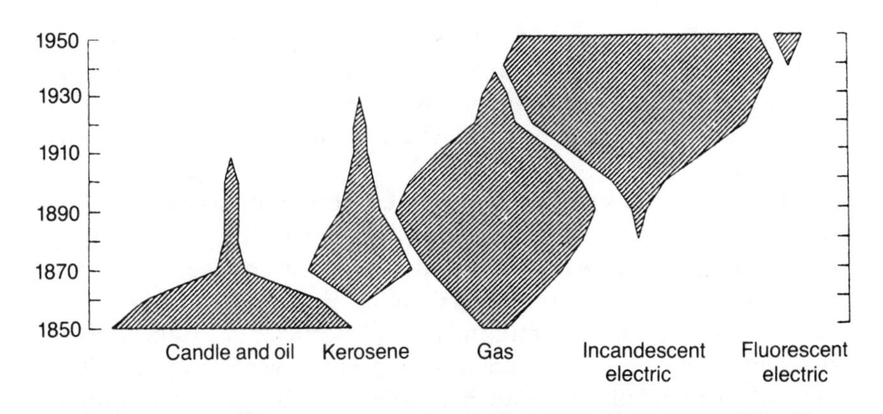

FIGURE 4.6 Clade diagrams show the growth and decay of artificial lighting devices. If there are as many bottom-heavy as top-heavy clades, diversification may not depend on which way time's arrow points. SOURCE: S. J. Gould, "Asymmetry of Lineages and the Direction of Evolutionary Time," *Science* 236 (1987):1437–41, fig. 6. Copyright © 1987 by the American Association for the Advancement of Science.

domness of nature's behavior, we ought to expect some clades to look bottom heavy. Such species achieve early success and die out relatively slowly. This kind of behavior would occur if a huge ecological niche were suddenly opened up, an event believed to have happened for the mammals when the dinosaurs became extinct over a relatively short period, allowing the former an unusual opportunity for relatively rapid diversification. On the other hand, some clades would appear to be top-heavy, thus indicating a long slow development and diversification. Now, let us look at large numbers of these clade development diagrams and focus on the symmetry of the distribution between top- and bottom-heavy ones. If there are as many top- as bottom-heavy clades, we could conclude that were we to run the clock of time in a backward direction, the diversification of species would take place in precisely the same way. If, on the other hand, the distribution is asymmetric, then at least one property of the evolutionary process suggests that things would be different in a hypothetical universe in which time ran in reverse. The fact is that the history of life as we know it, like the history of the universe, may be a one-way process. There does, indeed, seem to exist a bottom-heavy clade asymmetry—one particularly manifest early in the history of life. Life seems to be characterized by sudden breakthroughs and the slow demise of species rather than the reverse. And this likelihood in turn suggests that some deeper principle of organization yet to be discovered may regulate genealogical systems over the course of time. There is more to life than natural selection.

Expanding universe and asymmetric lineage both seem to suggest that time flows in a single direction—the way we live it. Supporting evidence on the one-way nature of time comes from nuclear physics and thermodynamics. The neutral K meson, or Kaon, a subatomic particle, decays into two separate and distinct particles in a time interval that can be shown to be different if time were reversed. The result is derived theoretically by mathematically reversing the basic atomic and nuclear properties, reflecting them into a mirror and running the time parameter in reverse. A well-known theorem from quantum mechanics predicts that the result should be indistinguishable from what would take place in the real world. But the results from the laboratory imply quite the opposite—a grain of sand in a nuclear oyster, small but too irritating to be ignored. One physicist regarded it as a "minor welding defect" in the construction of nature,[18] a remark reminiscent of the persistent belief in a mechanistic, if no longer fully deterministic, universe.

Entropy, a much overused and abused physical concept from thermodynamics, has been applied to everything from the initial fireball to the totalitarian state. Entropy is the measure of disorder, and the second law of thermodynamics states that entropy shall always increase. The classic example is that of billiard balls spreading out from their highly ordered state when first in the rack to the relatively disordered condition in which they later come to rest.

Today we think of chaos as the ultimate state toward which things tend if we assume our universe began in its most highly ordered state—in its lowest state of entropy with matter, both chemically and atomically; situated in its simplest, most singular condition; all packed together into a very small volume. An arrow of time seems to be built into all nature's processes, and it points from order toward chaos. Because there are an infinite number of states of disorder—that is, of positions the balls can take on the table—and only a single state of maximum order—wherein the balls are tightly packed together in a triangular configuration—chance favors disorder. In this light we wonder whether man created God to battle the inevitable and everpresent law of entropy—to undo what time wrought? After all, God's job description seems precisely contrary to the modern conception: in Genesis, He establishes order from chaos. And while all of these considerations of modern physics depend upon assumptions about the earliest stages of the developing universe, how do we know there did not once exist another set of microscopic laws that set up time's irreversibility in the first place, or principles that created a reversible sense of time that we simply have not yet observed? These are but some of the questions being asked at the frontiers of exploration of evolutionary time.

Conclusion

These ideas about evolutionary time are the foundation for the concrete time periods we in the West have constructed. Thus, the course of Western history—and concerns human and civil, religious and, lately, scientific and technological—have been responsible for our unique view of time. We have borrowed heavily from the Greeks: their logic, idealism, and reason lie at the foundation of how we think about time and its relationship to the natural world. The practical elements associated with time, the ones found when I unpacked the calendar unit by unit, seem to be descended directly from the Romans and their version of timekeeping, the ideas about time in the Middle Ages, especially its endowment with a linear quantity, having grown out of Christian beliefs propagated by the Eastern Holy Roman Empire, the formal bureaucratic structure that survived Rome. As mechanism replaced animism in the Enlightenment, and mechanistic determinism became the preferred form of reasoning, the world machine offered the most felicitous explanations about the history of the universe. Making good *ideas* to explain how things work was part of the success story of the Western invention of good *things* to simplify life. New technology was accompanied by new empiricism—a pair of compatible forces welded together by the desire for precision and regulation that had stemmed from medieval ecclesiastical concerns and was later fueled by the rise of the merchant class. Time's line became more continuous than discontinuous and, like the mechanical clock, was engraved with a carefully tooled set of fiducial marks to quantify its course, every indication on the measuring stick having been placed there as a result of direct experiment.

In the Renaissance, Greek idealism resurfaced in the form of the quest for underlying fixed and eternal first principles—the need to fit individual events into a general set of laws. The dialogue concerning these principles could best be conducted in the language of mathematics.

The spatialization of time, vividly expressed in the twentieth-century "world diagram" of figure 4.5, also started with the Greeks, who gave us logic in a most unusual and succinct form—one not practiced as extensively anywhere else in the world. It is Greek geometry that has allowed modern cosmology to be so spatial, so orbital and planar, and is the basis of "space science," the geocentric, heliocentric, and galactocentric coordinate systems, and the cartesian reference frame. In our visually oriented world view, time has been seen in reference to space; and since the time of Descartes, we have laid out space in concrete terms.

Our last few hundred years of thinking about time began optimistically—a decided upswing in the pendulum's amplitude from the protracted medieval pessimism over the failed Second Coming. Even the nineteenth century's last holdouts began to despair of that long-anticipated Resurrection and Assumption. The notion of temporal continuity coupled with optimism made for a progressive way of thinking about change that is still operative. And our passionate love of progress coupled with our insatiable appetite for technology may know no limits.

Today's Empire of the Clock

If we think for a moment about the differences between *ideational* time (the kind I have just been discussing) as opposed to *lived* time, it is easy to see that, of all the artifacts and ideas ever created, not a single one causes us to separate the two more than does the clock. The American social philosopher Lewis Mumford has said of the clock that it "disassociated time from human events and helped to create the belief in an independent world of mathematically measured sequences: the special world of science."[19] The hold the clock has upon Westerners cannot be understated.

"Time rules life" is the motto of the National Association of Watch and Clock Collectors—a statement borne out in the formal time units that make up our calendar as well as in the way everyday events in our lives have become organized and packaged. In many sports, play is controlled by the clock. Hockey has its three 20-minute periods; football its four 15-minute quarters; and basketball, a pair of precisely timed halves. We time our records in individual sports—except for tennis, golf, and baseball, of course—to the nearest hundredth, sometimes thousandth of a second; and athletes often speak of breaking time barriers—4 minutes for the mile or 10 seconds for the 100-yard dash. The sport of kings has its photo finish. We design our equipment and condition our bodies to meet the challenge time imposes upon us. In team sports, we speak of "fighting the clock"—that is, calling a temporary halt to proceedings at the appropriate spot toward the end of the game. Often the last 2 minutes of active play in professional football and basketball games are punctuated by numerous "time-outs," strategic sessions when time for the participant literally stops—though not so for the unfortunate TV spectator, who is assaulted by a banal barrage of precisely timed commercial messages.

Time scheduling also coordinates work activity. In the building industry, efforts are carefully time-oriented so as to avoid bottlenecks. We have become regimented under the empire of time. Think of an assembly line in an automobile factory, where human automation is barely distinguishable from the mechanical. Rather than being directly involved in fashioning an object, the worker is plugged into a fixed and immobile temporal framework and allotted a single task to repeat over and over again. An object arrives from the left in an earlier state—say, without a door latch—and departs to the right having been endowed with one. Speeded up to maximum flow, the worker's task becomes not only tedious and boring but also stressful. Like the quarterback running out of time, the efficient worker also must battle the clock—a situation memorably parodied in Charlie Chaplin's 1931 film *Modern Times.** First introduced into the United States in the 1920s, the assembly-line process of mass production reflects many

*A more recent parody of this situation was an episode in the 1950s of the popular television series "I Love Lucy." In the scene, a takeoff of the Chaplin film, the timing of the conveyor belt in a chocolate factory assembly line goes awry, forcing our heroine to struggle with temporal demands well beyond human capacity. Unable to coat them with their final layer of chocolate fast enough and not daring to let a single item pass on to the next stage unanointed, she tries to stuff the rapidly accelerating sweet morsels into her mouth as they whiz by. Later, in desperation, she begins to cram them into every available aperture in her clothing.

of the properties of scientific timekeeping I have discussed—sequentiality, consecutive change, and control—with emphasis on linear progression as much as upon product.

The journalist Max Lerner has said that our society cuts away the sensitivity to death and grief, suicide and mortality by emphasizing the here and now, youth and action.[20] Imbued in the youth culture, we act to prolong and extend the present by attempting to keep ourselves in the same constant condition, both mentally and physically. We try to think young and act young. We pace ourselves through the assembly line of Nautilus and Jacuzzi and ask the machines we have built to work wonders on our bodies—that material, tangible element of the human fabric. Often we tax ourselves with unreasonable physical demands like 10-mile daily runs, and hope that through changes in diet and exercise we can delay the sagging effects of universal gravitation or the inevitable onset of disease. We, the great advocates of change, all desire to remain suspended in time; not a single one of us does not abhor growing old. When we do think about the future, we look only to one that is compassable, one that provides for retirement or our childrens' needs.

The scientific establishment is no less than the layperson hung up on the denial of time advocated by the youth culture. The final years of our millennium have witnessed one popular book after another on the birth and early evolutionary stages of the universe, from Steven Weinberg's *The First Three Minutes* to Steven Hawking's *A Brief History of Time*.[21] Though both brief, neither really reaches much farther back than the past two decades of thinking on the matter of genesis—such is the progressive nature of science. One historian of cosmology has commented that, if modern physicists were as much interested in eschatology (the death of the universe) as in cosmology, then "somebody would already have matched the best-selling *The First Three Minutes* with *The Last Three Minutes*."[22] At all levels, our eyes remain fixed on the careful management of the tangible present.

One of the most interesting episodes in America's recent attempts to imperialize time is the strange story of the failed world calendar. Like those mechanists of old who sought to design the elusive perpetual motion machine, the dream of all timemakers is the perfection of a perpetual calendar, which would self-adjust the irregular months, equalize the quarterly divisions of the years, and fix the sequence of weekdays and month dates so that they would be identical from year to year. The United Nations seriously considered adopting such a model—"a scientific system of time measurement without sectional, racial or sectarian influence," as one proponent called it.[23] Spurred on by American leadership and imbued with the same post–Second World War attitude of universalism that activated the notion of a common language (Esperanto), various propositions for the "One World Calendar for One World" were devised. The rationale for each was the same as that which drove the old medieval mercantile class: "time is money." A business executive once described the existing calendar as a "smooth and subtle thief."[24] Consider, for example, the time required to determine on what day of the week the tenth of the next month will fall or whether Christmas will occur on a weekend next year. One news commentator estimated that it cost the taxpayers of the City of New York $5,322,866.25 a year to reckon time—and that was in the 1930s.[25] Vagaries in our calendar produce variable quarters (accompanied by erroneous estimates in quarterly contracts), variable overtime, variable time-payment periods—the list goes on and on; and all these uncertainties are hidden in the interstices of an irregular time grid upon which it has become difficult to rely for the regulation and planning of activities, the keeping of precise records and statistics demanded by an advancing world. This was the argument.

Our ancestors had proposed many perpetual calendars, some dating all the way back to the Enlightenment period (such as the calendar associated with the French Revolution [pages 144–45]). A utopian version of the calendar that captured the most interest in reform-oriented, post–Second World War America was the one that advocated withholding the 365th day, thus making a normal year 364 days long. This number, also employed for similar reasons by ancient Maya timekeepers over a thousand years ago, has the distinct computational advantage of being easily divisible, especially into 4 equal quarters of 91 days apiece. And these quarters can be segmented into identical month sequences of 31, 30, and 30 days. According to the plan, the first month of each quarter would become the longest; thus, January would be assigned 31 days, while February and March would have 30 apiece; the second quarter would start with the 31-day April, and so on—a much easier mnemonic than "thirty days hath September, April, June, and November, . . ." (see figure 4.7).

But the supreme advantage of using the number 364 as the year base is that it overcomes the bugaboo of the wandering week, for it is

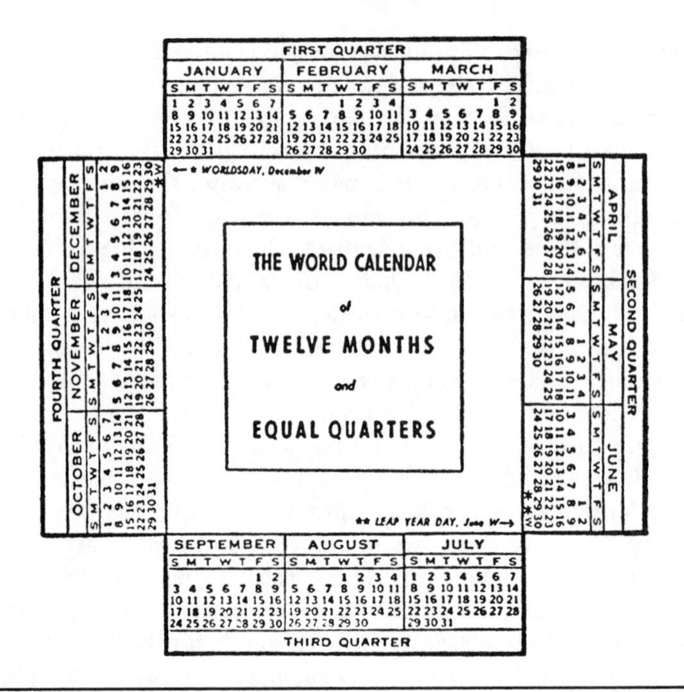

FIGURE 4.7 The perfect calendar. Equal year quarters each with months of 31, 30, and 30 days, and birthdays on the same day of the week—even an international "Worldsday" to cap off every annual cycle—a technically flawless world calendar that has repeatedly failed passage in the United Nations. SOURCE: E. Achelis, *Of Time and the Calendar* (New York: Hermitage, 1957), p. 161.

divisible exactly by 7. Thus, every year in the new calendar would have 52 whole weeks, and consequently every quarter would begin with a Sunday and end with a Saturday. Every 1 January would be a Sunday; every 1 February, a Wednesday; every 1 March, a Saturday (Monday, 2 January, would be the first business day of every new year). This same sequence would repeat for each of the 4 temporal segments that make up the year.

Now, because the year timed by the seasons is actually closer to 365 days (365.2422 days, to be precise), it would be necessary to add an extra day to every year and to intercalate yet another extra day according to the leap-year prescriptions described earlier. What could be more suitable than to call that extra day "World's Day"? This day, formally called "W December," but unnumbered in the usual sense, would follow the last day of December. It would be a day dedicated to universal harmony and unity, a day for bringing together all races and nations into a single band of fellowship. "Leap Year Day" would constitute a similar insertion in the calendar every fourth year. Placed at the halfway point of the calendar following Saturday, 30 June, it would simply be called "W June" and, like its December counterpart, given no weekday name.

The World Calendar was praised as a calendar for our time, and prospects looked bright for its adoption by the United Nations during the late 1950s or early 1960s. In 1954, the Vatican endorsed the World Calendar provided it could be demonstrated that "there were a general desire for reform, motivated by serious requirements of the economic and social life of the peoples of the world."[26] Thus, even the headquarters of the Catholic Church regards time management as primarily a civic rather than a religious concern.

Proponents of the calendar confidently predicted that it would be instituted in 1961, a year they pegged because its 1 January fell on a Sunday. But like the move to convert to the metric system in America by 1980 (also preceded confidently and optimistically by such gestures as highway sign conversion from miles to kilometers several years in advance), the perpetual calendar fell flat on its face.

It was never clear why. There have, of course, been objections to an imperial time system based strictly on Western Christian values—complaints from Orthodox Jewry and Seventh Day Adventists as well as countries practicing the Islamic faith, which still independently adhere to their largely lunar-based calendar. Other matters of far more important and immediate world concern have topped the agendas of legislative bodies, which so far have not felt it to be particularly worthwhile to waste time debating new ways to prohibit wasting time.

Thus, at least for the present, the effort to climb another step toward further regularizing and exactifying our formal system of time reckoning has been forestalled. Whatever the reasons for the World Calendar's failure, there may still remain within us a longing for the uncertain, the incalculable, the chaotic—that tiny little bit of the unknown we in the West preserve as the sacred turf of time to escape to and to explore freely in a life already too rigidly controlled by the clock.

PART III

Their Time: Following the Order of the Skies

Mythical Maya hero twins of the *Popul Vuh,* portrayed on a Classical Maya plate worshiping their ancestor. Courtesy Barbara and Justin Kerr Studio.

5

TRIBAL SOCIETIES AND LUNAR-SOCIAL TIME

> Though I have spoken of time and units of time, the Nuer have no expression equivalent to "time" in our language, and they cannot, therefore, as we can, speak of time as though it were something actual, which passes, can be wasted, can be saved, and so forth. . . .
>
> —E. E. EVANS-PRITCHARD, *The Nuer*

MOST comparative approaches to the study of time have an evolutionist ring to them: that is, they tend to look at what alien timekeeping systems lack in one or another quality relative to our own. We take this outlook because we are still experiencing the fallout resulting from the great explosion of ideas, emanating from the nineteenth century, about the time scale of life and the universe in which it resides.

Intoxicated by the success of the Darwinian rules that govern life systems, the new social scientists of the last half of the nineteenth century applied the evolutionary paradigm to create a developmental, progressive model of human society in general, a model that today we have begun to regard as misguided.

When we apply the notion of the ladder of progress to the classification of the timekeeping systems of other cultures, the results are fairly predictable. We find that all non-Western systems fall short for one reason or another: absence of technology, too little precision, lack of logic, or failure to apply sound reasoning. Such deficiency-oriented schemes always situate us at the top of a value-laden pyramid. But using our own way of knowing as a standard for studying time in other cultures robs us of the opportunity to confront the issue of whether and to what extent human knowledge is determined by the way a society behaves as opposed to objective forces or properties that lie outside of human culture.

In the following chapters, I shall explore the extent to which other peoples, from quasi-sedentary societies to bureaucratic states, can experience the same natural phenomena and, out of their own desire or necessity, devise an equally complex and useful time scheme as our own. In speaking of "their time," I shall mean how they both experience and reckon it—modes that can be quite different from those the British anthropologist E. E. Evans-Pritchard observed in the epigraph that opens this chapter.[1] The tribal societies I describe are relatively small groups of people who depend little on technology but strongly rely upon the seasonal cycles of nature, even to the degree that some of them do not remain in the same settled location the year round.

The Nuer are a tribe who reside in the upper Nile in what is now the Sudan, about 500 kilometers south of Khartoum. Back in the 1930s, when the British anthropologist E. E. Evans-Pritchard did the fieldwork for his classic study of their modes of livelihood and political institutions, they numbered nearly a quarter of a million and were relatively free of contact with outsiders. The young researcher found them to be a formidable people: tall, long-limbed, and handsome, with a complex and sophisticated kinship and political system and an economy immediately and directly centered on the raising of cattle. In fact, the best advice Evans-Pritchard could give anyone who wished to understand Nuer social behavior was "*cherchez la vache*," for they depended on their cattle for meat, blood, and milk. People follow cattle, and cattle follow people.

Anyone who leafs through the pages of *The Nuer* will discover a book that is basically about the structure of time. Reading it helps us to understand how time was conceived and comprehended by a pastoral seminomadic people who, possessing little in the way of material culture and technology, were directly and heavily dependent upon the local environment. The Nuer provide a stark contrast to the way time thought developed in the larger, more bureaucratically organized civilizations such as the Maya, the Aztec, or the Inca, whom I shall discuss in chapters 6 to 8.

Nuer people think about and make use of two different kinds of time rhythm. First of all, there is *ecological time,* which is decidedly cyclic in character. It connects people with the environment through changes in nature to which they react. "Eco-time" is made up of relatively short periods framed within the annual solar cycle. Then, there is *structural time,* a framework of much longer duration that connects people with one another; it seems to deal more with social rather than ecological concerns. Let us look at each of these schemes separately.

The Ecological Cycle

The basic time markers in the Nuer ecological cycle are given by the events that control the movements of people and their cattle within the environment, such as rains and water flow, changes in patterns of vegetation, or the movement of fish. Basically, like the oysters in New Haven harbor or the plant that raises and lowers its leaves on a daily basis, Nuer people seem to go with the flow of nature's basic rhythms. They move from permanent village to temporary camp according to the division of climate into the two extreme seasons of rain and drought. From March to September, the rain falls (torrentially toward the middle of that period), and the rivers in flat, clay-based Nuerland quickly flood the grasslands, turning them into one vast morass. Food becomes scarce. To prepare for the future, the people take to permanent villages, which are located on higher ground in between the flooded depressions, and engage in horticultural activity, raising corn and millet. Then, just as suddenly as it began, the rain ceases. The blazing, hot sun quickly dries out the land, the streams vanish, the river falls from torrent to mere trickle; and for the rest of the year, from October through February, the countryside becomes drought-ridden. Now is the time to move to outlying camps to engage in hunting and fishing during a time of plenty. Most important, it is time to bring the cattle to the natural source of food that nature provides for them. Toward the end of the dry period, they burn the brush and prepare for the next planting season (see figure 5.1).

The Nuer call these two seasons *tot* (wet) and *mai* (dry). Just as we recognize the transitional seasons of spring and autumn in between summer and winter, they refer to *rwil* and *jiom* as times of rapid change. *Rwil* is the time when they move from camp to village (about April–May) before the rains peak out. They think of *rwil* as part of the

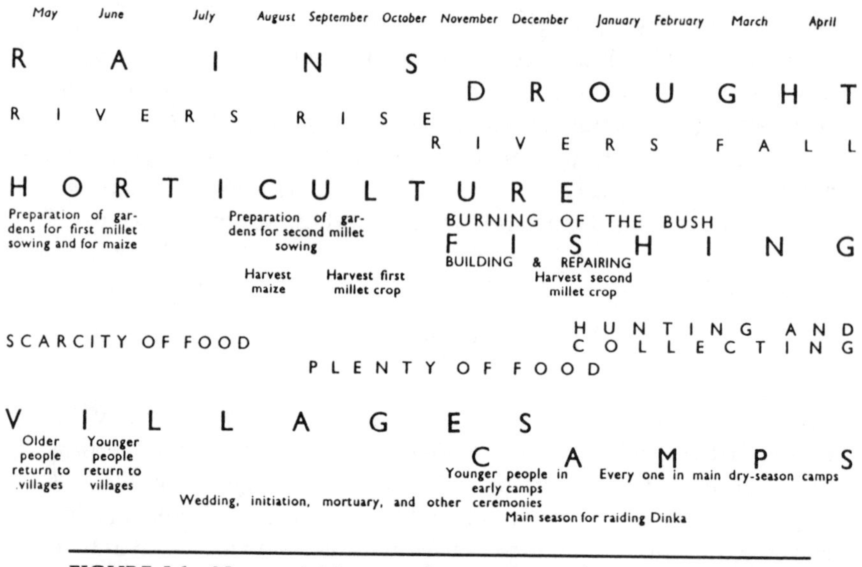

FIGURE 5.1 Nuer activities span the year. Except for details that we lack, their schedule of events in ecological time seems little different from the preclassical Greek calendar in Hesiod (see p. 42). After E. E. Evans-Pritchard, *The Nuer* (Oxford: Oxford University Press, 1940).

village-life portion of the year. On the contrary, *jiom* (which means "wind") is associated more with the winter or dry half of the year. It is the time when the north wind begins to blow (October–November) and people start to form the early camps. So the Nuer consider their ecological year cycle to consist of a pair of alternating opposite halves, with relatively short periods of drastic transition in between—a pattern encountered earlier when I discussed ancient Old World mythology.

Now, none of this time is marked out by equinox or solstice with the extreme precision of our own Western calendar. Instead, time for these people becomes the relation between activities on several different planes of existence. Reminiscent of Hesoid's *Works and Days*, particular celestial events that happen to be conveniently represented in the sky are selected as a kind of general forecast of the imminence of significant transitions in the agrarian phases of the Nuers' lives.

Conceptually, the Nuer anchor the cycle of ecological changes within a system of full moons, each of which has its own particular name. In the moon named *kur* (roughly November–December), they make the first fishing dams and build the first cattle camps. In *dwat* (May–June), they break camp and return to the village. Month names represent the activities that go with that month. Evans-Pritchard tells us: "Nuer do not to any great extent use the names of the months to indicate the time of an event, but generally refer instead to some outstanding activity in process at the time of its occurrence."[2] The temporal logic seems to be: if I'm going to church then this must be Sunday; or *since* people are on the move going between camp and village, *then it must be* dwat (or some time close to it). Activity supersedes time in the sense that we know it. Eco-time is the relation between activities, not a separate entity apart from the activities themselves. Evans-Pritchard calls it a series of conceptualizations of natural changes, the selection of points of reference being determined by the significance that these changes have for human activities.

Duration for the Nuer is also a relatively imprecise concept. There are no weeks, perhaps because there is no market cycle for people who spend much time on the move. Events several days back into the past or forward into the future are marked either by counting sleeps (if the interval is two or three days), or for longer intervals, by referring to the phases of the moon. "Sleeps," as time units unbroken by social activities, seem a more logical reference to the passage of days than "suns" or days themselves.

Unlike our fixed clock intervals, the value of time units also varies depending on when they occur. An hour is not always an hour; nor a day, a day. The dry season events, because they take place in a period of severe shortage of water and pasturage, are measured out in more precise terms, rather like the way a modern working parent cautiously speaks of the "quality" time spent with a child who has passed much of the day at a day-care center. This precious and critical time interval requires far more thought to coordinate a routine than the more monotonous events associated with the wet half of the ecological cycle. Longer time reckoning by the moon gets less attention than the daily time reckoning in the dry season.

Nuer eco-time in being both imprecise and human-centered is foreign to our way of reckoning time, confounding Western notions by using the name of a month to specify an activity, on the one hand, while at the same time suggesting that the name of the month is known *because of* the nature of the activity taking place. It sounds almost as illogical as the Anthropic Cosmological Principle! And, we ask, which comes first: the month or the activity? How can you have it both ways?

This same notion of eco-time is shared by the Mursi of southwestern Ethiopia, who argue about it incessantly, and by the Trobriand Islanders of Indonesia, who use worms to stay in tune with it. Both these cultures offer some answers to our questions about eco-time and, better still, force us to rephrase the questions.

The Mursi, like the Nuer, are seminomadic people who live in a wet–dry season environment between two rivers. They survive by a combination of rain and flood cultivation as well as by herding. Unable to sustain themselves by either one or the other, they make themselves physically mobile in order to survive. For the Mursi, a calendar is not a piece of paper to be hung on the wall of a hut, but a live interactive process filled with social dialogue and as much heated argumentation as a close call at home plate in a baseball game. Disagreement and retrospective correction quite naturally form a part of it; in fact, these people *need* to argue and discuss their calendar, even make things up as they go along, in order to survive.

The Mursi count 12 lunations in a seasonal cycle which they call the *bergu*. Every informed Mursi tribesman is able to recite the set of social and agricultural activities that goes with the lunation set. As in all environmentally based calendars, transitions are critical, and the most crucial part of the cycle occurs at the onset of the major rainfall period that causes the rivers to flood. Generally, the rain falls when the *bergu* is 8 lunations old, which, ethnologists have been careful to point out, does not mean the same as "in the eighth lunation." The activity is thought to occur at a stage in the annual cycle rather than in a specific lunation. The Mursi seem to be concerned more with lapsed time or duration than with time as the marking of events in a sequence.

At the end of the twelfth lunation, the cycle of activities is completed and then they say that the time is *gamwe*. Specifically, when the moon rises after *bergu* 12, it is *gamwe*. They think of this time period lying outside the cycle of seasonal activities; no *human* activity in the list is associated with it. But a *natural* event, the flooding of the river, is tied into the system and serves as *the* key event the Mursi use to restart the year. If the river crests during the lunation *gamwe*, they say the flood will be extensive; and if the peak is attained in *bergu* 1, the lunation following the flood will be small. One difference between us and the Mursi is that, at least since the time of Caesar, we relate seasonal events directly to the solar year; but they relate them to the passage of lunations. We might well conceive of *gamwe* in our heads as *bergu* 0, so that at the end of it, *bergu* 1 would indicate that the annual cycle is one month old, just as on my fortieth birthday, I can say that I have completed forty years of my life. But, how did my age in years get reckoned before my first birthday?

But there is still a problem here, as earlier in the Roman calendar: whether we count elapsed time or not, there simply cannot be a fixed number of months in every year. And yet the Mursi reckon a *gamwe* in every *bergu*. Doesn't time get out of joint? And how do the Mursi set things straight? The plain fact is that things *do* get out of joint in the Mursi eco-calendar, but it is nature and not man who is held responsible. The rains and the floods occur neither at the same time nor to the same extent in all of Mursiland (or of Nuerland, for that matter). Therefore, in the statement "(*a*) it is *bergu* 8 (*b*) when the heaviest rains begin," we must think of part (*b*) as the controller of part (*a*)—or else there can never be a calendar that fits nature. For example, if the rains begin when the count is at *bergu* 7, a local *bergu* expert will likely need to change his opinion. That change of opinion will be based not only on the occurrence of the rains but also on the recognition of a number of other seasonal indicators, such as the occurrence of the heliacal rising of the Pleiades or the positions of bright stars, all of these phenomena being given equal status in making the judgment. Thus, the Mursi say the sun enters his house (the solstice) during the first phase of *bergu* 5, when the rising sun approaches a particular gap or cleft along the horizon. The Pleiades set heliacally in the west in the second phase of *bergu* 9 and are overhead at sunset in *bergu* 6. Another *bergu* expert might, on the other hand, weigh all of these factors and decide not to change his opinion about what *bergu* it is. The decision becomes more like a judicial court case than an objective scientific judgment; and in the end, different times may well end up being kept in different locales. The important point is that, given all of the time checks used by the expert watchers of the environment all over Mursi territory, rarely will there be disagreement among all of the people that amounts to more than one lunation. The system of Mursi "flex-time" I have just described is not laissez-faire. They simply are not as hung up on precision, regularity, and absolute objectivity as we are when they deal with the natural world. Their time is as precise as it needs to be while still allowing them to keep a calendar that stays in close harmony with natural changes that occur at the local level. They need no all-encompassing system of the calendar, no compromise like the universal one we chose to make when we switched from local solar time to zone time.

Initially, the anthropologists who studied Mursi and Nuer time systems found people hard to pin down about which lunation it was, and misinterpreted, as laziness and lack of care, the attempts of a people who were acutely aware of their environment to preserve the flexibility and disagreement in the calendar that were necessary in order to keep their annual cycle truly adjusted to seasonal variations in a delicate and highly variable ecological region. Their system for dealing with time is only as complicated as what is required to meet the needs of the people, and no more. The Mursi had to learn to agree to disagree about how to keep time. Temporal knowledge is determined by the individual as a participant in organized society. Unlike our culture, there are no fixed, objective standards for stating what time is that can be transmitted to the individual layperson.

A similar, but slightly more complicated "eco-flex-time" exists among the Trobriand Islanders of northern New Guinea. Their system seems to involve a greater degree of regional social integration than that of the Mursi. Trobrianders are unconcerned about any solar year toward which the lunar year needs to be corrected. They list ten months, whose names they correlate principally with horticultural activities, and mark them by a series of astronomical and biological time checks.

The primary event in the Trobriand calendar—the one used to regulate the start of the seasonal cycle—is the appearance of a worm! If ever there were a case of man adapting his own clock to one of nature's most reliable biological rhythms, the relationship between Trobriander and palolo worm is it. This worm, an inhabitant of Pacific coral reefs, executes a "circa-lunar" rhythm. Once a year for three or four nights, the posterior parts of the worm wriggle dramatically on the surface of the water and let loose their genital products. To use the language of biorhythms discussed in chapter 1: the dim lunar illumination is the zeitgeber that entrains the animal to a lunar periodicity.

The spawning marine annelid is seen on the surface of the sea at the southern extremity of the island chain every year following the full moon that falls between 15 October and 15 November (our time). Trobrianders name this important month Milamala after the worm, and celebrate a great festival in its honor to inaugurate the planting season. (They also eat the worm roasted and prize it as a delicacy.)

Anthropologists had discovered that this same festival was held one month earlier in the main part of the island just north of the area, two months earlier in the outlying islands to the south, and three months earlier in the east. Though there are different time structures in these areas, the Trobrianders seem, unlike the Mursi, to have a somewhat more rigid sense of regulation. Like the Mursi, however, these people seem to be able to change the name of their month after they have completed it—a liberty I wish my society allowed me!

Trobriand space and time join in Milamala for it is considered, on the whole, as a set of four months that are broken down regionally among different island groups. For example, consider the month-naming sequence in the four districts A, B, C, and D in table 5.1. Suppose everybody agrees that district D will be the control or checkpoint, because the people in district D are the fishermen who actually observe the worm. Now, once the worm appears, the people of district D call the full moon "just past Milamala"—that is, they name it retroactively. But what if the worm is late, as we can expect it to be in a lunar calendar with a fixed number of months? In that case, the "D

TABLE 5.1
*Scheme of the Trobriand Calendar**

District A	*District B*	*District C*	*District D*
1			
2 Milamala	1		
3	2 Milamala	1	
4	3	2 Milamala	1
5	4	3	2 Milamala checkpoint
6	5	4	3
7	6	5	4
8	7	6	5
9	8	7	6
10	9	8	7
—	10	9	8
—	—	10	9
1 (same as 10 in D)	—	—	10
2 Milamala	1	—	—
	2 Milamala	—	—
		2 Milamala	1
			2 Milamala checkpoint

*This scheme shows how it is possible for each of four different social groups to mark out an annual calendar by naming only ten months. After E. Leach, "Primitive Time Reckoning," in C. Singer, E. Holmyard, and A. Hall, *History of Technology* (Oxford: Clarendon Press, 1954), vol. I, pp. 110–27.

people" simply name and celebrate an extra Milamala month. This would be a bit like those of us in northern climes celebrating another December if snow didn't arrive in time for Christmas! However, such action puts all the other districts out of line relative to D, as the table shows. This doesn't seem to upset the Trobrianders, for they respond later (even up to a year or two later) by doubling their own Milamala season. In due course, each district decides what it will do and when it will do it based upon their own separate time-checking schemes. Just as some states in our nation decide not to adopt daylight savings time, local autonomy reigns.

One curiosity about this system is that the whole territory completes a twelve- or a thirteen-month lunar cycle, and yet no given area actually counts more than ten months. As in the Mursi and Nuer calendars, there is a time that lies outside of the regular calendar. In this case, the "time-out" is a free-floating adjustable period tied to the occurrence in nature of a singular event that resets the calendar—the appearance of the palolo worm. This system is not unlike that of the Romans before Caesar, whose months, also ten in number, began with March, the month of the equinox (the reset mechanism) and ended with the eighth, ninth, and tenth lunations, as their Latin names betray—*Octo*ber, *Novem*ber, *Decem*ber. As for the rest of the year, following the tenth month there was a gap of two, maybe three moons until the next cycle, likely timed by an event that signaled the equinox. This uncounted interval corresponded to the temporal limbo when fields lay fallow, a borderland in which the farmers waited patiently for nature to signal the awakening of spring and thus rekindle the cycle. Although we have come to think of the solar year as somehow categorically *correct*, these lunar-based calendars (especially the four-district system of the Trobrianders) illustrate how a minimum of systematic knowledge about the passage of natural events can be organized into a complex and workable system.

Of course, none of these people ever needed to conceive of the problem of intercalation. The events that demarcate time as seen from their point of view occur not in a solar year of fixed length but in a seasonal year whose length goes undetermined and is of no real concern. Transition matters more than duration. They look for the events in the natural world that portend change—the temporal signposts to which they must react in order to survive. In each case, the time marker—whether flood, worm, or stars—is recognized to have a seasonal cyclic rhythm independent of human action, a property any time-correlation device must possess in a successful lunar calendar. And so to return to the question, Where is the logic and which comes first? The logic is there but is different from that of our calendar. Though tribal time-reckoning schemes are riddled with imprecision and subjectivity, and though there is no fixed universal time standard, the system works.

Though eco-time is cyclic, as I argued earlier, time that repeats itself is too complex to be exactly represented by a simple circle, for that would imply that time duplicates itself. A better model might be a wire hoop: sever it in one place, then twist it into a helix so that time's path curves back on itself in a succession of parallel planes, like that in figure 2.4B. It doesn't repeat exactly because the most recent wet season is never precisely the same as the last. Eco-time is not relived. Only its properties and qualities repeat. Nothing ever gets duplicated precisely. Next, make kinks in the wire to represent the points signaled by natural events that segment eco-time into each of its well-defined intervals. Run your finger along the helical hoop, and you get a tangible sense of the logic of eco-time: it recycles but still has a direction. Eco-time moves irreversibly forward along the upward coil of the helix. The breaks and kinks in the smooth path make the definition of one's place in time more precise. Without them, time really has no meaning—just as an unbroken sleep of any duration always seems the same.

Never forget that the hoop is only a metaphor—and *our* metaphor at that. I have resorted to it as a way of getting a feel for what another culture experiences in the general realm of the word *time* as defined by us in the West. Running our finger along our twisted, kinked hoop, we can sense duration by the speed with which we move it. We feel change and transition every time we hit a kink, and can recognize repetition after several trips back to the same side of the cyclic path. But let us not think for a moment that eco-time is universally conceived as an abstract grid onto which life's activities can be mapped. It sounds too Greek. To judge from the anthropological evidence, eco-time is the process of sequencing the activities themselves.

Structural Time

While eco-time tries to relate the response of human behavior to the cycles of nature, I also mentioned another notion of time practiced by many tribal societies, one that binds people to each other with little regard to what happens in the natural world.

To get a better feeling for what structural time is, think of the question, How old are you? And of some possible answers: Old enough to vote. Old enough to go to war. Old enough to marry. Old enough to know better. Unlike other conceivable answers to this question—such as I'm twenty-one, twenty-eight, or eighty-eight—in the first set of responses we think of time in relation to certain social activities, duties, or privileges we are accorded as members of an organized society as we move up through its chronologically based hierarchy. But the answers cast in the form of pure numbers, those we tend to give if posed the question on a job application form, are little more than abstractions. In giving such numerical answers, we all consciously associate numbers with tasks and liberties that define where we stand in life's pecking order. But aside from the voting age, isn't the time when someone can marry or bear children really a subjective matter? And, is twenty-one really old enough to know better? I used to think so!

Our society has few fixed rules that strictly relate age to social status or expectation. True, we do legally limit the age of a president to over thirty-five; a driver must be over sixteen (in most states); a salaried worker, under sixty-eight but over sixteen. But these legislated age restrictions have little to do with how we treat one another in the social hierarchy, with such questions as, When is a boy a man, or a girl a woman? or When did I become an elder? To these questions, other societies often have specific answers.

Structural time is reckoned among the Nuer by the *age-set system*. For this I want to suggest another kind of model: lay a piece of plain white paper over a bar magnet. Then toss several pinches of metallic powder across the surface. A physicist would say that the focused linear pattern traced by the iron filings represents the invisible lines of force that emanate permanently from the poles of attraction of the magnet; but these stationary lines of force were not revealed to us until the tiny bits of iron passed through them. So it is with age-sets. Structural time does not really move. Instead, like Shakespeare's seven life stages, people pass through its everpresent, invisible framework in endless succession. Age-sets are fixed, stratified, and segmented age groups through which every male member of Nuer society moves, the way iron filings slide over the fixed force lines of the magnet.

Members of a male Nuer age-set are initiated at the same time and seem to move in a progressive rather than a cyclic way, from one stage of life to the next, from juniorhood to seniorhood, after which they die and their set becomes a memory.* It is rather like imagining that as we age, rather than piling up years measured by the increasing numbers of candles on our birthday cakes, we glide from one stage of life to the next, from boyhood to fatherhood to grandfatherhood. Just as the physicist believes in the existence of magnetic lines of force, so the Nuer believe in the reality of the fixed age-set in structural time. The different age-sets are already there waiting for us to arrive, just like the station in Giorgio De Chirico's painting (page 324) which is always there waiting for the train to pull in.

Among the Nuer, the initiation from boyhood to manhood, which usually takes place between the ages fourteen and sixteen, falls at the end of the rainy season. (Indeed, eco-time and structural time are connected.) It begins with a bloody and painful surgical procedure in which the priest makes six long and deep parallel knife incisions all the way across the forehead between the eartops (see figure 5.2). A number of rites, including a period of seclusion, sacrifices, and festivity, follow the operation. All boys who are thus initiated over a several-year period belong to one age-set. The high priest in charge—the "Man of the Cattle" as he is called—then decides to close the initiation periods. Like some temporal sausage maker, he "cuts the sets." Then he "hangs up the knife" and performs no new initiations for three or four years. From his fieldwork, Evans-Pritchard estimated an age-set length to be about ten years, give or take a couple; in other words, about one or two age-sets generally intervene between a man and his son.

The boys of a given age-set are regarded as age-mates, a concept that seems to have no analogy in our timekeeping system. Perhaps "classmates" comes close, except we need to think of them spread over a few years. Their group is given a name that is generally held in common among different tribes across Nuerland, and each set is stratified internally with two or three named subdivisions like "junior segment" and "senior segment." Altogether, members of as many as six or seven sets might be alive at a given time; but when they die, it is like removing the powdered iron from the surface of the magnetized paper: the age-set to which they belonged quickly passes out of recognition.

How are age-sets linked to human behavior? The Nuer have three

*It is indeed unfortunate that few anthropologists in the first two thirds of the twentieth century had paid much attention to the transition points in the lives of native women. Today this gap is just beginning to be filled.[3]

FIGURE 5.2 The forehead of a Nuer youth reveals the mark of structural time: the banded scars are a reminder of his initiation into the age-set ritual. SOURCE: E. E. Evans-Pritchard, *The Nuer* (Oxford: Oxford University Press, 1940), p. 236, pl. xxvi.

distinct age-grades of boy, warrior, and elder, through which each of the sets pass. Each passage, like the major transitions in eco-time, is marked by an elaborate initiation ceremony. What unites the age-mates, however, is neither common ground on the battlefield nor politics, but rather the social activity that reflects the domestic and kinship ties that attach a man to his brotherhood. Change of status on the domestic front occurs, for example, when one passes from boyhood to manhood. He receives a spear from his father or uncle. He is given an ox and an "ox-name"; he is declared to be a herdsman; and, until he becomes a husband and a father, his principal activities consist of dancing and love making. In the next stage, he fights in the wars, he cultivates his gardens, and he marries a wife.

Age-set time lies at the core of social relationships among Nuer males. It incorporates a set of complex social rules and values that governs their behavior, that structurally defines the status of every male in relation to every other one in terms of equality: seniorhood or juniorhood. The son, say from age-set F, of a father of age-set C, thinks of all members of age-set C as his fathers. Likewise, he may neither marry nor have sex with the daughter of an age-mate, for she would be his daughter, too. But he may, with consent and an exchange of beasts, marry the daughter of one of his father's age-mates—but only if the father is deceased. Moreover, any set tends to think of the set immediately senior to it as equals in relation to junior sets and of the set immediately junior to it as equals in relation to senior sets. Thus, if a member of D kills an ox and there are members of C present, but not B, then D eats with C and E, while F and G eat by themselves. But if men of age-set B are present, then C dines with them because they are their father's set. Likewise, E must eat with F and D goes with C. These are but a small fraction of the age-set rules Evans-Pritchard discovered, but they are sufficient to illustrate how age-sets and the interactions among them are defined by family-like relationships and how the concept of structural time is used to regulate social behavior in the same way ecological time is employed to guide human action in the area of subsistence.

As wonderfully complex as the Nuer time sense might be, we can legitimately ask, Where is this people's sense of history? Their year seems to be the longest interval in ecological time; and in structural time, where all is relative, a year has no meaning at all. However, long distance between events seems to be reckoned structurally rather than in years. Thus, a particularly notable flood might be said to have taken place after the birth of age-set C or in the initiation period of age-set B. The concern is with *how many sets ago* something happened, not how many years ago or when within the year it took place. The structural system of time reckoning is based partly on points of reference in time that matter to locally or territorially related groups—that is, what gives them a common sense of history—and partly on the time distance between specific sets in the system. This combination gives rise to the concept of structural time as one form human history can take; for these people history is made up of a blend of space and time.

One cannot speak of Nuer history without referring to Nuer kinship. The Nuer system of lineage can be thought of as a fixed system, just as the age-set system, for there is thought to be a constant number of steps between a group of living persons and the founder of their clan. Four generation steps of blood-kin are marked in the male line: grandfather, father, son, and grandson. These give time depth to a group and provide reference points in a line of ascent. Persons with the same father who trace their descent to a common ancestor constitute a *clan.* Clans are segmented genealogically into lineages that can be traced, like branches on a family tree. Of course, the very words *clan* and *tree* are *our* conceptualizations of *their* kin relations; there is no reason to think they dream up these notions precisely as we do. In fact, as I have already hinted, there is evidence that the Nuer think of the system in a nonabstract, territorially based manner, merging kin relations and local groups. When asked to describe the relation between lineages among his fellow clansmen, one Nuer, instead of using the "tree" metaphor, drew on the ground a series of radial lines and labeled adjacent spokes of the diagram according to a combination of who lived next to each other and who were blood relations. The system is probably even far more complex than we make it out to be.

How far back does their history go? The anthropologist traces or maps out the lineages of a clan by asking the native some such question as, From whose seeds do you descend? By mapping out the structural pattern, Evans-Pritchard was able to discern that Nuer clans have a historical depth of about ten to twelve generations from the present, and no more, going all the way back to the ancestors who gave rise to them. Prior to that, events get scrunched together on a kind of mythic horizon; at least, they seem to be viewed always in the same time perspective. All the peoples and cultures of the world existed together at some time in the remote, structurally undifferentiated past. There is a strong tendency to merge branches of the lineage as one goes backward in time. For example, the descendants of one or two brothers of a flock of several tend to dominate; while weaker descendants either die out or attach themselves, usually for political reasons, to the stronger, more dominant line. A Nuer can generally trace a branch of his lineage, and conceives of his relationship to another member of the clan, defined by the lineage within the clan. Knowing the particular descent pattern of that other clan member is not important. To use a crude example from our culture: a female cousin may be separated from me by a certain socially defined structural distance which prescribes that we may be "kissing cousins," but we know that if we fell in love, we ought not marry. However, the detailed line of descent of my cousin does not matter in establishing our relationship.

The Mursi, like the Nuer, also do not express a person's age in years the way we do. To ask a Mursi how many *bergu* have elapsed since he was born would make no sense to him. Age *differences*, on the other hand, are important, for the same reason they matter among the Nuer: they define social status and all of the duties and obligations that attend it. Consequently, a Mursi will know, to better than a year, how much older or younger he is than other people in the neighborhood. Together with their age-mates, all children pass through the grades of their age-set beginning at about age seven. As the young person matures and moves up the social hierarchy toward increasing privilege and greater burden of responsibility, the stages of the set grow increasingly longer; but still within a given stage, there are substages that enable him to know distinctly where he stands in the pecking order.

Thus, these people do not believe in history the way we do, though they do have a sense of history. As in the events and relationships that comprise tribal life, there is a kind of immediacy to both cyclic ecological time and linear structural time among these tribal societies. Ultimate origins do not matter in their temporal time schemes. Interaction, with either nature or other people, is the real reason to keep time; and when things cease to interact or before they ever had interacted, there is no need of reckoning it.

Unkept time, alternating bursts of activity punctuating inactive durations structured and named after the human actions that take place, with variable hours, years comprised of different numbers of months: these are the earmarks of tribal timekeeping. A former student of mine who studied time keeping when he lived among the Bororo, a people of central Brazil, reported that after he and his wife had been in the community for several months, built their own house in the village, and partaken of all the routine chores and come to know the natives quite well, their quiescent, almost unrecognized existence was suddenly broken. One day the village chief abruptly confronted them and told them they would now be "given names." Two days later at sundown, there began an elaborate ceremony in which the pair were seated on a mat facing the sunset position. They were told their names by the adopting clan and then were sung over and twice led around the plaza in a counterclockwise direction. Elaborate rites continued throughout the night and to the middle of the next day. My student writes: "From that time on, we were expected to 'act' properly, use the appropriate terms of address for our fellow villagers, and participate as best we could in village activities."[4] Though both in their thirties, these outsiders had experienced the same initiation rite as native infants, who the Bororo say are "soft" for several months after biological birth. It is not until they are "hardened" by exposure to village life that they are ready for the far more significant event of social birth, whereby the naming process connotes that they are ready to participate in the life of the village. The absolute age and the historical background of my student and his wife were simply not a consideration for their Bororo clan adopters. Who they were and where they came from did not matter. The important point was that they did not exist socially until the time that they were named. Before that day, there was no time for them.

6

THE INTERLOCKING CALENDARS OF THE MAYA

> . . . the cult of the gods, the wisdom of the destinies, the basic duties of man in his belonging to his family, group, town, chiefdom, his activities as a farmer, warrior, artist, merchant or in any other profession. In brief, to exist for the Mesoamericans one had to observe the sky. Without skywatchers the ethos of this people, its distinguishing spirit, its own genius would not have developed.
>
> —M. León-Portilla

Introduction: Three American Empires

Now I turn to the development of timekeeping systems among bureaucratically organized societies, those commonly designated as states: that is, fully sedentary societies that develop extensive, highly organized, and specialized economic systems, and often expand and exert their influence, even their rule, upon adjacent peoples. State societies usually develop rigid patterns of social stratification, with kings and nobles at the top of the hierarchy dominating a large peasant class at the bottom.

I am going to look at three ancient societies, all of which developed into states of one form or another in the New World. First are the ancient Maya of Central America, perhaps best known for their celebrated obsession with time and the abstract system of writing by which they expressed it. Their civilization expanded during its classical period, roughly between the second and the tenth centuries A.D., and encompassed all of the Yucatan peninsula including the modern countries of Guatemala, Belize, and portions of El Salvador and Honduras. The Maya city-states were a bit like the Greek—autonomous units, each propagated by a royal bloodline; but they interacted with one another, intermarried, made war, drew up treaties, and, oddly enough, shared a common, complex, and sophisticated calendar. Moving westward into the highlands of central Mexico and forward in time by almost a millennium, I next look at Aztec timekeeping. In the popular vein, the Aztecs are often portrayed as fierce cannibals, devourers of human flesh. They worshiped the sun and keyed every festival, every tribute to the payment of their debt to the solar deity, Tonatiuh, extracting the hearts of the vassals of their tributaries in order to assure themselves that the Sun would continue to move and that time would keep on its course. While the Aztecs were exerting control over Mexico, the Inca were expanding their empire along the thousand-mile-long spine of the Andes of western South America, from the modern countries of Ecuador to Chile. Their military state, like that of the Aztecs, also created its own timekeeping system; however, the Inca must have encountered some difficulty in trying to impose it upon subjects who resided in remote areas where the agrarian climatic conditions were quite unlike those experienced in the high altitude of Cuzco, the Inca capital.

I chose Maya, Aztec, and Inca for two reasons. First, these civilizations were virtually free from outside contact before the sudden European invasion in the early sixteenth century. Hermetically sealed by two oceans, whatever ideas they may have developed about time would have taken place in a pristine condition from the very start. Hence, when we pry into pre-Columbian systems of thought, unlike Asiatic and African cultures, we can be certain that what we discover about how people make and model their time systems must have been invented and developed independently, devoid of borrowing through social contact, at least from Western civilization. This axiom gives me an opportunity to raise a big question about human nature: given basically the same stimuli from nature, whether it be the annual crosswise movement of the sun on the horizon or the up-and-down celestial slide of the morning star or crescent moon, do different people in different places all react to what they see in the same way? If not, then what factors might cause them to react differently?

There is another reason to study the Maya, Aztec, and Inca systems of timekeeping. They have all been described, at one time or another, as "classical": that is, they all fashioned great architecture, art, and sculpture, developed extensive and complex economics and trade, and possessed advanced systems of learning based upon precise knowledge. Even though they are not our Greeks (Western culture's classical civilization par excellence), we are forced to admire them. We cannot do without studying their systems of timekeeping. It is tempting to compare them with ourselves, for at least on the surface, they appear to have accomplished what our ancestors attained. Because we value and revere such accomplishments, we feel compelled to know how they dealt with some of the same problems our cultural ancestors confronted: how to reckon time by nature; and how to use it as a way of ordering religious, civic, and social activities. How did they deal with the problem of time's origin? Where did they think *their* universe came from? But I am not concerned solely with comparing each of these three pristine societies, taken together as a group, with our own Western society. Indeed, all three are different enough to warrant a comparison among themselves.

Were it not for the deliberate destruction of so much of the material remains of these first Americans by the expansionist sixteenth-century Europeans—people who lusted only after gold, and later collected lost souls—today there would be an abundant historical legacy about who these indigenous people were and what they thought. First came the soldiers like Bernal Díaz del Castillo, Spanish warriors with the fresh taste of victory in their mouths, having recently thrown off the yoke of the Moorish oppressors who had occupied their country for several centuries. Díaz was among the handful of equestrian soldiers of fortune, the conquistadors, who put the first ax and torch to the great cities of the New World:

> And we could not endure the rocks, stones, and javelins they hurled from the roof tops; they hurt and wounded many of us. I do not know why I write about this in such a tepid manner, for some three or four soldiers who were there with us and who had served in Italy swore to God many times that they had never seen such furious fighting in any they had encountered between Christians, against the artillery of the King of France, or against the Grand Turk, nor had they seen men like those Indians, with such spirit in closing their ranks as they advanced.[1]

Then came the priests, like Fray Diego Durán, so determined to convert the natives to the one true religion that he organized a disinformation campaign as a way of influencing their moral views:

> But my sole intention has been to give advice to my fellow men and to our priests regarding the necessity of destroying the heathen customs which they will encounter constantly, once they have received my warning. My desire is that no heathen way be concealed, hidden, because the wound would grow, rot and fester, with our feigned ignorance. Paganism must be torn up by the roots from the hearts of these frail people![2]

Were it not for this physical and mental plundering, not only would we be in a better position today to analyze the calendar systems as they were first encountered at the time of the Spanish Conquest, but also we might have a fair chance to trace the development of calendars as well, perhaps even acquire details about the origin of some of the intricacies that made them up. Sad to say, the wanton destruction of books and buildings, the looting of the tombs, still goes on today; and the suppression of ideas by the European invaders, while it does not rule out our attempts to understand native views of time and nature, nevertheless seriously hampers our task. We are left in the dark in too many areas. Still, one can learn something of the problem of how timekeeping changed in the Americas, with the advent of social contact between two worlds and the merging of their people, from valuable resource materials that have survived.

To begin with, these original Americans are by no means dead and gone. Descendants of Maya, Aztec, and Inca survive as native speakers in remote areas surrounding the sites of these great ancient cultures. Today two million people still speak more than a dozen dialects of the Maya tongue. Nahuatl, once spoken by the Mexica (popularly termed the Aztecs), is not quite a dead language; and Quechua remains the principal tongue of eight million people who live in the high Andes of Peru and Bolivia. More important, many of the native customs and beliefs survive, including parts of the (ancient) calendar—though as we might expect, they have been altered, not only by Spanish domination but also by a decline in their practice long before the Spanish arrived. It takes years to learn what indigenous people think, believe, and practice; but the ethnologist who lives in the native community can, with patience and understanding, bring back a legacy of ideas that connects up in a logical way with the ancient information we can retrieve separately from the historical written record.

That historical record consists of two parts. There are the post-Conquest written records: among the Maya, the books of *Chilam Balam* (a book of calendric prognostications) and the *Popul Vuh* (a story about creation) represent native attempts to preserve calendrical and cosmological traditions; but the bulk of the documents were written by the invaders, most of them Roman Catholic missionaries. They tell what the culture was like at the time of the Conquest as witnessed by Spanish eyes. For the Maya, Diego de Landa, for the Aztec, Bernardino de Sahagún, and for the Inca, Felipe Guaman Poma de Ayala (himself part-Inca) are among the handful who have given us gold mines of information that our modern inquisitive appetites have delved into for most of the past century. Isn't it ironic that today the descendants of sixteenth-century Europeans value what their ancestors worked hard to destroy a few hundred years ago? But we must remember that for the holy men from Spain, America was not a place to inquire into human potential; it was instead the battleground where they would help fight God's eternal war with the devil, whose ways seemed deeply embedded in the hearts of a native people who ate snakes, worshiped the earth, and sacrificed their children to graven images. Today the agenda has changed. We now want to know why indigenous Americans behaved the way they did before we ever came on the scene, and are concerned more with the working of their minds than with the few material things they left behind that we did not destroy.

In addition to post-Conquest writings, a few records from pre-Conquest times have survived. In Mexico and the Yucatan peninsula, there are the codices, painted picture books made out of tree bark and filled with numerical and calendrical symbols. Most of what has survived can be piled on a modest-sized desk top: the missionaries burned the rest, declaring it to be the stuff of devil worship. Among the Maya, inscriptions carved in stone and inscribed or painted in other media such as ceramic pottery, bone, and wooden door jambs and lintels testify to a diligent preoccupation with keeping the right time. Thanks to significant advances in the decipherment of Maya hieroglyphic writing, the past two decades has constituted something of a revolution in our understanding of Maya timekeeping and its connection with social life; for the Inca, the native written record is embedded in the *quipu,* a complex of knotted cords that looks like a colorful string mop. Though *quipus* still elude real decipherment, nonetheless they offer the potential of revealing a detailed knowledge of precise timekeeping which is sure to square with what we have learned both from the chroniclers, and to some extent from the archaeological record, about the way time and space were marked out in ancient Cuzco, the capital.

The Maya and the Body Count

There was a time before the revolution in decipherment when Maya researchers operated under a different paradigm. The scholars of fifty years ago garbed the Maya in relatively peaceful, intellectual clothing characterizing them as an almost mystically religious people who became captivated by time and events in the heavens. They were thought to have pursued knowledge for the sake of knowledge. Wrote one author: "The great men of Athens would not have felt out of place in a gathering of Maya priests and rulers."[3]

Alas, we now know the Maya, like the Greeks, were "real people," constantly bickering over their time and their turf—even making war about it. They wrote their inscriptions not so much to glorify time as to attract attention to themselves, to write their own histories. Recently, in Maya studies, there has been a coalescence of the old theory (which says the Maya inscriptions were only about astronomy) and the new theory (which implies that their writing is only about the history of the rulership). Contemporary Mayanists believe the ancient Maya were no less concerned with numerology and nature than investigators in the 1920s thought. Time and number simply were a means. The revolution lies in our changing understanding of the concept of the ends. Strictly speaking, the Classic Maya seem to have been concerned not with writing only nature's history, or only their own, but rather, with the task of weaving the two together. The Maya ruler seems to have been using time and all of its natural indicators as a vehicle for legitimizing and validating his rulership and authority. This is not to suggest that either the king or the commoner tilling the fields possessed no fundamental set of deeply revered beliefs underlying a Maya philosophy of time. The rulers were not simply manipulating time to hoodwink the people.

Imagine the Classical Maya king seated on his inscribed throne: he is immediately surrounded by the royal lineage, as he proudly uses the spine of a stingray to let blood from his genital member in penitence to the gods as an assembled public throng looks on. The written record suggests he might even perform his penitential act as the very celestial body from which he believed he drew his powers of durability and dependability moved over temple and throne in the sky above. There was grandeur in the rites connecting the phenomena of nature with courtly events in the life of a ruler, whether it be initiation of a battle, a marriage, sealing a pact or peace treaty, a capture, a burial, or the fragile moment of transfer of power from a deceased leader to his offspring. For the Maya, life's events needed to be regulated and timed precisely. This was the Classical mentality: the order of nature made manifest in the governance of society, at least as we now *think* we understand it. But it is full of far richer detail.

The first bits of information in any Maya inscription are about time. What sets the Maya apart is not the number of time units they devised, or even their complexity; rather, it is their preoccupation with "commensurateness"—perfecting the way time cycles interlock and fit together. Where did these Maya ideas about time come from? What need did they fulfill, and how was timekeeping catapulted to a lofty level in ancient America a full millennium before Columbus arrived on these shores?

The Maya carved their earliest chronological records in stone; but unlike the Babylonian economic motive for developing numeration in written form, they kept permanent chronological records principally for political reasons. The Maya seem to have been trying to encapsulate historical events in a closed chronological network of time loops, a network that grew to staggering proportions as Maya civilization became more segmented and stratified.

The Maya got some of their ideas about time from the Olmec, their predecessors, who flourished on the Gulf Coast west of the Maya zone as early as 1200 B.C. They also borrowed from the Zapotecs who lived in the highlands of central Mexico around the region of Oaxaca, somewhat before 600 B.C. The pre-Maya monuments shown in figure 6.1 exhibit some of the earliest New World references to timekeeping. Their rudimentary hieroglyphs are a far cry from the ornate forms of the Classic Maya period (figures 6.7 and 6.8).

The numbers in figure 6.1 that appear on the pair of stone stelae from the ruins of highland Monte Alban were carved about 275 B.C. The way they are written provides a clue to how these people used the parts

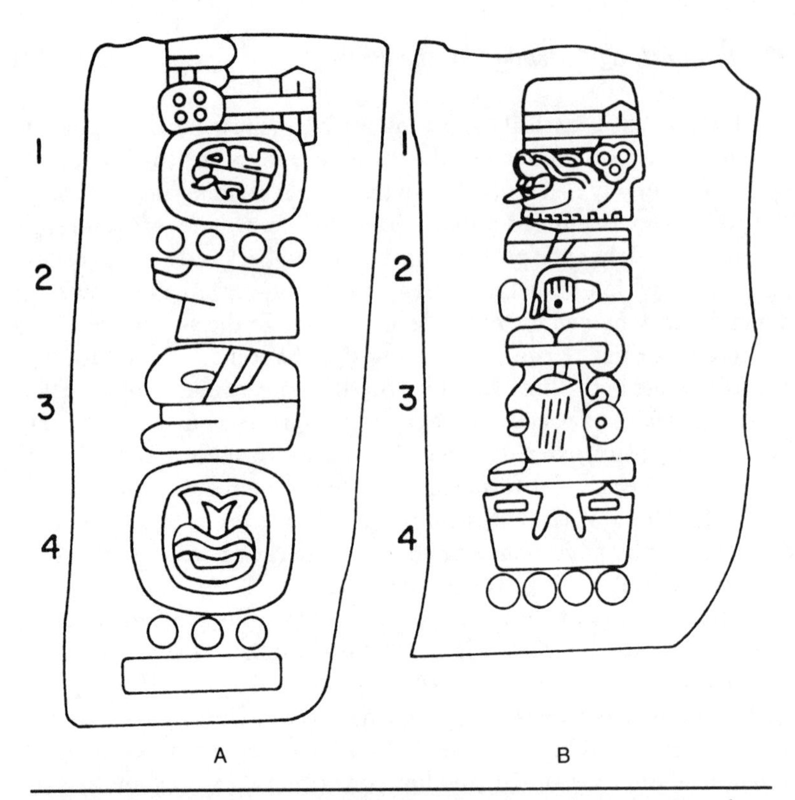

FIGURE 6.1 Antecedents of Maya timekeeping. Stelae 12 and 13 from Monte Alban, Oaxaca, offer graphic clues that connect Maya abstract numbers to a count taken on the fingers and toes. SOURCE: P. Drucker, "Ceramic Sequences from Tres Zapotes Veracruz Mexico," *Bureau of American Ethnology Bulletin* 140 (1941); and J. Marcus, "Origins of Mesoamerican Writing," *Annual Review of Anthropology* 5 (1976), figs. 3 and 4. Copyright © 1976 by Annual Reviews, Inc., courtesy of the author.

of their bodies to count the days. The round circles or dots are 1s and the bar (position A4 on the left) represents 5. The dot symbols are abstractions that probably represent the tips of the fingers, which once were used for tallying (see figure 6.2). For example, begin with the little finger of the left hand (call it day 1), then count across both hands through 10, and then across the toes to 20. If you do, you literally will compile a "person full of days"—about three of our weeks, an easily recognizable and sensible duration. Indeed, we still speak of a handful of days in our language. On the Monte Alban stelae in figure 6.1, a bundle of five days is represented by the finger itself held horizontally or by the clenched fist with the thumb at the top extended (position A2 at the left). The Mesoamerican counting system operates the same way as our decimal (base 10) system, except that it takes all the fingers and toes, rather than just the fingers, to fill a position in a number sequence. (Might the entire history of Western mathematics be dependent on whether my ancestors wore shoes?) There is no question that 20 is the basic unit in the Mesoamerican counting system; at least, no combination of dots and bars in a single position ever exceeds that amount. Among the numbers in the Maya vigesimal system that can be recognized in figure 6.1 are 8 (a bar and three dots in A4) and 6 (a stylized finger and dot in B2 on the right).

The fundamental unit of time for all Mesoamerican people, and the one counted in dots and bars on all Maya inscriptions, was the day. They still call it *kin*, an all-inclusive term that also means "sun" and "time." But they conceived of the day, not as one of a multitude of time units, but as a manifestation of the cycle of the sun. In other words, time *is* the sun's cycle itself—not an entity apart, not an orbit, not a course or a schedule that the sun follows. The hieroglyphic symbols for *kin* are among those most frequently displayed in Maya writing, and their images carry the power of this concept: each *kin* glyph marks out the four directions of the sky-earth; and the tips of the floral symbols at the center of each cartouche or frame, which symbolize procreation, also map out the extreme positions of the sun at the horizon, where it rises and sets at the winter and summer solstices. The names of the parts of space are time-related, too. The direction of east, *lahkin* in the Yucatec Maya dialect, still translates as the "sun accompanying"; while *chi-kin* literally means "the sun is devoured," an appropriate name for the west, where the sun is swallowed up each night. Thus, the spatial directions are keyed directly into time: for in Maya, space also means time (figure 6.3).

But Maya time is also motion, as Aristotle once posited. East and

FIGURE 6.2 We find the beginning of ancient Maya writing and time counting in simple hand gestures. The four hand gestures (*on the right*) used by contemporary Maya farmers signify (*from left to right*): the moon is going down; the young moon is rising; the moon retains water (the dry season); the moon lets water escape (the rainy season). *At left:* Some of these signs appear in hieroglyphic form in the ancient Maya writing. SOURCE: Drawings by P. Dunham; after H. Neuenswander, Summer Institute of Linguistics, Guatemala City, Guatemala.

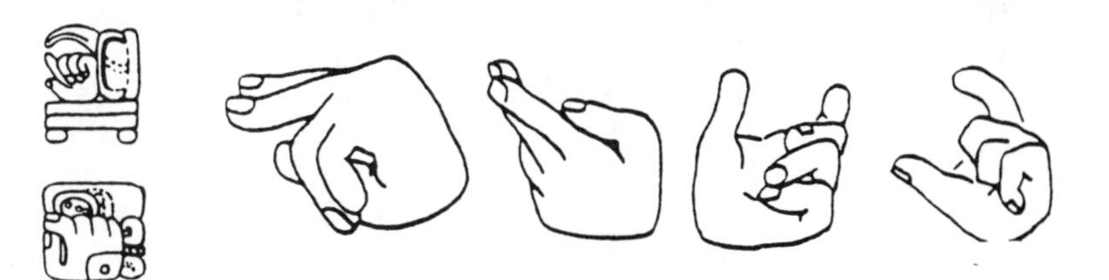

FIGURE 6.3 Maya *kin* glyphs symbolize sun, day, and time. The points of the floral cross may represent the extreme positions of the sun at horizon. *Far left:* The ancient hieroglyphic sign for "day is done," a closed fist (signifying completion) beneath *kin*. SOURCE: Drawings by P. Dunham. From A. Aveni, *Skywatchers of Ancient Mexico* (Austin, Tex.: University of Texas Press, 1980), fig. 6la. By permission of the University of Texas Press.

west are not points in space, as we might refer to them, but are the instants when time reverses its movement. In the present Mam Maya (southwest Guatemala) dialect, the directions east and west are derived from the verbs "to enter" and "to exit," respectively. East, the rising place of the sun, is thought of as the principal cardinal point—where all things enter the world. The moment the sun sets and leaves the world in the west, it reverses its direction and moves beneath us so that it can reappear the next morning in the east. The Maya believed that these cosmic transitions in the sun's cyclic risings and settings governed their fate as well as the fate of everything in existence. The luck of the day held sway over all inhabitants, whether noble or peasant, of the Maya universe. A day's name and number told when the time was right to marry, to plant, to make war, to be interred in the earth. This example of a day series from a lengthy Maya book of prophecy from colonial times reads like Hesiod's "Days" (see page 50):

3	Ben	Bad for those who go through the forest
4	Ix	The queen bee is fertilized
5	Men	Bad
6	Cib	Bad for walking in the forest
7	Caban	Bad. The deer's cry is imitated
8	Eznab	Bad for the people of worship
9	Cauac	Good for bees
10	Ahau	Good. The burner begins the fire
11	Imix	Bad for leaders
12	Ik	Bad. Good wind

13	Akbal	Bad for those who keep watch
1	Kan	Bad. Calm
2	Chicchan	Bad
3	Cimi	Bad
4	Manik	Good
5	Lamat	Good
6	Muluc	The passage of the sun is measured
7	Oc	Bad
8	Chuen	Bad
9	Eb	Good
10	Ben	Bad[4]

Why do people see fit to ascribe powers to the days? Why do they organize days into groups and give a special name to each one? Cultures directly dependent upon nature can see an element of predestination; the environment provides anticipatory cues. Nature really *does* deal with human beings in alternating patterns of good and bad, of ups and downs, as Hesiod has said. Isn't it logical to try to find a pattern in the interaction that takes place between ourselves and the rest of the world? Often we use the order we find in nature to structure human culture, elaborating upon models of nature already inborn in us. In this sense, the Maya differed little from the Greeks or the Babylonians.

Recall that in our calendar, each day of the week is named after the planetary god who governed the first hour; and to each cog in the wheel of 30 days that counted off the phases of the moon, Hesiod gave a name and a prognostication. These formed an oscillating pattern that detailed what a good farmer should or should not do. The Maya calendar is really no different, except that the root of their principal cycle, the number 20, is ordained not in the lunar principle of the heavens but in the human body. The Maya attached a named deity to each unit in the 20-day cycle, for like the counting of fingers and toes, the days and the behavior of the ruling gods all repeat themselves.

The Maya did not believe that a ruling god *represented* each one of the 20 days. Rather, the day revealed the god it showed his attributes plainly and visibly to both eye and mind. As with sun-day-time, the Maya conceived of the day as a *manifestation* of the god itself. The auguries that go with the days behave like the days themselves—some fabulous, others nasty. On the maize-god day, manifested by a youthful god who always seemed to be kindly disposed toward humans (see

FIGURE 6.4 The names of the Maya days of the week also are gods: Cimi is death; Kan, maize. Animal deities include jaguar (Imix is a jaguar's paw), monkey (Chuen), and serpent (Chiccan). SOURCE: Drawings by P. Dunham. From A. Aveni, *Skywatchers of Ancient Mexico* (Austin, Tex.: University of Texas Press, 1980), fig. 53. By permission of the University of Texas Press.

his glyphic symbol in figure 6.4), one could go forth with considerable confidence that all prospects would be pleasing—but beware the death-god's day, for then evil is cast over the land. They responded to such omens with resignation, as their cultural upbringing had accustomed them to do. Curious, isn't it, that a people so disposed toward the celestial fates would begin with their own bodies rather than those in the sky to count the days.

Given the relatively uncomplicated agararian life the early Mesoamerican civilizations led—a life of basic hard work in the fields—the average person would have needed to know little more than what the fates had in store for a few or perhaps several days in advance. But as Maya societies became more hierarchically organized states, and people became more specialized and interests more diversified, their calendar became more structured and formalized. Small periods were built up to create lengthier ones. Where in our calendar months become bricks in an edifice of years, there is sound archaeological evidence that by 200 B.C. the Maya had developed a system of counting the days in units of 260. The complete cycle, called the Maya *tzolkin*, or sacred day count, was probably invented by pairing number coefficients 1 through 13 with the rotating cycle of the 20 day names (listed vertically in figure 6.4), in much the same way that we lay the 30 (or 31) number days of the month alongside the cycle of 7 day names. (The glyphs placed next to the dot and bar coefficients in the inscriptions in figure 6.4 are the day names associated with those numbers.) Now, if we run through every conceivable match-up of number and name, we return to the beginning of the 260-day cycle after 13 times 20 pairings, or 260 days. There is nothing like the 260-day divinatory cycle anywhere else in the world. This calendar (of which I excerpted a colonial survival on pages 194–195) is *the* centerpiece of the Maya calendar system, the single most important block of time they ever kept, and still do keep in areas remote from modern influence. But why 260? And where did the 13 come from? While we have a pretty good idea of when it was instituted, no one really knows exactly why the idea of a *tzolkin* was thought up in the first place, though a number of clever theories have been conjured up to try to solve this mystery, as we shall see.

The secret lies in the numbers.* Like the days they, too, were gods.

*We find it difficult to think of a number having any meaning by itself. As the Greeks have taught us, we know a number by what is numbered. Yet for some people, 7 means good luck, all good things come in 3s, and airlines and hotels still manage to omit 13 from the count of their basic occupation units.

FIGURE 6.5 For the Maya, every number taken by itself possessed identity, each being a god. Note the similarities between "ordinal" and "teen" gods. Twenty, or "completion of a body," is symbolized by a closed fist placed alongside the face. SOURCE: Drawings by P. Dunham. From A. Aveni, *Skywatchers of Ancient Mexico* (Austin, Tex.: University of Texas Press, 1980), fig. 49. By permission of the University of Texas Press.

Though a number can be written by dot and bar combinations, each one also had its own particular face (see figures 6.5 and 6.7). For example, the head of the deity who is the number 8 was a youthful maize god—as is denoted by the maize plant growing out of his head; often he has a chain of maize kernels draped over his ear. Number 10's countenance is a skull with a bared jawbone and a nose devoid of its flesh: it evokes the image of death. Now, the 2 number symbols I just chose from the list of 13 look the same as the 2 day signs among the 20 mentioned earlier; and like the numbers 8 and 10, the day names Kan and Cimi also happen to be 2 days apart. Indeed, the order of the head number glyphs in figure 6.5 matches quite well in most attributes with a sequence that can be traced out among 13 of the 20 day names in figure 6.4; but the 7 day-name glyphs that follow all look different from any of the number glyphs. Furthermore, whenever the Maya scribes wrote the head variant numbers 14 to 19, they represented them always as a merging of the two digits that make up the number. In other words, 14 looks like 4, except that the god 14 has a fleshless jaw; 15 looks like 5; and so on. This may represent a translation into glyphic symbols of the spoken forms of these words, as it would work in our language. Thus, the Maya word for "eight" is *uaxac;* for "eighteen," *uaxac-lahun;* and the "eighteen" glyph *looks* like the "eight" glyph, except for a few alterations. Likewise, when we say the word "fourteen" and symbolize it by putting 1 in front of 4, what we really mean is "four-and-ten." Two facts, then, make it clear that one key to unlocking the mystery of the origin of the *tzolkin* may reside in the number 13. First, the singling out of the form of the coefficients 1 through 13 from all the others (13 is clearly the transitional number in the set; sometimes it looks like a variation of 3; but in other cases it takes on a totally unique form, though it is usually spoken as a derivative of 3, the way we say "thirteen"); and second, the matching of these 13 forms with thirteen consecutive forms of day glyphs.

Can it be a coincidence that the distinct array of 13 recognizable nonrepeating glyphic forms should also come to serve as the complete cycle of numbers used to match up with the 20 day names? Not likely. The 13 cycle must be very old. Perhaps at one time the Maya thought of it separately from the cycle of 20 day names. Maya cosmology featured a thirteen-layer heaven with each layer assigned to a ruling power. The thirteen gods of heaven were said to have dueled with the nine lords of the underworld for possession on our earthly region. If the Lords of Heaven constitute the most ancient Maya symbols, the notion of assigning gods and names to each of the 20 days may have been an extension of the original 13. If several more gods were added to the list of 13, every entry in the vigesimal (or body-count) system would then have its own name. A distant memory of the expansion of the series of the 13 days to 20 may be echoed in a passage in the book of *Chilam Balam* of Tizimin, which reads, "Thirteen makes seven then, said the word of the Devil to them."[5]

Was the 260 daytime count born out of the intermingling of 13 and 20? We are fairly sure about where 20 came from. And as we have seen in our own history, time reckoning is a build-up rather than a break-down process: the idea is to anticipate the future; and as we become more organized, we tend to look further and further ahead, building cycle upon cycle as we go. If the Maya recognized the 260-day period by itself early on, then maybe the number 13 is derived, the result of dividing 260 by 20. Take an analogy from our calendar: our long base unit is the year, while the week is one of its principal short duration intervals. Now, we did not set out to design the year to contain 52 weeks, but that is the ultimate outcome. The same is true of the number of months in the year. The lunar and solar bases are well reckoned from celestial observation. From these we derive the number of constellations in the zodiac. Thus, we can think of 52 and 12 as numbers resulting from the design of our calendar.

Still, there is something missing in these explanations for the *tzolkin*. People do not invent large units of time simply to derive intervals that are the product of two smaller numbers. Did 260 denote some universal characteristic then? We return once again to biorhythms for an answer. The average duration between human conception and birth is 266 days, and today Maya women still associate the *tzolkin* with the human gestation period. They time it by the lunar cycle, counting 9 months of the phases—265.77 days by modern calculations. While we cannot prove that the *tzolkin* was originally derived from a birthing cycle, we do have a practical human motive for this cycle. Recall that an archaic form of the old Roman calendar was adapted to the gestation period of cattle.

In pre-Hispanic times, one took one's name from the day name in the 260-day count on which one was born. Modern diviners in highland Guatemala pass through a 260-day initiation period before they are allowed to practice. Furthermore, the birthing cycle is a fair approximation to the length of the basic agricultural cycle in most areas of Mayaland, a factor that may have led to the habit of associating one of nature's fertility cycles with another. In some remote areas the moon is still thought to take away "nine bloods" from pregnant women to give to the lives of the newborn.[6] If human gestation and counting on the fingers and toes were the motives behind the earliest and most fundamental Maya time units, then theirs is truly a quintessentially human-oriented timepiece, as opposed to our annual cycle.

In practical terms, the Maya planting cycle was tied to the onset of the rains, which can vary drastically from year to year. This was the temporal milestone, and all calculations were made from the rain event. The time at the end of the year after the harvest mattered less. But, if gestation time were used to count agrarian time, then why did the Maya fail to hold the 260-day count fixed in the year? Why did they record the days of the *tzolkin* the way we tally the days of the week—in a continuous never-ending fashion, paying no regard to the season whatever? Did they grow so devoted to the ruling lords of time that they were willing to give the godly-time cycle precedence over the practical agricultural calendar, with its gap of "dead time" at the end of the planting season? Or, did they employ an entirely separate but unwritten agricultural calendar to regulate agricultural activities, a calendar out of which the formal *tzolkin* was born?

If you lift your eyes up from your body toward the heavens, you find other natural events that transpire over approximately a 260-day period. First, there is the average interval of appearance of the planet Venus as morning or evening star at 263 days; second, the average duration between successive halves of the eclipse season, at 173½ days, fits into the *tzolkin* in the ratio of 3 to 2. If this seems contrived, there is evidence in the inscriptions that the Maya used the *Tzolkin* to predict where Venus would appear and when eclipses would occur. For example, certain named days in the 260-day count seem to have been designated as inauspicious because they were the ones on which it was possible for eclipses to occur. For a society so carefully attuned to anticipating celestial events that marked temporal transitions in their lives, we can only imagine how dramatic an unscheduled plunge into darkness at noon would be. This means that certain days that are particularly vulnerable to the occurrence of eclipses can be found in clusters at intervals set one third of a cycle (about 120 days) apart in the *tzolkin*. In the Maya Dresden Codex there is even a table of eclipse predictions that names these inauspicious days and associates them with declaratory statements like: "Woe to the maize!", "Woe to the pregnant women!"[7] Because the numbers do not mesh exactly, the eclipse warning days have shifted considerably from the time the prediction table was written.

A third celestial rhythm with a 260-day beat, one that has meaning only in tropical latitudes, is connected with the interval the noonday sun spends north as opposed to south of the overhead position. These intervals vary depending on the latitude, but in latitude 14½° N, close to the locations of the great Maya city of Copan and the pre-Classic city of Izapa, the annual cycle divides up neatly into 105- and 260-day periods. Given the present-day archaeological record, the required latitude lies a bit on the periphery rather than at the center of the area where archaeologists have unearthed the earliest calendrical inscriptions.*

And so, like the *tzolkin* itself, round and round goes the argument about where it came from. Which part came first? I believe there is no single answer to this question. Perhaps the Maya sacred time cycle acquired its importance when Maya timekeepers discovered that the number 260 signified many things—like our gravitational constant or the speed of light, numbers that crop up in many modern physicists' mathematical equations. The discovery of the coming together of many of nature's phenomena—birth, the moon, Venus, and eclipses—may or may not have arisen in the Maya's number-oriented heads all in a flash. We will probably never fully sort out the parts of this puzzle. What is important is that in the Maya mentality, nature and number joined together. Once a Maya genius may have recognized that somewhere deep within the calendar system lay the miraculous union, the magical crossing point of a host of time cycles: 9 moons, 13 times 20, a birth cycle, a planting cycle, a Venus cycle, a sun cycle, an eclipse cycle. The number 260 was tailor-made for the Maya and the *tzolkin* count, interwoven by so many temporal threads, became the grand fabric of Maya timekeeping.

The 260-day divinatory cycle was not the only cycle the Maya used to keep time. Inscriptions dating from a few hundred years later than the first recognized *tzolkin* count also display dates reckoned in a 365-day year (or *haab*), which the Maya divided into 18 "months," each of 20 days (not to be confused with the 20-day round of the

*Furthermore the 11-August-date equivalent in our calendar for the creation of the present sun, which can be calculated by an extrapolation of the Maya Long Count (page 208) back to its zero point, corresponds closely to one of the two dates in the year when the sun passes exactly overhead in this special latitude.

tzolkin). Unlike most early civilizations that broke down the year of seasons interval into 12 units as a result of the lunar influence, the Maya seem to have remained with their body count unit of 20. At the end of the 18-times-20-day run, they added an extra 5-day month, an unlucky period thought to reside outside of the regular year, reminiscent of the "time-out" period at the end of the cycle discussed earlier in the Trobriand calendar and in our old 12 days of Christmas. We shall encounter this idea yet again in the calendar of the Inca of South America.

While the Maya may originally have tacked on these 5 days at the end of the 18-times-20-day period to bring it into line with the seasonal year, the alignment, as their astronomers would later discover, was not perfect. But, unlike us, they did not intercalate days into the *haab* in order to make it fit more precisely with the tropical year of 365.2422 days. As a result, their year cycle, like that of the Egyptian calendar, gradually slid out of joint relative to the seasons by a quarter of a day in 4 years. Consequently, the month and day count did not return to its original position in the seasonal year until about 1,460 years had elapsed.

Now, it would bother us enormously to see Christmas retreat into autumn or the Fourth of July back up into the winter season. It certainly bothered Julius Caesar and Pope Gregory—but then, that's *us*. The Maya, however, would likely have seen intercalating days as violating one of their cardinal rules of Maya timekeeping: always keep absolutely fixed the points of initial dates of various cycles that mesh together. The *haab* (we call it the "vague" year because it is only an approximation to the astronomically determined year) is a fixed 365-day unit that divides precisely and with no remainder into other longer cycles the Maya employed to mark lapsed time. Though there is no evidence they had leap years, there is every reason to believe the Maya kept track of the accumulated error between vague and tropical years, and that they could calculate where the months stood in the seasons at a moment's notice.

In spite of the inexact nature of the *haab* as a seasonal indicator, the names of the months of the Maya year and the hieroglyphic symbols that represent them convince us that they instituted the 365-day year basically as an agrarian calendar. For example, like our zodiacal constellations, their so-called watery months and those with earth-based designations seem to cluster together in one part of the list of month names. These are followed immediately by a month thought to

FIGURE 6.6 The Mesoamerican fourfold divisions of time and space. (*A*) A New Year's calendar from early colonial Mexico shows early European influence. Time winds its way from the sun-centered universe outward in a counterclockwise direction. A full spiral turn completes four years, and thirteen turns make up a sacred cycle of fifty-two years. SOURCE: D. Durán, *Book of the Gods, Rites, and Ancient Calendars,* trans. F. Horcasitas and D. Heyden (Norman: University of Oklahoma Press, 1971). (*B–D*) Other quadripartite dividers in ancient Mesoamerica include the pecked quartered circles carved in the floors of buildings at Teotihuacan (*B, C*) and the quartered universe with each of its associated entities in a pre-Conquest picture book. SOURCE: (*B*) Drawing by H. Hartung. (*C*) Photo by L. Aveni. (*D*) Codex Fejerváry-Mayer p.1, Free Public Museum, Liverpool, 12014/M (HMAI Census no. 118); Codices Selecti, vol. XXXVI (Graz: Akademische Druck-u. Verlagsanstalt, 1968).

symbolize the maize seed they garnered for sowing. Then come the months associated with the deities who presumably could be seen in the clear skies of the dry season.

The divinatory *tzolkin* and the agriculturally based *haab:* two different calendars with two different purposes, each represented by units of days and numbers thought to be the gods themselves. The cycles tripped along concurrently, the same way we time our work/rest patterns by the seven days of the week and manage our long-term fiscal affairs by two half-year periods. The *tzolkin* appears alone in the Middle Formative Period of Maya civilization (about 600 B.C.), and not until a couple of hundred years later did the *haab* become written into the inscription alongside it, the two then being organized together as a combination. We cannot say why the Maya combined them, but since the Maya society had grown into a set of organized city-states, it is not unlikely that a rulership strongly represented by the noble class had realized the wisdom of juxtaposing urban and agrarian interests. If society is to be organized on a broader, more complex basis, then its time also must be organized; a unified calendar is better than a lot of separate ones. Thus, we find inscriptions that read 1 Ahau, 18 Kayab; or the day 1 Ahau in the *tzolkin* cycle which is also the 18th day of the month of Kayab in the vague year. Joined, the *haab* and the *tzolkin* form a longer cycle of 52 years, called by Mayanists the Calendar Round.

The number 18,980, which represents the lowest common whole multiple of days in both *tzolkin* and *haab* ($52 \times 365 = 73 \times 260 = 18{,}980$), records the interval over which name and number combinations begin to repeat themselves. A little arithmetic shows that only certain name and number combinations are possible in the full 52-year calendar round. For example, only 4 of the day names in the 20-day cycle can be matched with the first day of the 365-day year—because 365 divided by 20 gives a remainder of 5. Therefore, in successive *haab*, the name of New Year's Day must be either the 5th, 10th, 15th, or 20th name in the list of 20. Some examples of the quadripartite nature of Mesoamerican space and time are depicted in figure 6.6 (see

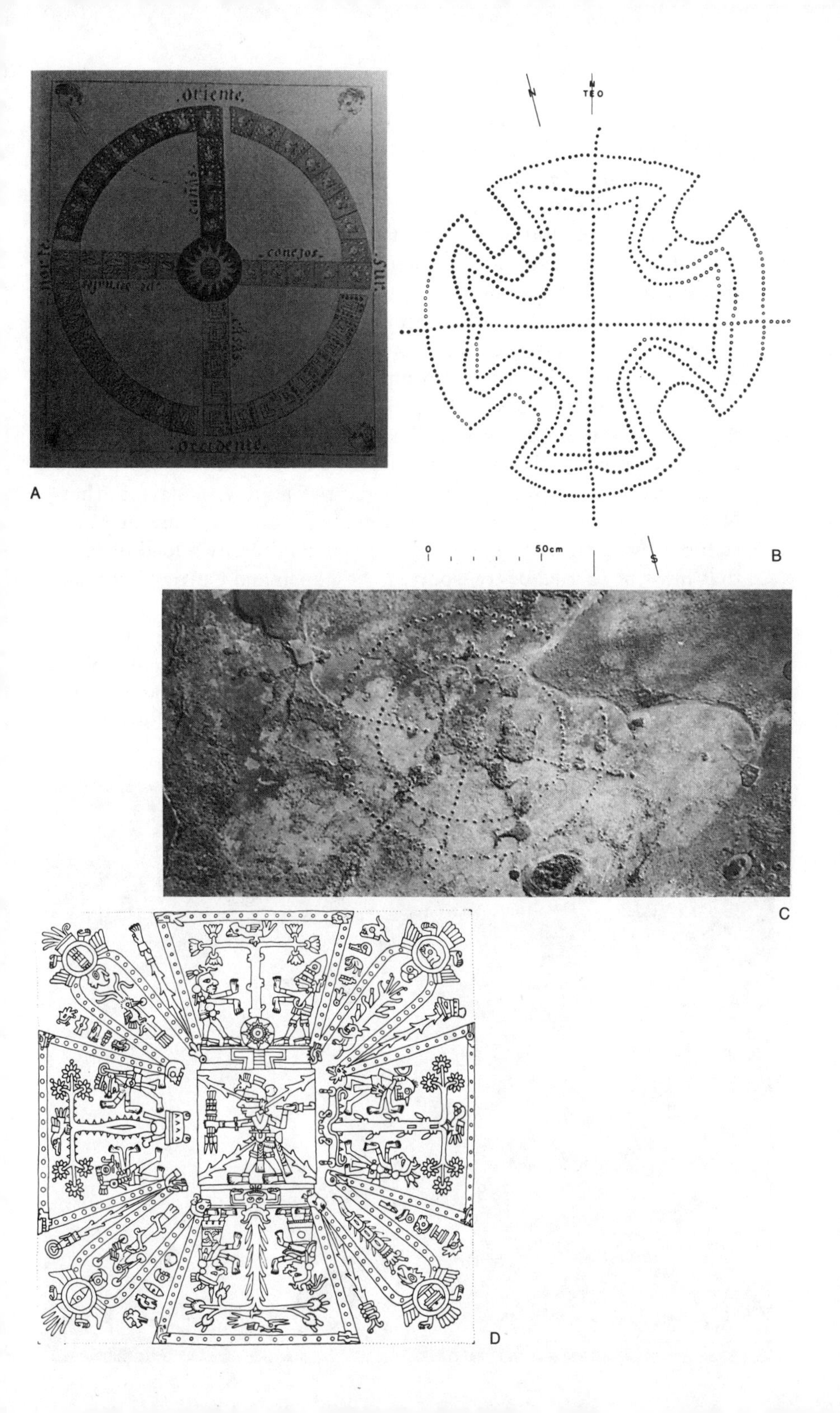
.oriente.
.norte.
.sur.
.occidente.
.cañas.
.conejos.
.casas.
A
N
N
TEO
0
50cm
S
B
C
D

figure 6.6A for the arrangement for the four New Year day names). The Maya called these four day-gods the alternating "year bearers" because, by being the initial day of the year, they were delegated the task of carrying the burden of the full year of time upon their backs, the way the planet accorded the first hour bore the name of the day in the old Babylonian week-day naming system (page 103). Figure 6.7 graphically displays the ties that bind Mesoamerican time. This notion of carrying a burden also applies to the other time units, as we shall see. Evidently the year bearers become fatigued after undertaking their great temporal journey; and so, at the resting point, once they complete their travels, they pass the burden on to the next set of gods in the temporal relay, who then carry the bundle of days farther. Nowhere but in the carved monuments do we see time in Maya thought more clearly depicted as a quantity, literally a load of stuff that must be carried or transported. At Copan and Quiriguá, we can see the neck muscles of the Maya gods of number straining at the pull of the cargo on their tump lines as they are about to pause for a rest before exchanging loads (figure 6.7A). This discontinuous concept of time, an interval of rest-completion following a period of bearing the burden, is central to an understanding of the Maya philosophy of time.

FIGURE 6.7 The Maya burden of time. (*A*) These full-bodied hieroglyphs from Copan depict the facial characteristics of the anthropomorphic time-bearing gods of number who carry burdens of time upon their backs (compare figure 6.5). The inscription, which celebrates the completion of a *Katun*, reads "9.15.5.0.0 [long count] 10 Ahau [*tzolkin*] 8 Ch'en [Haab]." (*B*) Time bound. A hand-held rope tightly binds together an Aztec date ("eleven monkey" *at right*) with a date possibly of Maya influence ("nine house" *at left*) on the Temple of the Plumed Serpents, Xochicalco. SOURCE: (*A*) After A. P. Maudslay, *Biología Centrali-Americana* (London: Porter & Dulau, 1889–1902), vol. I, pl. 48. (*B*) Photo by L. Aveni.

The Long Count and the Origin of Time

By the second century A.D., Maya royalty had mastered cultivation of the land, expanded the state, built great cities, monumentalized architecture and ultimately established one of the great civilizations of the ancient world. About this time, the rulers also made a fundamental revision in their calendar, one that we might well anticipate, having seen it happen before. They created a skyscraper-sized time cycle, a brilliant invention built upon an amalgam with a profusion of subcyclic levels that held the potential to catapult them all the way back to the gods' creation itself. Eight hundred years after the first appearance of the *tzolkin* the Maya began to reckon time in the Long Count.

Similar to our Julian-day scheme, the Long Count has been likened to the way the odometer on an automobile functions, by clicking off one day at a time in endless succession. (A misleading model, for there actually is no evidence the Maya ever used gears or machinery to keep time!) There is one essential difference, however, between your automobile and the Maya universe: when the odometer turns over, thus signaling the resting point on the longest Maya time cycle of all (some five thousand years by our way of reckoning), then the universe will be destroyed and reborn anew.

When does a culture begin the process of seeking the origin of time and when does it resort to telling the story in word and picture? We might think that the making of an origin myth on this scale is possible only in highly stratified, literate, bureaucratic societies—those we often refer to, perhaps somewhat mistakenly, as advanced. We do know that the Maya became passionate about giving the most precise material expression to their long-term history during the Classical period when they erected most of their buildings, did most of their writing, created most of their art and sculpture, and expanded their influence over a sizable region of Mesoamerica. They did it by expressing the past in Long Count inscriptions in a base-20 system: in other words, each place in a number series contained 20 times the quantity of the previous one, except for the third place upward in the hierarchy; this they called a *tun*, and it held 18 times 20 instead of the logical 20 times 20 days, probably because 360 was a closer approximation than 400 to a year. Twenty *tuns* made up a *katun*; and 20 *katuns*, a *baktun*, conventionally the highest number in the chain.

Take, for example, the Long Count number 9.15.5.0.0., which translates as follows:

(*kin*)	0×1
(*uinal*)	0×20
(*tun*)	$5 \times 20 \times 18$
(*katun*)	$15 \times 20 \times 18 \times 20$
(*baktun*)	$9 \times 20 \times 18 \times 20 \times 20$
Total	1,405,800 *kin* since the last creation

The root word *tun* means "stone" and signifies the carved stone stelae, the time pillars that the Maya faithfully erected every time the odometer turned—but especially on *katun* endings. In our culture it would be like erecting a monument every time a decade or a century elapsed (a *katun* is about 20 years). Like the days of the 260-day cycle, each *katun* carried its prophesies while at the same time it also harbored the record of what had already transpired. This curious admixture of time-past and time-future concocted within a cyclic framework suggests that the ancient Maya did not segregate past from future as we do. For them, the past could and, indeed, did repeat itself. If you paid close enough attention to time, you could see that the past already contained the future. Read the curious blend of past and future tenses in one segment of a *katun* wheel taken from the Maya colonial Book of *Chilam Balam* of Tizimin. The year was 1607; the fourteenth year of a civil war, the *katun* had begun with day name 1 Ix in the *tzolkin*. Strife and suffering had taken hold of the people. The priests made predictions by divining shark entrails and by giving attention to meteorological and astronomical events:

On the fourteenth measure
  On the first of Pop,
On the fourteenth tun is the time
  In the katun period.
There remains being made to fight
  oneself.
  The fighters arrive with the East
  priest Uayab Xoc
At the time of seeking fire,
  Of seeking shark tails.
That is the return of seeking things,
  When one seeks then

In the sky,
  In storms,
Sun phases,
  Far seeing,
At the time of covering of the face of
  the sun,
  Of covering of the face of the moon,
(Which recurred)
  On the fourteenth tun again.
Destroyed is the year
  By pleasure:
Suffering mouth for the mother,
  Suffering mouth for the father;
Suffer girls,
  Suffer boys.
Destroyed is the residue
  of the governor.
Already past is his change;
  Already past is his change of office,
Destroying the town
  He had seized,
The East priest
  Uayab Xoc
On the day of the pyramid of pain,
  Of the stone pyramid.
No one goes
  When it is given
On the seen day,
  On the return of the katun period.
Already past is the moon of the remainder
  Of the judgment of the Itzá
To return to the north,
  To return to the west.
They join together the descendants of
  the tun,
  The stalks of the hills,
Which recurred
  In this fourteenth tun again,
The day of suffering,
  The katun of suffering.

Descended are the stingers
  Descended are red were bees,
Attracted to the wells,
  To the springs,
Occurring
  According to what is in the
  arrangement
Of the writing
  And glyphs.
It was to return in this fourteenth tun
  Or was to occur at the need of the
  chiefs,
The renewal of the governors
  In this fourteenth tun again. (2409–67)[8]

The Maya traced creation all the way back to 11 August 3113 B.C. Like Bishop Ussher, who sought to follow the Old Testament lineage back to Adam and Eve by counting generations in Genesis, some astute Maya priest must have sat in his temple scribbling calculations concerning how his rulers' lineage might be tied into some already canonized information about the gods. Since most carved stelae date from *baktun* 9, with a few 8s and 7s sprinkled in, we are led to believe all this calculating went on early in the Classical era, in the same way our B.C.–A.D. system of counting was instituted well after the death of Christ.

The Long Count with its 18s and 20s seems to have grown out of the Calendar Round. As I hinted, the Maya may have conceived it as a year or *tun* cycle of 360 days obtained by lopping off the unlucky days at the end of the vague year. Later, some clever Maya mathematician extended the cycle by multiplying each successive order by 20. In a sense, that genius flattened out a portion of time's circle, thus giving it a more linear appearance. It is only when we contemplate the Long Count over a considerable duration, like thousands of *years* instead of days, that time becomes cyclic again. The effect is rather like looking out over a distant horizon: experience tells you the land before you is really flat, but if you go far enough, your trajectory will curve back on itself.

The lengthening of durational sequences in Maya timekeeping early on in the Classical period clearly was done with a motive, by a person or class of persons desirous of propagating the notion that the present can be stabilized by projecting it much further back into the past than anyone had ever contemplated.

Time and the Maya Ruler

What kind of people dare to write their history alongside that of the gods—to use time as a vehicle to legalize and canonize the authority of their rulership of the state? What were they like? Their effigies first begin to appear on carved monuments in Maya cities shortly before the beginning of the Christian era in the Old World. Today there is no doubt the figures sculpted on Maya statues were real people; we have even given them nicknames like Bird Jaguar, King Chocolate, 18 Rabbit, Lady Zac-Kuk, and Shield Pacal, after the pictorial aspects that make up hieroglyphs that represent their names. Most of them were members of various royal bloodlines, though a generation ago we believed they were gods or other mythical characters conjured up by mystical metaphysical Maya who were concerned only with astronomical knowledge for its own sake—those New World Platos and Aristotles I mentioned at the start of this chapter.

The Tablet of the Cross at Palenque (figure 6.8) reveals one of the most graphic attempts made by the Maya to link genealogical with mythic time. This Classic Maya city is located southeast of Vera Cruz, where the humid Gulf Coast plain meets the mountainous inland rain forest. The engraved stucco tablet, with its nearly life-sized figures, lies in a recessed chamber in the Temple of the Cross, so called because the cross-shaped symbol at the center of the sculpture was thought by early explorers to represent a Christian cross (more likely it is a tree of life that symbolizes the axis connecting heaven with the underworld). The cross is adorned with a double-headed serpent and topped by a huge bird; both represent sky deities, the ultimate source of royal power. King Shield Pacal is the smaller figure on the left. He has just died and, like the setting sun, is about to pass into the underworld. The lid of the sarcophagus in his tomb, which lies only a hundred yards away, shows him making the descent. In the Cross tablet Pacal passes the symbols of royal power on to his son, Chan Bahlum, the larger figure who stands to the right of the cross-tree.

The Tablet of the Cross depicts a great moment of tension—the point of transition in the succession of rulership; he had erected the temple in celebration of this important event. Much of the imagery in the scene is concerned with the vestments of power. Notice the sacrificial stingray spine, the instrument of penitential bloodletting, along the axis of the cross. Chan Bahlum holds the same symbol in his hand just below the *kin/sun* floral symbol in the depiction on the outer left panel; we can even see the sacrificial blood flowing from it. And on the inner right panel, he receives the serpent-footed jester god, an instrument connoting rulership.

The inscriptions (not shown) on the tablet artistically express the symmetry the Maya envisioned between natural and supernatural worlds, laced together by the thread of time. They tell of events in the lives of the king that are combined with happenings from the past, many of which refer to astronomical events that actually occurred in Palenque. The inscriptions on the left half of the tablet, flanking the dead father, date from *baktun* 1, a time preceding Pacal's life by three and a half millennia. They offer a chronology of legendary events—the birth of the gods at Palenque from whom Pacal is said to have descended, gods who existed even before the beginning of the present era. In stark contrast, in the text to the right of the living descendant of Pacal, the scribe has recorded dynastic historical events—what we would call real time events. In tune with the *katun* and *tzolkin* prophecies, the Maya rulers advertise these events on the right side as re-enactments or consequences of the godly events carved on the left side. Quasi-historical happenings bridge the birth of the gods with statements about the immediate past lineage of Pacal and his son. For example, one name recorded at the bottom of the left panel as well as at the top right is thought to be that of an Olmec ruler born in 993 B.C. (our time), who acceded to the throne in 967 B.C. Since Pacal died late in the seventh century A.D., shortly after which the Temple of the Cross and its commemorative tablet were erected, this luminary must have preceded the Palenque king by nearly seventeen hundred years, a greater interval than the one separating Moses from Christ. Surely he

FIGURE 6.8 (*overleaf*) Dynastic and deistic time. The Tablet of the Cross at Palenque commemorates the accession of Chan Bahlum (*right*) who receives the instruments of office from his deceased father, Shield Pacal. A glyphic text to the right of this panel is essentially family history, but a matching text to the left ties the ruling family's roots to the birth of the gods at Palenque. SOURCE: L. Schele, "Accession Iconography of Chan Bahlum in the Group of the Cross at Palenque," in M. Greene Robertson, ed., *The Art, Iconography and Dynastic History of Palenque*, part III (Pebble Beach, Calif.: R. L. Stevenson School, 1976), fig. 6.

must be legendary, at least in part. This same juxtaposition of real and mythic time scales, with legendary interfacing, is repeated in other tablets at Palenque. In each, the principal focus is the passage into the afterlife of the great lord Shield Pacal. To judge from all the Palenque monuments dedicated to him, he must have been as influential in the New World as Caesar or Alexander in the Old.

Maya kings tell their dynastic story not by writing out a chronological list of dates, but by recording elapsed time. The inscriptions always begin with an event, then an interval, then another event follows, then another interval, another event, and so on, most of the human events in the inscription denoting the birth and death of his ancestors. Epigraphers call these intervals "distance numbers"—with, I think, the deliberate intention of implying that the Maya really thought of time as distance, a road traveled. Things happen in rest periods or breaks separated by these distance numbers. All events are pegged to a Long Count date at the beginning of the inscription. During these intervals, the gods carry time's burden from one happening to the next. With inventions like Long Count and distance number, a ruler like Pacal could proclaim the longevity of his bloodline in concrete terms. His rulership could acquire new depth, and the monumental carvings would demonstrate his permanence in the public eye.

The change of presidents and premiers, the death of a ruler and the succession to office of his replacement—what a critical time in the history of the state. Such an abrupt discontinuity, so unsettling; for the Maya it would be like a kink in the circle of time. How would the people know that once their old leader was gone, his son would possess the same powers? What guarantees it? For Americans, continuity of rulership is promulgated by faith in the democratic process; for the Maya, it was by faith in the bloodline. Blood is the answer. Kinship and dynastic bloodlines are proclaimed the principle of immortality in the Tablet of the Cross. Pacal tells the whole world that Chan Bahlum is his legitimate extension, and uses celestial analogy, as well as specific sky events like eclipses and planetary conjunctions, to validate the principle of continuity in his rulership. The Maya creation myth says that when the old gods died they re-emerged in the east as the sun and Venus. So it was for the kings. After Pacal undergoes his apotheosis in the underworld, his powers are reborn in his son. Like the coming spring after a long dark winter, his life will re-emerge invigorated in the body of his son. In one text from Tikal, a dead ruler is actually shown being transported through the underworld in a large canoe paddled by a group of mythical animals who represent the underworld gods. The text is carved on a series of bones discovered in the ruler's burial crypt.

Today in Palenque you still can see the sun at winter solstice, the year's most prolonged period of dark time, diving into the place where Pacal is interred beneath the Temple of the Inscriptions, a short distance west of the Temple of the Cross. In this architectural hierophany, the plunging sun *is* Pacal beginning his descent into the underworld, as he takes the first steps on the temporal road to his resurrection. He is clearly represented falling into the underworld on the elaborately carved lid that caps his sarcophagus beneath the stairway to his crypt. The message plays on the Maya fascination with cyclic time. Death is necessary to life: it is the ultimate creation force that produces birth. Chan Bahlum not only lives on as a continuation of his father; he literally *is* his father, a re-emerged, re-envisioned version of his predecessor by blood in the cycle of courtly life. Pacal's death is not an event to mourn; rather, it should be celebrated, for it is the first step in a great transformation that gave human insight and access to the process of renewal and rebirth.

Maya people and Maya culture live on through the immortality of Maya rulership. The events carved on the Tablet of the Cross assured it—events that extol the unbroken connection between mundane time and eternity. Through the Long Count, Maya royalty could vastly lengthen time, give themselves room to breathe, room enough to hitch their impressively long bloodline to that of the very gods who created the universe and time itself. How much time did they need? Enough to assume that if they extrapolated into the future they could be assured of the same long-term stability and permanence their society had experienced in the past. Maybe it isn't so different for us. We care little about the "imminent" destruction of the world the modern cosmologists tell us will surely occur several billion years from now when the sun violently blows away its hot outer atmosphere. While we may be concerned about our children and our children's children, we are hardly moved to think and act in anticipation of events forecast millions of generations ahead. Like the Nuer, age-sets far away from us in either direction of time seem to dissolve into an equidistant past/future that seems unreachable. We occupy ourselves only with those strips in either direction along time's road that have not vanished in the impenetrable fog as we try to steer our way with a minimum of disorder through that portion of the highway we can keep clearly in focus.

Just how far back into the past did the Maya calculators delve?

How deep did they seek to penetrate that fog? We know they produced a recyclable table for predicting eclipses that was 11,960 days long and a reusable Venus ephemeris of 37,960 days—over a hundred years. These tables appear in written manuscripts used by the priestly élite to make astrological predictions. Each table had been updated and recycled over a period of at least a few centuries. But some of the cycles they played with were far longer than this. Some Maya inscriptions take us four places higher than the *baktun*—that is, to 64,000,000 years; and one inscription may involve a calculation going as much as half a billion years into the past. The use of such vast quantities of time by Maya astronomers would have staggered our imaginations had we found out about them before we made the geological discoveries of the last two centuries that have drastically expanded our own historical time scale.

The numbers the Maya wrote down emphasize completion and the harmonious intermeshing of cycles. Take the Long Count date 9.9.16.0.0 for example, which is recorded in one of the Maya codices, where it is thought to mark an appearance of the planet Venus because it opens a tabular ephemeris for that planet. Epigraphers call it the "super number of the codices" because of its unusual capacity for swallowing whole time units. The super number is not only an integral number of *tzolkin*, Haab, and Calendar Rounds, but also an exact multiple of the cycles of Venus and Mars as seen in the sky. Clearly, the Maya élite in their Classical heyday somehow became as fixated on commensurate numbers as they did on the matter of coordinating religious activities with astronomical events. Consider the enormity of the task of choosing a Venus event that fell close to the day 9.9.16.0.0 for the starting position in a calendrical table. Imagine mathematically searching for a coronation day or an inauguration date in our own calendar whose Julian-day number must be divisible without remainder by a host of other important magic numbers. But at the same time the celebration must take place at the actual time of occurrence of a major celestial event! Calendrical coordination of this order, with ritual restrictions underlying purely astronomical calculations, has been attempted only rarely in the history of the world. There must have been a Maya Kepler or Newton at work among the calendrical specialists who walked the halls of the royal palaces mulling over these very serious problems of their temporal art. In our world, too, the clergy were also the first to become possessed with how to gain a foothold on eternity.

The innovative keepers of the Maya calendar who made those laborious calculations must have been members of the noble class, close advisers to the rulers, persons of high rank:

> They had a high priest whom they called Ah Kin Mai [literally, "he of the sun"]. . . . In him was the key of their learning and it was to these matters that they dedicated themselves mostly and they gave advice to the lords and replies to their questions. . . . They provided priests for the towns when they were needed, examining them in the sciences and ceremonies, and committed to them the duties of their office, and the good example to people and provided them with books and sent them forth. And they employed themselves in the duties of the temples and in teaching their sciences as well as in writing books about them. . . . The sciences which they taught were the computation of the years, months and days, the festivals and ceremonies, the administration of the sacraments, the fateful days and seasons, their methods of divination and their prophecies, events and the cures for diseases, and their antiquities and how to read and write with the letters and characters, with which they wrote, and drawings which illustrate the meaning of the writings.[9]

This description of the calendar-priest and his duties by a sixteenth-century Spanish clergyman documents the specialized nature of the calendrical scribe's work. He was skilled not only in mathematics and astronomy but also in matters of calendrical divination. Descendants of such experts still exist among some of the Maya groups. These calendrical shamans retain a knowledge of the 260- and 365-day cycles and know all about the year bearers that tie the two together.

Modern anthropologists have observed that day counting and tallying is but a small part of the astrological repertoire of these specialized calendar-priests, who still are much revered by the indigenous community. A shaman's task is not simply a matter of consulting a list of days and naming their properties. The whole process of making calendrical prognostications operates more like a dialogue between priest and client, and much of the outcome depends on their social rapport. A modern ethnologist describes one such dialogue:

> Then the diviner announces that he or she is taking hold of the divining bag and borrowing the health of the particular day (of the 260-day cycle) on which the divination is taking place: "I am now borrowing the yellow sheet-lightning, white sheet-lightning, the movement over the large lake, little lake, at the rising of the sun (east), at the setting of the sun (west), the four corners of the sky (south), the four corners of the earth (north)." At this point, sensing that the "blood" and the days are ready to respond, the diviner, after saying "one is now giving clean light," then proceeds to frame the divination in a formal way. For example, the first formal question in the case of illness would be, "Does the illness have a master, an owner?" etc.[10]

We have no reason to think the calendrical divining process was basically any different in the past, except that in Classical times the priest probably carried his book of computations along with him and may have been accompanied by considerable fanfare. Sadly, practically all the books were destroyed in huge bonfires by the Spanish priests who believed they promoted idol worship, which would seriously impede the invaders' plan to establish the Roman Catholic religion among their subjects.

The Cycles of Venus

The Dresden Codex

Unlike the carved inscriptions, which dwelled mainly on the deification of the nobility, the codices held all the valuable factual data of observation and tables on astronomical phenomena, as well as the basic rules of procedure for the diviner and prognosticator. Also unlike that on most stelae, this information was private rather than public. The Dresden Codex, portions of which are shown in figure 6.9, must have been intended for priestly eyes only (a portion of another Maya codex is depicted in figure 4.2D.). In these documents, the Maya calendrical mentality reaches its loftiest heights. We cannot even begin to comprehend the depth and precision of Maya timekeeping without touching on at least a portion of one of these books.

One of four surviving Maya books, the Dresden Codex dates from eleventh-century Yucatan. Made of lime-coated tree bark and painted with vegetable dyes, it corresponds precisely to descriptions of Maya books given over four hundred years ago by eyewitnesses to the Spanish conquest of Yucatán:

> Their books were written on a large sheet doubled in folds, which was closed entirely between two boards which they decorated, and they wrote on both sides in columns following the order of the folds. And they made this paper of the roots of a tree and gave it a white gloss upon which it was easy to write.[11]

To get at the substance of the Maya timekeeper's knowledge, I want to dismantle the calendar on these pages and focus upon its component parts the way I dissected the intervals in our own calendar in part II. These particular pages are devoted to the Maya expression of the complex movement of the planet Venus, though this is not the only celestial body the Maya paid attention to; there certainly were others. Because this is one of the few completely deciphered passages in the codex, the example of Venus gives us a rare opportunity to glimpse the detail, the precision, and the complexity that engrossed Maya calendar makers.

What did Venus symbolize in Maya thought? We know that for the post-Classic Maya, the planet was a male god: the light-skinned Quetzalcoatl, as he was called by the Toltec invaders of the Maya world, or Kukulcan by the Maya. The god may be based in part on a real person who was exiled in the tenth century from Tula, the Toltec capital of central Mexico, from where he is said to have journeyed east to Yucatan where he founded a new empire. He predicted his own return, which coincided with Cortés's invasion of Mexico in 1519—an item the Spaniards were quick to capitalize on. Motecuhzoma (Montezuma), king of the Aztecs, who inherited the Quetzalcoatl myth from the Toltecs, laid down his arms in anticipation that the god had returned to claim his kingdom. The familiar resurrection myth is given celestial expression in the Annals of Quauhtitlan, a post-Conquest book from Central Mexico:

> They said that Quetzalcoatl died when the star became visible, and henceforward they called him lord of dawn. They

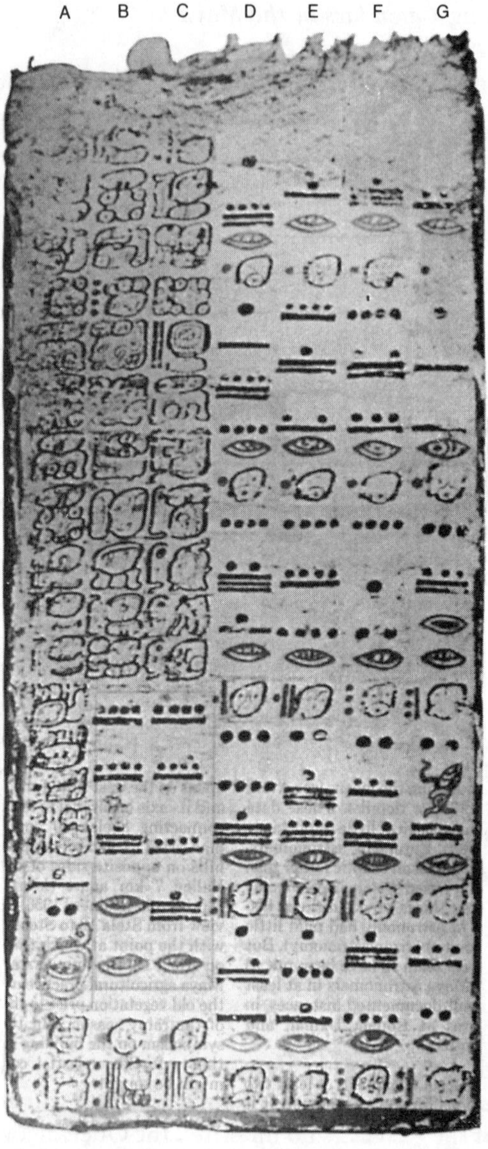

A B C D E F G

FIGURE 6.9 The Dresden Codex. Portion of a Maya Venus calendar painted on lime-coated tree bark. The correction table for the Venus calendar (*left*), which immediately precedes the page shown on the right, may be regarded as a multiplication table with a base equal to the length in days of five Venus cycles (columns DEFG). Higher multiples toward the top of the table (sections 1 and 2) deviate slightly from a whole number of cycles, a clue that has led epigraphers to believe that the priest-astronomers applied small corrections to keep the ritual table in step with the actual events observed in the sky. The five-place dot-bar numbers toward the bottom of columns B and C are Long-Count dates for Venus events counted from the beginning of the creation epoch; the ring num-

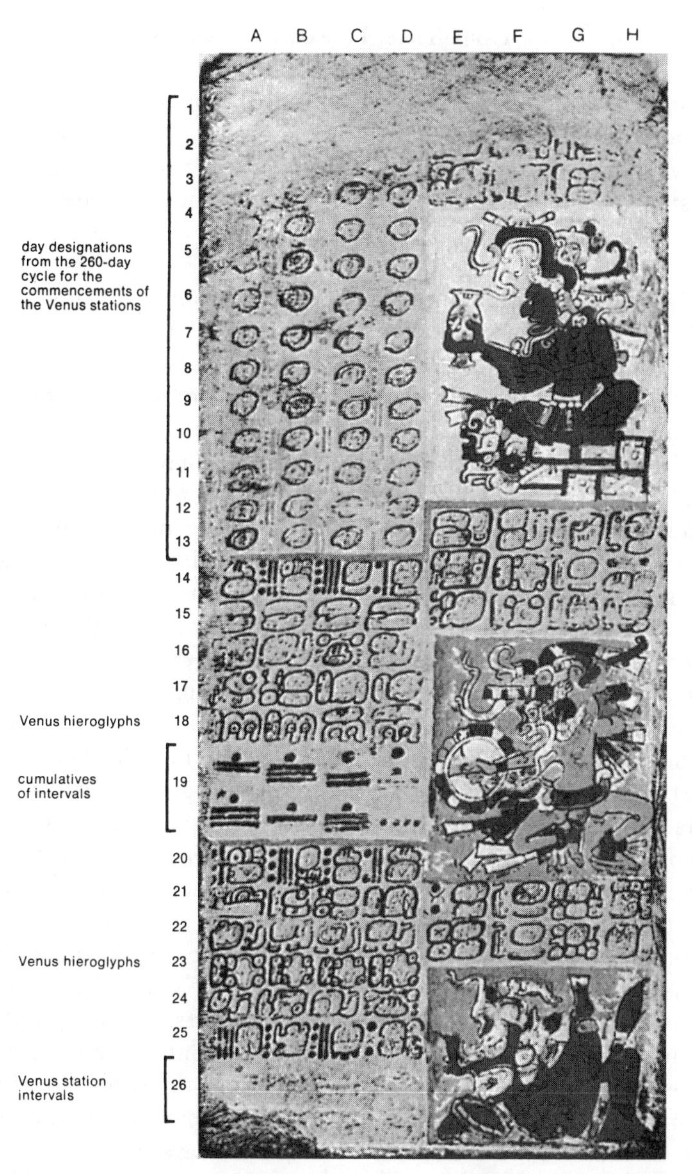

ber is the difference between the two Long-Count numbers. Page 46 (*right*) of the Dresden Codex is the first page of a five-page Maya Venus calendar, which is accurate to two hours in five centuries. The top few lines have been worn away. The four numbers in line 26, representing the number of days in the four Venus stations, are 236, 90, 250, and 8 (*reading from left to right*). The numbers of line 19, which are 236, 326, 576, and 584, represent accumulated totals. Lines 14, 20, and 25 comprise day designations in the solar year cycle (Haab); lines 1–13, in the 260-day ritual cycle (*tzolkin*). Note the Venus hieroglyphs. SOURCE: J. Eric S. Thompson, "A Commentary on the Dresden Codex," *Memoirs of the American Philosophical Society* 93 (1972):24, 46–50.

> said that when he died he was invisible for four days; they said he wandered in the underworld, and for four days more he was bone ["dead"]. Not until eight days were past did the great star appear. They said that Quetzalcoatl then ascended the throne as god.[12]

One European-sounding variation on the resurrection myth states that Quetzalcoatl became excessively drunk. For such sin he was forced to abandon his country and travel to the east, where he raised a funeral pyre to destroy himself. Out of the fire his heart rose into heaven to become the planet Venus.

What physical aspect about the celestial behavior of Venus were the Maya getting at? The underworld journey of Quetzalcoatl corresponds to the 8-day disappearance of the planet in front of the sun. After it vanishes from the western sky in evening twilight, Venus is next glimpsed in the morning sky prior to sunrise approximately 8 days later. Gradually, from night to night, its dazzling light, exceeded only by the sun and moon in luminance, increases to maximum brightness, as Venus begins to rise earlier in the morning twilight and can be viewed for a longer time each day. As it pulls farther away from the sun on the sky, it fades in brilliance. After about 9 months (263 days) of celestial meandering, it closes back in on the sun, and finally disappears in the solar light. Venus is absent from the celestial scene for approximately 50 days while it passes behind the sun. Then it comes back into view in the evening sky for an additional 9-month period, its brightness waxing until, toward the end of the evening-star viewing period, it once again achieves greatest brilliance. This is the round of Venus as the Maya would have seen it and as we still observe it in the sky today—a closed cycle divisible into four parts: two unequal disappearance intervals separated by a pair of 9-month periods when its bright white light can be seen hovering above the sun's glow in the morning or the evening twilight. Venus completes the full cycle in a little less than 20 of our months, or about 584 days, and this is exactly the way the Maya record it in the Dresden Codex.

Since the great scientific Renaissance, Western culture has acquired the habit of thinking of the planets in a sun-centered, space-based, orbital framework, Venus and the earth playing the role of solar attendants that mutually revolve about the sun, though in different periods. At regular intervals, when Venus passes between the earth and the sun, it is lost from view as it gains a lap on our orbit. As Venus moves ahead of us, it treks among the stars across the morning sky. Its orbit takes it to the opposite side of the sun, then into the evening sky. Algebraic manipulation reveals that a Venus orbit is completed in 225 days as conceived in a framework fixed on the sun. Though, of course, we do not live on the sun, we have somehow come to think of this interval as the "true" Venus period.

But that is *our* explanation of the Venus cycle. To know how it was envisioned by the Maya, we must divest ourselves of the heliocentric posture we have acquired since the Renaissance, and learn that the 584-day Venus cycle, as far as an earth-based spectator is concerned, is really far removed from the sun-centered Venus year of Western astronomy. Figure 6.10 depicts a part of the Venus cycle as it unfolds directly on the sky.

Venus is unique among the nine planets in respect to its motion (Mercury comes closest to following Venus's pattern but, because of its proximity to the sun, is usually hidden from view). For example, Mars, Jupiter, and Saturn range over the entire sky. Unlike Venus, they can take up points opposite the sun on the sky and therefore can be seen high in the sky at midnight. But Venus really seems to be hitched to the sun by an imaginary elastic thread upon which it bobs back and forth like a yo-yo, never receding more than an arc of 45 degrees from the sun on the sky. When it cannot be seen closely following the sun as it moves downward in the western evening sky or announcing the rising sun in the morning in the east, it is obscured by the blinding solar light. The completed cycle of 584 days became a magic number for Maya prognosticators. And given what can really be seen of Venus's aspects, we might well have expected the Maya to have split its time cycle into four periods or "stations": one for the "evening star," one for the "morning star," and one for each of the two unequal disappearance intervals. Given the pan-Mesoamerican penchant for quadripartition (see figure 6.6), how satisfied the Maya must have been with themselves!

The Maya discovered another divinely harmonious property of the Venus number: its seasonal quality. The 584-day cycle intermeshes with the 365 days of *haab* in the perfect ratio of 5 to 8. Translated into visual reality, that means a Venus station, once recorded, is bound to recur after 8 calendar years. Suppose, for example, that a heliacal rise of Venus takes place today; then astronomers can be assured that the event will recur on nearly the same date 8 solar years from today (a day or two correction would be necessary because the seasonal year is actually a trifle longer than 365 days).

The Maya saw something quite different in Venus from what we

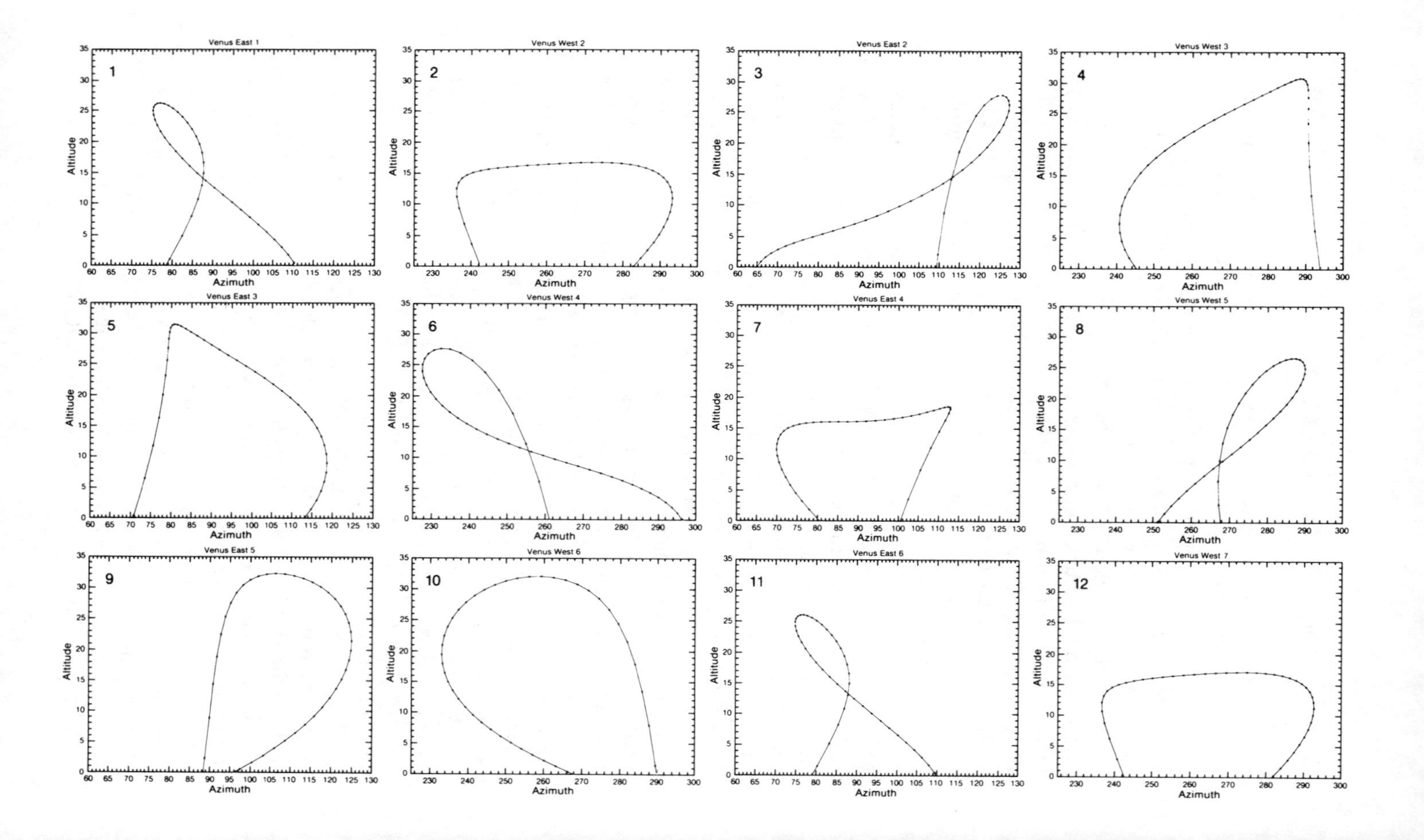
Venus East 1
1
Altitude
Azimuth
Venus West 2
2
Altitude
Azimuth
Venus East 2
3
Altitude
Azimuth
Venus West 3
4
Altitude
Azimuth
Venus East 3
5
Altitude
Azimuth
Venus West 4
6
Altitude
Azimuth
Venus East 4
7
Altitude
Azimuth
Venus West 5
8
Altitude
Azimuth
Venus East 5
9
Altitude
Azimuth
Venus West 6
10
Altitude
Azimuth
Venus East 6
11
Altitude
Azimuth
Venus West 7
12
Altitude
Azimuth

see, as this brief nontelescopic astronomy lesson helps us to understand: because it always attends the sun, they transformed it into the mythical god who rose out of the sun's ashes. With these visions of how Venus moves over the sky, we can turn the pages of a Maya Venus calendar and begin to understand how it worked. The right-hand picture in figure 6.9 shows the first of five consecutive pages (numbered 46–50) that make up the Dresden Venus Table. Each page follows the same format, a common one in Maya calendrics. On the left side of that page we find a series of time intervals recorded in dot-bar notation, and on the right, a set of pictures representing celestial events. Certain hieroglyphic symbols interspersed among both pictures and time intervals are equatable with Venus because they occur with unusual frequency on these particular pages of the Dresden Codex.

The four light-colored numbers across the bottom of our representative page 46 add together to make 584. It is no coincidence that four subintervals are recorded: these are Venus's four basic stations. The right-most number, written vertically as "08" (note the stylized shell which substitutes for the clenched fist as zero symbol above it), tabulates the 8-day disappearance segment, the most critical one in the Venus cycle. The black numbers located higher on each page are interval accumulations. Each is obtained by adding the previous black number to the light-colored (red) interval recorded below in the same vertical column, another rendering of the punctuated time sense as on the Tablet of the Cross at Palenque: interval–event–interval–event. This program continues across each of the five pages up to the final cumulative number on page 50, totaling 5 times 584, or 2,920 days. Then, like Maya time itself, the table recycles, and we return to page 46 to begin another run through the table, still in perfect step with Venus in the sky.

The lion's share of each of the 5 Venus pages is occupied by the first 13 horizontal lines of glyphs at the top of the table. Each line contains 4 symbols representing the commencement days of the 4 Venus stations expressed in the language of the 260-day cycle. Note that each

FIGURE 6.10 The Venus cycle as it is played out on the theater of the sky. Alternate pairs depict views of Venus in the sky at twilight near the horizon. Each dot is separated from the next by a 5-day period: (1) first morning interval; (2) first evening star interval; (3) second morning star interval; and so on. Note that after five pairings (5 × 584 days) the patterns repeat; thus compare 1 with 11, 2 with 12, and so on. SOURCE: Courtesy O. Gingerich and B. Welther, Smithsonian Institution, Astrophysical Observatory, Cambridge, Mass.

symbol has 2 parts: a number from the sequence of 13, and a hieroglyph from the cycle of 20. Thus, line 1, read across all five pages, represents five consecutive passages of Venus through its 584-day cycle; line 2, the next five passages; and so on. The length of the entire table, then, is 5 pages times 13 lines times 584 days per page, or 37,960 days. This interval also possesses that special property of commensurateness with other cycles of importance. It is a whole multiple of (1) the 584-day Venus period, (2) the 260-day *tzolkin,* and (3) the 365-day vague year. This habit of creating almanacs consisting of bigger and bigger cycles each exactly divisible into one another seems to be a calendrical preoccupation with the Maya. The same theme is echoed in the lunar-eclipse table in another part of the Dresden Codex.

Just what is supposed to happen when Venus makes its great transformation and is born again into the morning sky after the critical 8-day absence? The vertically arranged sets of three pictures on each page give us an answer. In the middle illustration of page 46 we find one manifestation of the Venus god. He holds an *atlatl* in his left hand, having just used it to hurl the spear of dazzling light rays evident when Venus becomes visible upon the morning heliacal rise. His spear is destined for the sacrificial victim below, who appears with it already driven through his bowels. Between the pictures, glyphs detail the omens attending the Venus stations. Today these compound hieroglyphs, which contain numerous prefixes and suffixes, are only partially deciphered but the message is clear: the key turning points in the motion of the celestial deity portend significant junctures in the future of men and women. They can be translated roughly as "woe to the maize seed" or "presence of disease." In the upper picture, a toothless old god squats on a throne, offering sacrifice to the Venus god in honor of his reappearance.

The whole document gives the impression of a very exact system of Maya astronomy tightly bound to astrology. The Maya needed a precise knowledge of the Venus cycle for practical reasons. Because they believed every facet of daily life was regulated by the heavens, the performance of a civic, religious, or agricultural function required precise timing. Conformity to nature superseded ennoblement of the mind, at least in the traditional Western way the Greeks have passed it on to us. The very notion that true astronomical knowledge lies buried beneath ritualistic constraints rings a dissonant chord in our ears. Again the problem is one of perception. We regard astrology as a mystical drag force that would damp out any illuminating results the Maya might have envisioned about the way things behave in the heavens—just ask any modern astronomer what he thinks about astrology! Rather, religious thought seems to have been the motivating force that spurred on the Maya sky watchers. They were always looking for better ways to meet the challenge of harmoniously mediating the affairs of nature, society, and the gods.

Paradoxically, while Maya timekeepers were concerned about accurate prediction, they still seemed to be getting away with grossly distorting the Venus dates for religious purposes. Take the 236-day interval written down at the bottom of page 46. In reality, this interval is a long way from the 263-day average interval of appearance of Venus in the sky. Why canonize a fictitious interval? This is but one of a host of Maya Venus mysteries that have yet to be solved. Oddly enough, three of the four tabulated Venus intervals are either whole or half multiples of the lunar month (236 days is close to eight lunar synodic periods). They may have been making an attempt in their table to tap out the Venus rhythm with a lunar beat. In fact, a clever user can even employ the Venus table to predict lunar eclipses. But this carries us far into the remote and not yet fully explored recesses of Maya calendrical astronomy. If they tabulated semifictitious Venus intervals, then how could they use their ephemeris to predict when Venus would reappear? To answer this question, we turn to the left-hand page of figure 6.9, which immediately precedes the 5 consecutive Venus pages. It emerges as sort of a user's instruction manual. Columns D, E, F, and G of this page comprise a table of multiples of 2,920 (5 × 584) days, a handy device for any Maya priest who might spend a lot of time calculating bundles of Venus cycles. But some of the higher Venus multiples, those in rows 1 and 2 at the top of the page for example, deviate from a whole number of Venus revolutions by just a few days. Venus watchers used these "aberrant" multiples as correction devices so that, after many passes through the table, the priest still would be able to correlate omens, drawn at the proper time from his table, with the passages of Venus in its true cycle on the sky.

When did the Maya priests become aware that the real Venus period (actually 583.92 days to one who looks carefully) was ever so slightly less than 584 days, the whole number they had fixed to represent it? (The Maya made no use of either decimals or fractions; their basic unit was the whole day.) This means that after one trip through a given line of the table, the real Venus would lag 0.4 days (5 × 0.08) behind that tabulated for the planet in the codex. After 65 Venus revolutions (one complete trip through all 13 lines of the table), Dresden time would be ahead of real Venus time by 5.2 days. Now, the scheme on the user's page of the Venus table suggests that to get back on schedule, the priest must not trek all the way to the 65th revolution at the end of the ephemeris. Rather, he must stop at the end of the 61st revolution, deduct 4 days, and then return to the beginning of the table. Such a little prescription has the effect of bringing the table back into closer alignment with the actual timing of Venus; it is still 0.88 days short since $(61 \times 584 - 4) - (61 \times 583.92) = 0.88$. More important, it enables the user to obey another one of those incontrovertible rules of Maya calendrics: the table must be re-entered at its beginning always on the same number and day name—in this case, the day 1 Ahau of the 260-day ritual cycle, or Venus's name day. In our calendar this would be a little bit like requiring that new moons at the start of a year must be celebrated only on a Wednesday!

Another portion of the correction table implies that on a different occasion the priest must stop progress through the main table on the 57th instead of the 61st Venus revolution. At the end of this, he must subtract 8 days and return to "go"; this also results in precisely the correct day in the *tzolkin* for a new entry into the beginning of the table. Furthermore, the second correction must be applied only after the first had been set into operation on four consecutive occasions. The results have astonishing implications about precision in Maya timekeeping: after $61 + 61 + 61 + 61 + 57 = 301$ passages through the table, the accumulated error would amount to $4\,(61 \times 584 - 4) + (57 \times 584 - 8) - 301\,(583.92) = 0.08$ days. Expressed in our terms, the difference between the real and the tabulated position of the planet amounts to a mere 2 hours in 5 centuries! A veritable "Maya Institute of Advanced Studies" must have been at work.

Like our old calendar reformers and computists, the Maya chronologists must have struggled arduously to achieve a satisfactory solution to the Venus problem. In fact, there are striking parallels between what they did in the tenth century and what our own ancestors did on at least two occasions (in 45 B.C. and A.D. 1582) to correct our year calendar. The Long Count number 9.9.9.16.0, written in the user's page to the Venus table, had long been thought to be the starting date of the table—a date on which Venus rose heliacally. It corresponds to 1 Ahau in the *tzolkin* (as we might expect) and to the eighteenth day of the month of Kayab in the *Haab*. Now it turns out that 9.9.9.16.0 (6 February A.D. 623 in our calendar) is fully 17 days before a Venus heliacal rise;

if, however, we skip forward from this date 3 Great Cycles, or 3 passages through the Venus table, we land on another 1 Ahau 18 Kayab, and it corresponds exactly to a real Venus heliacal rise—the one of 20 November A.D. 934. Had the Maya instituted their table on this latter date, when ritual and natural time so miraculously coincided? The anthropologist Floyd Lounsbury thinks so. He argues that since the Dresden document (probably a copy of an earlier codex) dates from about two centuries later than this, astronomers may have been trying to establish a precise Venus table for quite some time.[13] It was only well after the A.D. 934 date of formal installation of the Venus calendar that Maya astronomers recognized that their scheduling had its shortcomings. They needed to shift the base of the table (that is, drop days) in order to keep the all-important 1 Ahau date of the *tzolkin* in step with the real events in the Venus cycle.

Still another pass through the table takes us to 10.10.11.11.17 (25 October A.D. 1038) by which time the difference between the canonic and true Venus heliacal risings had swelled to 5.2 days. Apparently Maya astronomers had not yet achieved enough of an observational base to be sure the Venus error would still accumulate; at least they appear to have undertaken no corrective action at this point.

If we follow the accepted correction scheme, as expressed through the aberrant multiples in the user's table over the next three successive passages through the Venus table, the calendar round bases shift first to 1 Ahau 18 Uo (6 December 1129), then to 1 Ahau 13 Mac (15 June 1227), and finally to 1 Ahau 3 Xul (17 December 1324). Each of these shifts brings the table almost precisely back into line with the true times of occurrence of Venus events, as figure 6.11 shows. In the first base shift, Maya priests appear to have exacted a double correction (8 days) thus reducing the drift of 5.2 + 4.2 = 9.4 days down to only 1.4 days. In the second base shift, a single (4-day) correction was called for. This reduces the error from 6.2 to 2.2 days; and finally on the third occasion, a 7.2-day error got reduced to 3.2. This takes us to a time only a handful of generations before the Spanish Conquest. But by then the Classical period had long since come to an end. Archaeologists don't really know why, but difficulties with agricultural management may have led to a bureaucracy that became further detached from the people. Whatever may have contributed to the collapse, the Maya gave up reckoning time by the Long Count about the same time as they quit carving stelae and building pyramids. No longer could the rulership lay claim to time and the birth of the gods. No longer perhaps

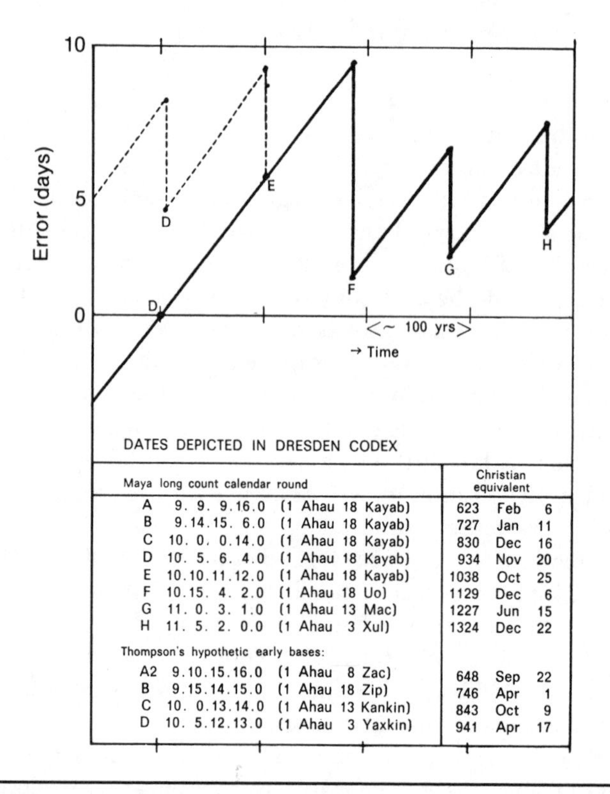

	Maya long count calendar round		Christian equivalent		
A	9. 9. 9.16.0	(1 Ahau 18 Kayab)	623	Feb	6
B	9.14.15. 6.0	(1 Ahau 18 Kayab)	727	Jan	11
C	10. 0. 0.14.0	(1 Ahau 18 Kayab)	830	Dec	16
D	10. 5. 6. 4.0	(1 Ahau 18 Kayab)	934	Nov	20
E	10.10.11.12.0	(1 Ahau 18 Kayab)	1038	Oct	25
F	10.15. 4. 2.0	(1 Ahau 18 Uo)	1129	Dec	6
G	11. 0. 3. 1.0	(1 Ahau 13 Mac)	1227	Jun	15
H	11. 5. 2. 0.0	(1 Ahau 3 Xul)	1324	Dec	22
Thompson's hypothetic early bases:					
A2	9.10.15.16.0	(1 Ahau 8 Zac)	648	Sep	22
B	9.15.14.15.0	(1 Ahau 18 Zip)	746	Apr	1
C	10. 0.13.14.0	(1 Ahau 13 Kankin)	843	Oct	9
D	10. 5.12.13.0	(1 Ahau 3 Yaxkin)	941	Apr	17

FIGURE 6.11 Portion of a graph showing errors between actual and predicted times of appearance of Venus according to the Venus table. The Maya corrected their Venus ephemeris (at points *F*, *G*, and *H*) in essentially the same manner as Julius Caesar and later Pope Gregory XIII reformed the Christian calendar. SOURCE: After F. Lounsbury, "The Base of the Venus Table of the Dresden Codex and Its Significance for the Calendar-Correlation Problem," in *Calendars in Mesoamerica and Peru*, A. Aveni and G. Brotherston, eds., (Oxford: British Archaeological Reports, International Series 174, 1983):1–26.

could a leader call himself divine. By the fourteenth century, there was no astute astronomer-priest around to apply the correction scheme written in the Venus table, much less deal with its mounting ineffectiveness—no Pope Gregory to improve upon what an earlier Caesar had done to repair the faulty time line. Given the slow drift in the Venus position (now about a 1-day error per century) under the clever correction program written in the Dresden, we can only wonder how long it might have taken Maya astronomers to realize it was time for another Venus calendar reform. And we can only imagine what other rules they might have devised once they finally recognized their empirical shortcomings.

I have suggested that the Venus Table in the Dresden Codex can be thought of as an accurate ephemeris—a device based upon past observation devised for the purpose of making accurate predictions about when significant Venus events would take place in the future. For more than four hundred years, Maya priests were increasingly careful about getting the table right. I have also implied that the Maya developed their Venus calendar in exactly the same way Old World astronomers worked at patching up their solar timekeeping system. Both were forced to change what they had already canonized—by the two-step process of resetting the calendrical mechanism to fit with the event, and by devising a more accurate prescription to slow down the future rate of slippage between real and representational time. In both the Old and New World, the script was the same; only the actors on the celestial stage were different.

Architectural Images of Time

THE RUINS AT BONAMPAK

One of the outcomes of events like those set forth in the omens in the Venus table is elaborately depicted in a mural painting found in the ruins of Bonampak deep in the rain forest of the central Yucatan peninsula. When first discovered over a generation ago, these colorful images provided some of the earliest evidence that the Maya were not the loving pacifists our scholarly predecessors once made them out to be. The Bonampak wall paintings are about war, not peace. Calcareous deposits that had dripped down onto them have sealed and preserved the delicate thousand-year-old record of a battle said to have taken place exactly when Venus was in its most auspicious aspect for the conduct of war. Accompanying glyphs state that the date was 2 August A.D. 792, and Venus was about to burst into the morning sky with its

FIGURE 6.12 Part of a wall painting from the Maya ruins of Bonampak suggests a schedule of war-making time by the arrival of Venus in different constellations of the zodiac. SOURCE: Drawing by A. Tejéda, Courtesy Peabody Museum, Harvard University. Insert after F. Lounsbury in A. Aveni, ed., *Archaeoastronomy in the New World* (New York: Cambridge University Press, 1982), p. 167.

penetrating daggers of light bearing omens after its 8-day absence; moreover, the sun would also pass directly overhead at noon on the same day. Here was a coincidence of two celestial events of cardinal cosmic importance. At the top of the room, the Venus glyph—the same one we see in the Dresden Venus table—is depicted in different houses of the Maya zodiac (the peccary at the left; the tortoise, probably Gemini, at the right [figure 6.12]). Below, captured lords from an alien city supplicate at the feet of the ruler of Bonampak, King Chaan-Muan. He stands bigger than life at the center of the scene holding his staff, while royal patrons flank him on either side and his Bonampak warriors look on from below.

The whole set of murals painted across the walls and ceilings of the three rooms of the temple of Bonampak celebrate the victory in battle, the capture of the enemy, and the institution of a new heir to Chaan-Muan's throne, together with the attending festivities—all of it couched in a celestial context with Venus as the cosmic centerpiece. The Maya hero-king had commissioned monuments to commemorate the successes he had achieved. His priests' secret codices indicated when the impressive power of the heavens could be unleashed on his behalf, while the great carved monuments recorded everlastingly and for all to see the outcome of his actions and activities. What perfect timing! Just as godly Venus appears with his light spears attacking both plants and animals, so, too, the king of Bonampak strikes down his enemies; his painted palace, his source of power, is clearly allied with the Venus sky deity.

Codices, ceramics, sculpture, and wall paintings: the Maya élite used various forms of expression for the passage of time. They even encoded the Venus time record into the architecture of their ancient urban centers.

The Maya city—unlike New York City or Washington, D.C.—looks like a helter-skelter jumble of buildings with no apparent order or arrangement. A Maya archaeologist once commented that, even in their Classical heyday, Maya masons were too sloppy to construct a right angle; they failed even to line up their buildings in the cardinal directions.[14] But in the ground plan of a typical Maya city like Uxmal (figure 6.13), a pattern begins to emerge: the buildings seem to be grouped about a north-south axis, twisted slightly clockwise from perfect north. This long axis connects a pair of major pyramid-temples which look as if they are responding to one another across a half-mile of open, scrub-covered terrain. Most of the structures along the axis face inward, toward the civic center. All around there are smaller clus-

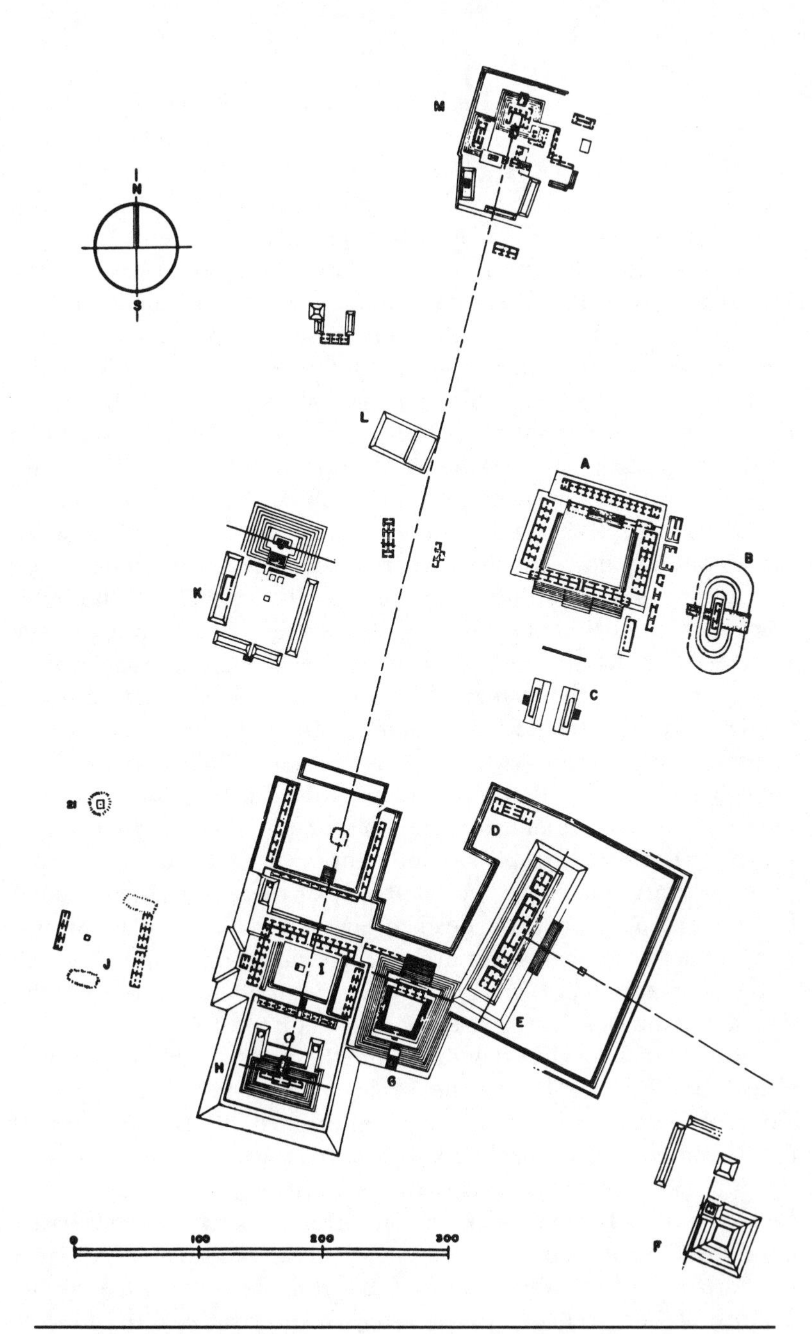

FIGURE 6.13 Maya royalty embedded their calendar into the space of their ceremonial centers. Maya centers from the Puuc zone in west Yucatan have in common curious orientational properties that may relate to the place of the sun at the horizon. Uxmal offers a good example. Note the skewed orientation of the House of the Governor (*E*). SOURCE: S. G. Morley and G. W. Brainerd, *The Ancient Maya*, 4th ed. (Stanford, Calif.: Stanford University Press, 1946), fig. 11.58, p. 342. Copyright © 1946 by the Board of Trustees of the Leland Stanford Junior University.

ters, most of them quadrangular in shape, with each of the component buildings looking inward toward the enclosed plaza. Were each of these clusters separate royal or perhaps priestly family units?

Exactly the same pattern occurs in a few other cities from the Puuc-Maya zone of West Yucatan, which flourished during the Terminal Classic period (A.D. 700–800). Here most sites are still in relatively good condition with a lot of solid standing walls even a full millennium after they were erected: Labna, Kabah, Sayil, Oxkintok—more than three dozen sites make up the list—all exhibit the same elongated axis with an east-of-north skew and those inward-facing palace groupings. It is as if they were cast out of a standard urban mold. In most cases, the long axis is marked out by a *sacbe*, or "white road," a raised paved walkway along which ritual processions likely once took place.

A time record is built into the stone walls that make up the architecture of these abandoned cities—a calendar that transcends the Maya written documents and carved stelae. These old urban centers may even contain the history of one of the earliest calendar reforms, for when we measure building alignments in a large number of cities, a significant orientation pattern begins to emerge. Puuc-Maya site orientations seem to be distributed about two basic directions along the horizon—14 degrees and 25 degrees east of north (or south of east and north of west if one happens to look out of the palace doorways that face crosswise relative to the main axes of these sites). The second direction is an astronomical dead giveaway. It marks the place of sunrise on the December solstice in the east or sunset on the June solstice in the west; each represents one of the key turnabout positions of the sun in its annual oscillatory trek along the horizon.

But what about the other direction, the one that fits a majority of cities? One explanation for the 14-degree skew is that it points to where the sun rose or set at a whole multiple of 20 days (2 times 20 days) counted from the day when it passes overhead. Because nearly all the Puuc sites hold steadfastly to this orientation, I think they were all following a unified calendar, an unwritten seasonal calendar based upon the division into 20-day units of the yearly motion of the sun at the horizon, just the way the ancient Babylonians had divided the circle into 360 degrees, one for each day of the year. A closer look at Uxmal, which has more buildings standing than any other site in west Yucatan, reveals that some palace doorways and sight lines between principal buildings point to other key sun positions—the equinoxes, the solstices, and the zenith passages as well as to the twice-twenty-day deviation at either side of the solar zenith passage.

Now, something very interesting happens when we take the two-bump distribution in the orientation axes of the Maya cities I have discussed, and wipe out all but the Terminal Classic (A.D. 700–900) data. When we remove the alignments of the earlier buildings, the 25-degree maximum vanishes. This may mean that in the Early Classic (A.D. 200–300) the calendar operating in the north of Yucatan was based on a solstice starting time; but when later dynasties came to power, the old tradition underwent some revision. New rulers switched to an annual cycle pivoted about the passage of the sun across the zenith. Indigenous writers tell us, shortly after the Conquest, about the ancient calendar their ancestors used to employ: "To this day the Indians . . . commenced [their year] . . . from the precise day on which the sun returns to the zenith of this peninsula on his way to the southern regions."[15]

Why the change? If only we had documentary evidence of the kind that has become available concerning the Gregorian or Julian reforms of our calendar, we might have some sense of what palace intrigue might have lain behind this Maya effort to control time. Did they, too, have jealous rulers, and were their feelings akin to the anti-Semitism that played a role in the Gregorian reform? Were they as vexed about when to start their cycle as our ancestors were about commencing it precisely on the equinoctial downbeat? While we can only guess what the royal motives were, we do have a pretty good idea of where they came from. At just about the beginning of the Terminal Classic, the whole of the southern Maya lowlands had rapidly begun to fall into disarray. A combination of failures to control both crops and population together with a host of other factors saw Tikal and all of its sister "skyscraper-sites" in the Petén rain forest of present-day northern Guatemala lose control of the trade economy of the whole Maya region. Now the cities of the northern lowlands became the great ruling centers—the inheritors of artistic and sculptural styles, writing, and the keeping of time.

This general flow of styles and ideas to the north also seems to have delivered in its wake a new way to think about time. Not only does the orientation shift in city planning from solstice-based 25 degrees to zenith-based 14 degrees take place at this delicate time of cultural transition, but also the new orientation scheme makes use of a curious sun-time principle that works only around the latitude of Tikal. If you go there and carefully mark the position of sunset at multiples of 20 days counting from zenith passage, you find that twice 20 days backward from one zenith passage takes you to an equinox and twice 20 from the other zenith passage lands you on a solstice. A count of 20 days forward from either zenith passage places you at the other equinox-solstice pair. In other words, the southern lowland sites are, by a curious quirk of astronomy and geography, situated at that "magic latitude" in which the year can be perfectly segmented into multiples of 20 days, with each segment ending on a highly visible pivot of the annual solar cycle. It is the ideal place to develop a year cycle based upon the principle of tying solstice and equinox to zenith passage in neat 20-day bundles, a principle I believe migrated northward in late Classic times along with practically every other aspect of Maya high culture. At least this is the story the building alignments seem to be telling us.

Not only do the axes and plans of whole Maya cities carry standard sun time locked within their walls, even the Venus phenomena that symbolize royal authority in the Dresden Codex and the Bonampak murals have been carefully fitted into the architecture of a number of pyramids and temples that we find deliberately deviated from the rest so that they could be fixed upon the key points along the horizon where Venus appears or disappears.

Windows on Venus

If we watch the daily position of appearance and disappearance of Venus relative to the horizon, two facts become obvious. First, Venus fluctuates like a pendulum over an 8-year period between two great horizon extremes centered on the east and west points of the horizon; and second, the seasonal recurrence of these great extremes is correlated with the length of the disappearance intervals before morning heliacal rise, which can vary several days either way from the well-known 8-day mean. One explanatory note about each of these observations: on the first one, we might well expect a Venus 8-year period because of the way the 584-day cycle interlocks harmoniously with the seasonal 365-day year in the ratio of 5 to 8; and on the second, the 8-day disappearance period quoted in dots and bars in the Dresden Codex is only an average. Depending on the season of the year, Venus can be absent from the sky for 20 days or more; and at the other periodic extreme, it can return to the sky on the very next morning following the evening on which it disappears. In both cases, 8 emerges as the quintessential—or we should say "oct-essential"?—Venus number, at least for those who follow its course across the sky with rapt attention.

These observations may seem a bit esoteric. Indeed, they were of little importance to European astronomers, who paid little attention to what happens in the region of the sky close to the horizon: our astronomers were simply too busy looking up. But both these Venus precepts were not only carefully watched by Maya astronomers, but also put to good use by the patrons they served to create an atmosphere for worshiping the gods in which aspects of both the environmental and social world merged in the space of the ceremonial center. There are at least three well-documented instances of such Venus timings in Maya structures at Copan, Uxmal, and Chichén Itzá.

Temple 22 at Copan has at least two attractions for the modern astronomer. We call it the Temple of Venus because of the presence of Venus symbols (exactly the same kind as in the Dresden Codex) in an elaborate sculpture adorning its entrance (figure 6.14A). It also possesses a window, an unusual attribute for a Maya building. The slotlike aperture, measuring about 70-by-15 centimeters, opens on the west side of the building. Its axis is parallel to the baseline connecting a pair of carved stelae (numbered 10 and 12 on Copan maps), two bigger-than-life pillars perched on hills 7 kilometers apart on opposite sides of the Copan Valley where they likely served as territorial markers. Now, the view from stela 12 to stela 10 aligns with the sun set point on 12 April, the annual start of the Maya agricultural season. Today farmers still initiate the growing season by burning the old vegetation to prepare for the coming of the rainy season. The Venus symbolism on Temple 22 suggests that it too may be involved in the sun-rain connection.

Today the view out the narrow window is blocked both by intervening vegetation and by another building the Maya added at a later date. But even if we could peer through the slot on a really dark night, we still would not see what the Maya saw when they built it; the sky has changed. With modern computer graphics, however, we can effectively climb back into that darkened chamber over a thousand years ago and replace our eyes with theirs. When we do, we discover that Venus passed through the narrow window at regular intervals that marked important points in the agricultural cycle. Its first appearance in the slot coincided with the coming of the rains in the year before the

FIGURE 6.14 Maya expressions of their knowledge of Venus. (*A*) Detail of the bicephalic (two-headed) serpent hanging over the doorway of Temple 22 at Copan shows in the middle a Venus symbol (*inverted*) like ones found on p. 46 of the Dresden Codex in figure 6.9. A sun symbol (*not shown*) lies higher up on the serpent. (*B*) The House of the Governor at Uxmal (see figure 6.13). (*C*) The view from its central doorway aligns with a distant pyramid as well as with one of Venus's temporal turning points on the horizon. (*D*) Carvings on the building attest to Venus worship. They include over 300 Venus glyphs, and (*E*) the magic Venus number "8," shown as a bar from which three dots are suspended. (*F*) Windows in the Caracol Observatory at the ruins of Chichén Itzá capture the rays of setting Venus at both its horizon extremes. SOURCE: Photos by H. Hartung.

D
E
F

planet reached one of its horizon extremes. And the last appearance of Venus in the window marked the coming of the rains in the year following an extreme. Imagine a cuckoo clock, whose pendulum climbs to its greatest height every time the little bird inside chirps. Venus swinging and rain falling: the two beat rhythmically together. Like bird and pendulum, they constitute a single entity when viewed from the proper perspective. This correlation could mean that one of the principal purposes of the temple was to host the rites related to the annual coming of the rains and the growth of the new maize crop. The window, like the secret Dresden document, could have served as the observatory through which the Maya priests sighted and determined the celestial event.

Uxmal is the site of a second example of an ancient Maya building tied to the celebration of Venus events. Typical of the ancient Maya ruins of northern Yucatan, the ruins of Uxmal overlook a virtually flat horizon dotted with distant sites which often are visible from the tops of its remaining pyramids. Most of them were erected about A.D. 800, and they all line up on the same east-of-north axis, except for the House of the Governor (figure 6.14B). This imposing structure, so nicknamed by modern travelers because it looks regal enough to have housed a ruler, lies perched on an artificial square mound about 10 acres in extent. Its visual aspect is remarkable even to the casual visitor, for it is skewed out of line nearly 30 degrees, its doorway facing south of east.

Standing on the "porch" of that building, you can sight along a perpendicular line from the great central doorway of the building over a solitary man-made mound 6 kilometers distant—the only blip on a featureless, flat horizon. The sight line from Uxmal to the distant pyramid also points to one of the Venus turnaround points or extremes. If you stood in the doorway of the palace at the right time in about A.D. 800, you would have seen Venus hovering above the distant pyramid when it made its first appearance in the southeastern sky at the time of arrival at one of its extreme points. Did a great ceremony celebrating cyclic closure take place on the porch of the Governor's House on that occasion? Was a ruler's accession to office being commemorated? Perhaps a marriage between rival city-states sealed the completion of the social side of the time cycle. Or was it the taking of captives in war for which the Maya of Bonampak already had invoked Venus, or perhaps an agricultural event as at Copan? Not many inscriptions survive on the fragile limestone carvings that make up

Uxmal's literary corpus. But the masks of the rain god Chac that adorn the building may hold part of the answer. Looking across the cornice of the Governor's House, we can see carved Venus symbols—hundreds of them—positioned beneath the eyelids of the Chac masks. And on the palace cornerstones is found the number 8, the Venus number carved in the form of a bar and three dots carved onto the stucco masks (figure 6.14D, E).

As for the distant pyramid, an arduous trip through the rough brush flanking the Pan-American highway near Uxmal reveals the aligned distant mound (C) to be the largest among an array of buildings that rival Uxmal both in size and in extent—the ruins of another ancient city. Once it was connected to Uxmal by a *sacbe* (an elevated "white road") that began just across from the Governor's House. The cosmic axis that ties Uxmal with the neighboring site raises an interesting question which cannot be answered without further study: Did the Maya notion of geography employ cosmological ideas that included a set of geometrical and astronomical rules for the placement and orientation not only of individual buildings like the House of the Governor but also of entire ceremonial centers relative to one another?

The Caracol of Chichén Itzá (figure 6.14F) offers a third example of Venus-aligned architecture. Chichén Itzá is a sprawling Maya–Toltec site in north-central Yucatan, and its Caracol symbolizes the joining of these two cultures, an event that took place in Yucatan sometime around the tenth century. According to Spanish chroniclers, the Caracol's round shape symbolized Quetzalcoatl-Kukulcan, the Venus deity. We know it was built at about the same time and roughly in the same area where the Dresden Codex and its Venus table originated. It may even be the almanac in stone that incorporated the Venus directions and astronomical observations that gave rise to the Venus calendar written in the codex. The building looks like a cylinder resting atop a two-layer quadrangular platform, one tier being misaligned by 5 degrees from the others. (Misalignment always seems to indicate that a building may be astronomically oriented—as is true even of our modern solar-heated houses.) Four entranceways on the perimeter of the cylinder give access to four more doorways inside. These lead to a narrow passageway that coils upward like the shell of a snail and leads to a rectangular room at the top of the structure (this coiled shape gives the building its modern name "Caracol" or snail in Spanish). Three windows are all that remain venting outward from the chamber; they open to the south and west and can only have been made to look at the heavens through. Modern studies have revealed close correlations with astronomical events at the local horizon for most alignments that have been measured on the building. These include sunset at the equinoxes and sunset on the days the sun passed the zenith.

Most significantly, four of the most important directions match the Venus horizon extremes. The base of the lower platform and the principal altar, which support a pair of round columns set into the stairway of the skewed upper platform, marked the Venus northerly extreme around A.D. 1000, about the time the building was erected. A diagonal sightline through one of the windows of the upper chamber is exactly parallel to this direction. Another window alignment matches with the Venus set position when it attained its southern extreme. In these orientations lay the priestly power of prediction, for by sighting where Venus disappeared at the western horizon, astronomers situated in the Caracol could predict precisely when it would reappear in the morning sky. Thus, they could anticipate whether the great luminary, the god Quetzalcoatl-Kukulcan, would arrive before or after the 8-day average they had calculated in the codices. Building alignments are a reasonable way of permanently encoding significant astronomical information in a landscape devoid of natural peaks and valleys that might aid in the task of marking time. As we look more deeply into the role that architecture played in Maya cosmic expression, we might well expect to find more structures astronomically aligned, for only recently have we come to view the Maya as a people who saw time and space connected in this concrete, physical way. Apparently this form of expression was allowed to develop and flower in different ways in different Maya cities. In the Caracol, the west-facing windows in a single building caught the essence of Venus's motion; while at Uxmal, time was a factor in the spatial relations among different buildings.

For the Maya, keeping time by the stars did not simply entail writing a bunch of numbers down in books. Cosmically timed cyclic events needed to be piped directly into the environment where the ritual took place. For us, it is the closed interior space of cathedral and synagogue; for the Maya, it was the open exterior space of the ceremonial center. Venus, priest, and people—all had to arrive in the theater at precisely the same instant. This is why the priests of Uxmal distorted the layout of the House of the Governor, why the Maya skewed the axes of their cities all across Yucatan. Keeping the right time was an intimate part of the Maya notion of city planning.

Both Maya public art and architecture seem to focus upon celestial power as the ultimate creative force in the universe. At the same time, these expressive media offer a public display of the power of kinship in Maya social life. The mixture of kinship and divine authority evokes values remote from the way we run our cities today.

Popul Vuh: The Maya Genesis

The special relationship between Venus and the sun found in the calendrical writing in the codices, carved on two-headed serpents, and expressed as architectural metaphor in the orientation of Maya pyramids, also is voiced in mythic terms in the *Popul Vuh*, the so-called highland Maya Book of the Dawn. Written after the Conquest, it has been interpreted as an alphabetic version of the pre-Conquest hieroglyphic codices—precise timekeeping housed in a simple myth, as in Homer and Hesiod. The principal characters in the creation myth are hero twins who represent Venus and the sun (see figure on page 166 where they are perched at opposite ends of a tortoise). The peregrinations are embedded in the *Popul Vuh* in a language almost as complex and precise as that of the Venus table. In fact, the poetic movements of Venus and the sun provide a kind of musical score for the enactment of the myth.

According to the Maya genesis story in the *Popul Vuh*, the gods, who come across as far more bumbling and humanlike than Jehovah of the Old Testament, fail three times at the act of creation. It is only when they fashion people out of corn that they finally produce a successful race—one that will talk, walk, and worship them. The gods' action upholds the universal maxim "You are what you eat." Even older than these makers and molders of men is an elderly couple whose calendrical names correspond to the day names in the *tzolkin* when Venus appears and disappears. The epic is too detailed to recount here, but it goes on to tell of the adventures in the underworld of their sons and grandsons, from whom the Maya lineage is said to have descended. The translator Dennis Tedlock has noted that the story offers clues along the way—some direct, some indirect—that link great moments of tension and climax to important days in the Venus cycle.[16] The thread that weaves the mythic fabric together is the repeated rising in the east of Venus, of dawn and life. The birth of Venus becomes the microcosmic analogy of the ever-recurring starting point in the cyclic creation of the entire universe.

Tedlock has demonstrated that the episodes in the *Popul Vuh* come in groups of five, just the way the Maya saw the motion of Venus as divisible into five repeatable though slightly different cycles that take place over the course of eight years. This Venus five-ness is the leitmotif behind the layout of the Dresden Venus table, with its five pages and five Kukulcans in different costumes, each of whom spears a different victim. Architecturally, there are five layers of rain-god masks on the House of the Governor—the same as the number of completed Venus cycles that fit perfectly into eight solar years.

The *Popul Vuh* reminds us in many ways of the time-related myths of the Old World, like the Babylonian *Enuma Elish* and the Greek *Theogony.* The Maya creation story is filled with acts showing parallels between celestial and human life—acts that continually repeat themselves. Maya creation seems to consist of a two-step process—a sowing and a dawning. Seeds are sown in the earth; the sun, moon, and stars in the sky; and humans, in the womb. The chaos that precedes the creation is a condition much like that found in the Old World creation stories. Sky and sea are separated by a vast calm that lies in between them. The episodic wanderings of the twin sons and grandsons of the ancient creator pair take the form of cyclic happenings at two levels: those that occur on the surface of the earth, and the ones that take place in the underworld. Time, like the creation, is made up of the continuous process of alternation between two opposing conditions: descent and ascent, conflict and resolution, sowing and dawning. Mimicking the movement of things in the heavens, the action that takes place in mythic time is repetitive; in a sense, it is cyclic, but it never repeats itself exactly—the upwardly coiling helix model I proposed in figure 2.4B comes to mind. In each attempt to create viable human beings, for example, the gods get a slightly different result. And when, in the final chapter, the heads of the lineage who descended from the hero twins make their pilgrimage in search of divine elements in their own lives, they do not repeat precisely the route their mythical cosmic hero-twin ancestors took; however, their actions leave no doubt about their blood ties.

> And then they remembered what they had said about the east. This is when they remembered the restrictions of their fathers. The ancient things received from their fathers were not lost. The tribes gave them their wives, becoming their

> fathers-in-law as they took wives. And there were three of them who said, as they were about to go away: "We are going to the east where our fathers came from," they said, then they followed their road. . . .
>
> There were only three who went, but they had skill and knowledge. Their being was not quite that of mere humans. They advised all their brothers, elder and younger, who were left behind. They were glad to go. "We're not dying. We're coming back," they said when they went; yet it was these same three who went clear across the sea.[17]

These are the princes from whom the Quiché Maya, for whom the *Popul Vuh* was written, had claimed their descent, a class of nobility influenced directly by the gods and commanded to repeat and re-enact past divine action. Believing themselves to possess blood ties with such princes, modern-day Quiché Maya still recognize their special nature; it is *they* who were chosen to be the one true people of the gods.

The Maya Decline

Eventually the exalted rulers of the Maya Classic period fell from power. Post-Classic times are notable for the trend away from sculptural and architectural monumentality. Carved stelae still appear, but in drastically reduced numbers; one look at them tells us that laboring in stone as a way of propagating the dynastic record was no longer one of the foremost Maya concerns. At least in the north of Yucatan, the Spanish chronicles tell of a not-so-distant past when dissenting city-states were constantly at war. Of the great city of Mayapan, one bishop said: "The quarrels between the Cocoms, who said that they had been unjustly expelled, and the Xius, lasted so long that after they had lived in that city for more than 500 years they abandoned it and left it in solitude, each party returning to his own country."[18]

To be sure, the priests still recounted the *katun* prophecies, but no longer were the predictions of the future contained in the past elaborately written down. By the eleventh century, the Long Count was dropped. The shorthand form of the calendar adopted around the time of the collapse consisted simply of supplying the name of the day in the 260-day cycle that corresponded to the beginning of the *katun*, or score of *tuns*, along with either the prophecy if one were talking about the future, or the history if one were dealing with the past; thus:

> 8 Ahau. They destroyed the governors of Chichén Itzá by the sinful words of Hunac Ceel.

Or:

> 6 Ahau. Completed the seating of the lands of Champoton.[19]

This would be like specifying that an important historical event took place "on Friday, the 13th," without giving the year. Mathematically, we can figure out that *katun* endings that match 8 Ahau occurred in 1204 and again in 1461; but, given only this information, it becomes very difficult to match dates in the Maya and Christian calendars. Even when we find post-Classic dates carved in stone, there is a much higher frequency of mistakes than in the Classic. By the late Classic the scribes were no longer rigorously trained in precise date keeping. After seven hundred years Maya attention had been diverted to other matters.

The inland people, who had developed and nurtured the art of writing and the science of timekeeping through its florescence, withdrew from prominence. A new merchant class of seafaring people from the west coast of the peninsula took over control of much of the economy. They had little interest in celebrating time.

By the time the first Spaniards arrived on the scene, the grandeur of Maya civilization—at least as measured by the profusion, quality, and monumentality of its sculpture, and the great love affair with time, all of which had unquestionably been fostered by a powerful élite rulership—was well on the wane. Nonetheless, the belief the descendants of these people expressed in the *Popul Vuh* that time carried omens within itself remained unswerving. One of the Maya sacred books written well after the Conquest tells of an Itzá tribe that retreated, in the face of the advancing Spanish wave, into the rain forest in the interior of the peninsula. They said that the *katun* was not right for the prophecy that called for their calendar to be fully Christianized. But when the correct *katun* 8 Ahau finally approached (a century and a half later!), they promptly sent an ambassador to the Spaniards to tell them that they were now ready for conversion.

Characteristics of Maya Timekeeping

Though the ways of expressing the calendar in post-Classic and Conquest times bear little resemblance to the elaborate undertaking one thousand years before the Spaniards first made contact with the natives of the American continents, time still held the same meaning for these people. We wonder what the Spanish really conquered. Surely not the mind, for the Maya philosophy of time endures today and continues to display distinct qualities.

Above all, cyclicity was paramount. Like the movement of the sun and other astronomical bodies, which must have served as one of nature's sources for such thinking, time's cycle was limitless. There are three cycles in the pilgrimages undertaken by the Quiché lords of the *Popul Vuh* as well as in the wanderings of the hero twins. There are cycles in the *katun* prophecies of the books of *Chilam Balam*. Unlike the linear movements our eyes make when scanning across the horizontal lines of a wall calendar, when we look at the *katun* wheel in figure 6.6A, the round-and-round movement of the eye in a circle becomes the metaphor for time's flow. Counting from one *katun* to the next, we move counterclockwise in an ever-widening circular spiral.

In the codices, the cycles of Venus and other celestial bodies resonate the same message: time repeats itself. Each page of the Venus table in the Dresden Codex foretells what will befall the people on the occasion of Venus's reappearance in the sky; but, as we also have seen, the omens as well as the planetary aspects that bear them change slightly with time. Great extremes at the horizon are not exactly the same, nor is every battle and subsequent capture. The Maya wrote history the way a musician composes variations on a theme: the themes that portray the actions of the gods and their distant ancestors are not repeated; they are re-enacted. Past deeds already harbor the seeds of the future.

The idea of time going round and round, of events being conceived as marks on a circle thematically repeating themselves, is totally at odds with our way of thinking of time. Our temporal model is like a long thin wire. We stand poised at one position upon it. One direction we face is the future, and we can sight along the wire in that direction only to a certain extent. Over our shoulder lies the past—the history of civilization, preceded by the history of *Homo sapiens*; before that, the creation of the earth; and farthest back along that one-dimensional continuum, the creation of the entire universe, where time's string starts. To be sure, our time string is much longer than one trip around the biggest Maya time circle—a few million times longer. But the Maya were calculating and contemplating previous creations; they could project themselves as far backward or forward as they wanted because, for them, time was something without end or beginning. As far as we can judge, they believed in *ultimate* creation; and for them, there would be no *final* destruction.

Though we tend to focus on history's linear aspect—we all remember dates like 1492 and 1776, and no schoolchild is taught to think the events that happened in those years will happen again—still the idea of time as oscillation, of successive destructions and creations, lies submerged beneath the fabric of our own calendar. If we care to dig down and look, history really does repeat itself. Even though we throw away our wall calendar at the end of the year, the new one issued for next year will contain the same repeatable intervals. We all believe there will be a lovely July (at least in part) to match this year's. We once believed in a world that existed before the present one—the world before the Flood. That world was destroyed, the Old Testament tells us, because of the evil wrought by mankind in breaking the covenant. Western Christendom once held the belief in Armageddon, a theme that has been revised and recast today in the form of impending nuclear disaster. And we focus attention and concern on the Jewish Holocaust because somehow we all have the feeling that it might just happen again. Some modern cosmologists envision another universe that existed before creation; but by the rules of the scientific game we play, they must offer material evidence to argue their case—a difficult requirement, for how can one collect data on the galaxies, stars, atoms, even the DNA molecules that might once have comprised a cosmologist's brain, if all the constituents of that previous universe were crunched into an archaic, ultra-hot cosmic fireball?

I have tried to suggest that the historical element of formal Maya timekeeping developed out of political concerns when a series of rulers, likely related by blood, came to power. They skillfully regulated agricultural production and trade among the competing city-states they governed. The calendar became a part of their ideology. They expanded the time scale and manipulated the time packets that comprised it, and expressed their temporal ideas in space-bound metaphors in the monumental art and architecture of the ceremonial centers they erected to commemorate themselves. Time became the transcendent principle—the set of rules used to justify their special place in Maya society. The more power they sought, the greater the need to justify it. Inscriptions became longer; temples of worship, larger.

In this respect, the Maya are no different from us. We all find strength from our identity with the past. Today's American political leaders will go to great lengths to enhance their status by boldly stating that they stand for the same principles that Jefferson, Lincoln, even Roosevelt stood for. Russian leaders appeal to Lenin; Indians, to Gandhi; Mexicans, to Juárez;* and so on. And most religious leaders also plant their values firmly on moral foundations written in the far distant past—in the books of the apostles and the prophets.

Astronomy, astrology, history, politics, and pure numerology—this seemingly unlikely set of bedfellows conspired to create one of the most unusual timekeeping schemes ever devised by any ancient civilization. In the eyes of the people, most of whom neither read nor wrote, the mathematical edifice of the Maya calendar was the product of religion and astrology—a structure designed to uphold the proposition that differential social ranking was the natural order of things. The people really believed their ruler was special: he was closer to the gods. His understanding and use of the power inherent in celestial phenomena proved it, just as any prophet's vision of the almighty demonstrates his special relationship to the deity. The Maya priest, perhaps in some instances the ruler himself, must have labored hard to forge precise temporal connections among these seemingly diverse aspects of human affairs—war, politics, agriculture, climate, the environment—and the movements of the celestial bodies. This strange-sounding mixture of careful observation and chronicling together with astrological prognostications has long dissuaded Western scholars from attributing intellectual worth to the Maya timekeeping system. But religion can be a positive as well as a negative force in the progress of science. More than the desire for pure mental exercise and curiosity drove these people. The Maya had real motives for devising a Venus calendar accurate to the day over several centuries: the social and spiritual livelihood of the community demanded it.

*Benito Juárez, a Mexican of Indian parentage, freed Mexico from its European oppressors in the nineteenth century.

Their cosmology lacks the kind of fatalism present in our existential way of knowing the universe, one in which the purposeful role of human beings seems diminished. These people did not react to the flow of natural events by struggling to harness and control them. Nor did they conceive of themselves as totally passive observers in the essentially neutral world of nature. Instead, they believed they were active participants and intermediaries in a great cosmic drama. The people had a stake in all temporal enactments. By participating in the rituals, they helped the gods of nature to carry their burdens along their arduous course, for they believed firmly that the rituals served formally to close time's cycles. Without their life's work the universe could not function properly. Here was an enviable balance, a harmony in the partnership between humanity and nature, each with a purposeful role to play.

7

THE AZTECS AND THE SUN

When we saw so many cities and villages built both in the water and on dry land, and this straight level causeway, we couldn't restrain our admiration. It was like the enchantments told about in the book of Amadis [a sixteenth-century romance of chivalry] because of the high towers . . . and other buildings, all of masonry, which rose from the water. Some of our soldiers asked if what we saw was not a dream. It is not to be wondered at that I write it down here in this way, for there is so much to ponder over that I do not know how to describe, since we were seeing things that had never been heard of, or seen, or even dreamed about.

—BERNAL DÍAZ DEL CASTILLO, *Chronicles*

THOSE who cast only a passing glance at the history of America before Columbus may think the Maya and Aztec one culture, just as someone unfamiliar with east Asia might regard Vietnam, China, and Korea as the same, or Argentinians, Mexicans, and United States citizens—all of whom call themselves Americans—as part of the same culture. In fact, there was great difference between Maya and Aztec, for their cultural zeniths came several hundred years apart, and the ecologies that sustained them were as dissimilar as night and day. The Aztecs, or Mexica, as they called themselves (from which we derive the modern name of Mexico), were mountain people whose roots we can trace to migrant tribes that came from the north to settle in the Valley of Mexico. Before they rose to prominence, they were but one among a large number of tribes bound together by the Nahuatl language. Once established, they came to regard themselves as the bearers of the great traditions of Tula (Tollan) and ancient Teotihuacan, whose pyramids stand like those of ancient Egypt to make their impression on today's tourist in Mexico. There in the Valley of Mexico they established their capital city of Tenochtitlan on an island in Lake Texcoco in 1325.

Today most of the remains of the ceremonial precinct lie submerged under modern Mexico City. But what the Mexica built there, once they came to dominate the tribes all along the lake shore, must have been impressive: When, in 1519, Cortés and his *conquistadors*, foot soldiers like Bernal Díaz, clambered over the last mountain pass between their landing point at Veracruz on the Gulf Coast and the mile-high valley enclosing the capital, they could not resist expressing their astonishment at the sight of the place, as in the epigraph that opens this chapter.[1]

As remote as the Aztecs were from the Maya in both time and space, the two cultures nevertheless possessed similar calendars and compatible philosophies of time. The Aztecs kept a 260-day calendar, which they called *tonalpohualli;* and a 365-day year (*xiuhpohualli*), which they, too, combined into a 52-year calendar round. Perhaps they favored this interval because 52 years is approximately equal to the length of a human life; thus, everybody could experience, at least once, every conceivable omen the calendar might have to offer.

Closure of the 52-year round was vital. In the New Fire Ceremony that marked the binding of the years, they extinguished all fires, destroyed their pots, mats, and other utensils, and reinstalled everything anew. These human actions formally completed one cycle and began the next. As the chronicler Sahagun tells it:

> Behold what was done when the years were bound—when was reached the time when they were to draw the new fire, when now its count was accomplished. First they put out fires everywhere in the country round. And the statues, hewn in either wood or stone, kept in each man's home and regarded as gods, were all cast into the water. Also (were) these (cast away)—the pestles, and the (three) hearth stones (upon which the cooking pots rested); and everywhere there was much sweeping—there was sweeping very clear. Rubbish was thrown out; none lay in any of the houses.[2]

They timed the end of the 52-year cycle precisely by the occurrence of a celestial event. The passage of the Pleiades overhead provided the signal for renewal. The priests went to a special place to observe it, an eminence called the Hill of the Star, located on a prominent peninsula that jutted out into the lake. When the Pleiades crossed the overhead point, it was a sign to the people that the gods would not destroy the world, that people would be granted a new age.

Yet, unlike the Maya, there is no evidence the Aztecs ever kept a long count. Furthermore, they expressed their mathematical notations only in dots; they did not use bars: that is, a 7 in Aztec notation was simply seven dots, rather than a bar and two dots. (Compare the juxtaposed Maya and Central Mexican dates tied together by a rope in figure 6.7B.)

Contrasting a Central Mexican book (figures 7.1 and 7.2) with a Maya codex (figures 6.9 and 4.2D), we find the contents of the former to be more pictorial and far less abstract than its Maya counterpart, which is laden with compound hieroglyphs; the overall content, however, is much the same: in both cases, these sacred books are filled with omens and prognostications interlaced with historical and legendary information. Their themes seem geared to demonstrate that the nobility are the direct descendants of the gods. Some texts contain long genealogical narratives, the kind of information the Maya rulers disseminated through the stela cult, or even like the long lists of "begats" in the Old Testament.

A Cycle of Celebrations

In one such book, the Codex Nuttall, a "folding screen" text from the Mixtec region south of Mexico City (dated about A.D. 1200), time literally flows like a wave across the folded-out pages of text (figure 7.1). The scene depicts a group of aristocratic individuals who probably represent the ancestors of local rulers. Like the heroes of the *Popul Vuh* (in the Maya ceramic plate on page 166), each figure, labeled by his or her calendar name, proceeds from one event to the next in a sinuous journey across the pages of time. Begin at the top right, where the two seated personages—the man, 6 Crocodile (right) and the woman, 9 Eagle (left)—point their fingers at one another. Allying himself with the forces of nature, the man wears the headdress of Tlaloc, the rain god, and has a sun disk perched on his back. The vague-year date of the event being described, probably their marriage, is 6 Flint, which appears above the figure; and the date in the 260-day cycle is 7 Eagle, shown in the middle of the blue background between them. Below are

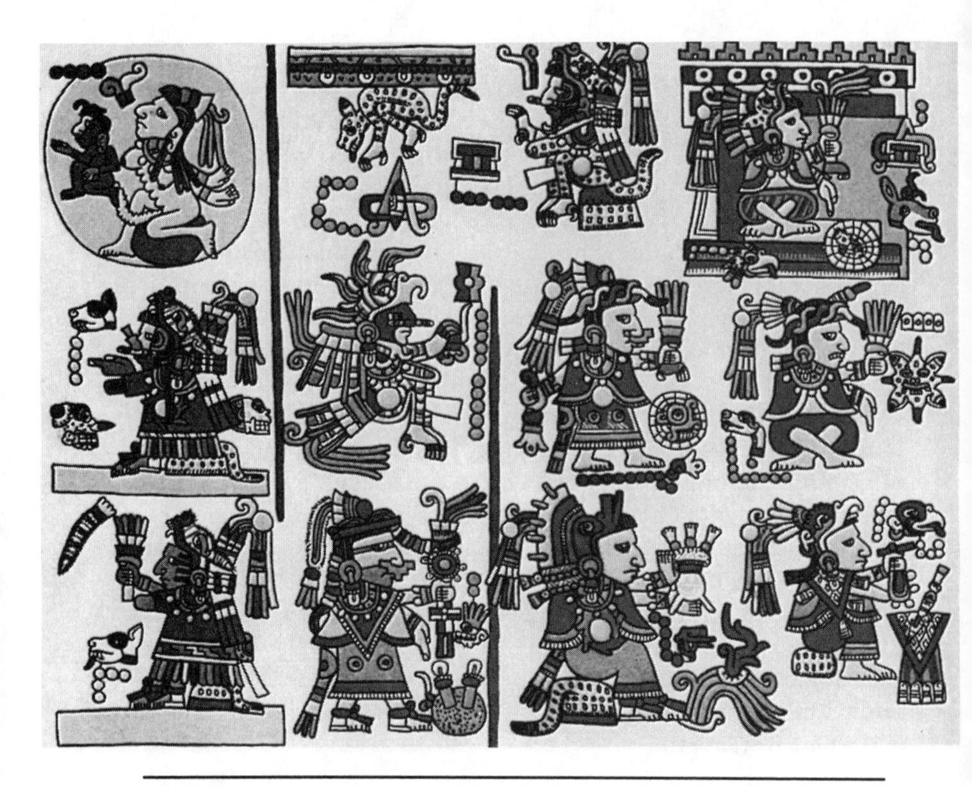

FIGURE 7.1 Time takes its bends and turns as it flows from right to left in this quasi-mythic history book about the genealogical descent of a Mixtec dynasty that ruled highland Mexico before the Conquest. SOURCE: "Codex

their three children, also named according to their days of birth in the 13-times-20 count. As you move down and turn to the left in time, the next event is recounted below the red boundary line. Enter the lady 11 Water. She is the second wife of 5 Crocodile, whom he married 17 years later. If we slide back upward again, we find in chronological order the three offspring of the second couple. Follow the spaces between the partitions (passing downward), and you will come to the next event—another marriage, this one between the man 8 Deer and the lady 13 Serpent. She offers him a ceremonial cup of chocolate to seal the pact. On and on the text of lineage events goes, most of it about as interesting to us as reading someone's family tree; but this time-embedded genealogical history served the purpose of establishing both continuity and depth in time to the noble class—for a sound ruler could not portray himself as just a latter-day conqueror who burst upon the scene exhibiting no familiarity with the local culture.

Still other Central Mexican texts, again like some of the Maya

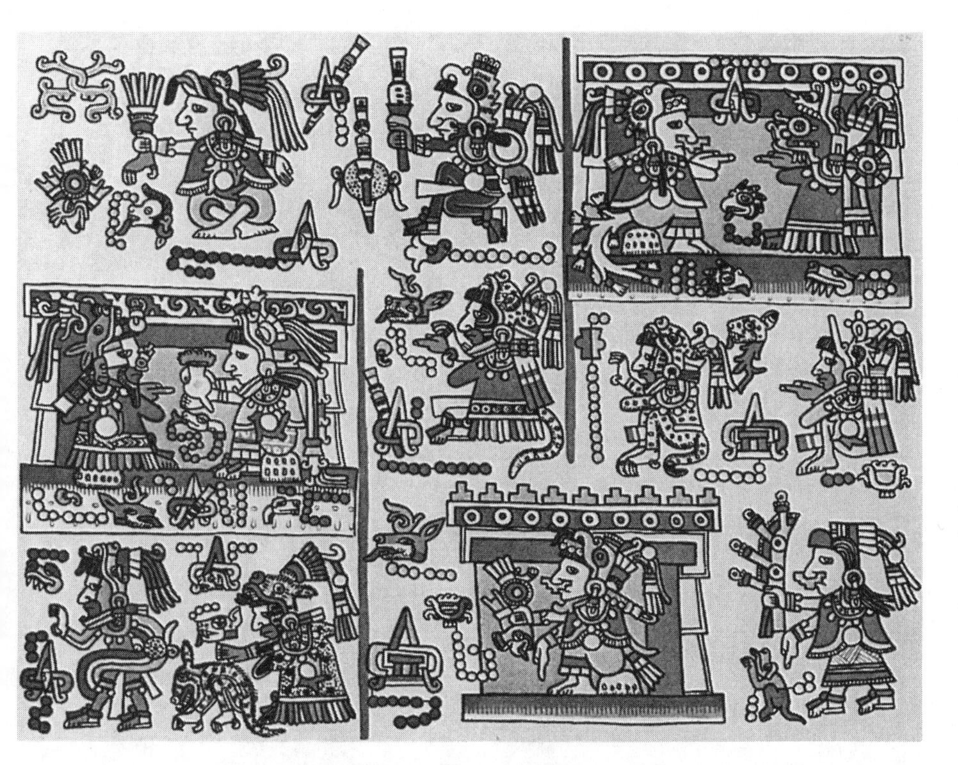

Nuttall, facsimile of an Ancient Mexican Codex Belonging to Lord Zouche of Harynworth, Cambridge, England." Peabody Museum of American Archaeology and Ethnology, 1902. Courtesy, Peabody Museum, Harvard University.

books, seem to have been intended exclusively for the use of priestly specialists: these holy instruction manuals detailed exactly how to carry out worship and celebrate the feasts in the various temples. Each temple in the pantheistic Mexican religion represented the center of worship of one of a vast number of cults, each housed by a tutelary or protective god; and each such god was a guardian of some form of power that emanated from the natural order: Tlaloc for rain and water; Huitzilopochtli, the sun and war; Coatlicue, the earth and fertility. Each god had its own temple and a unique set of specialized cultic practices that the expert priest needed to master down to the finest detail.

The illustrations in some of these codices touch the extremes of human imagination. Take the page of the Borgia Codex, reproduced in figure 7.2, which pertains to secrets of the cult of the earth. Once hidden away in the temple, and intended to be consulted only by the priests who served as specialists of this particular cult, it graphically

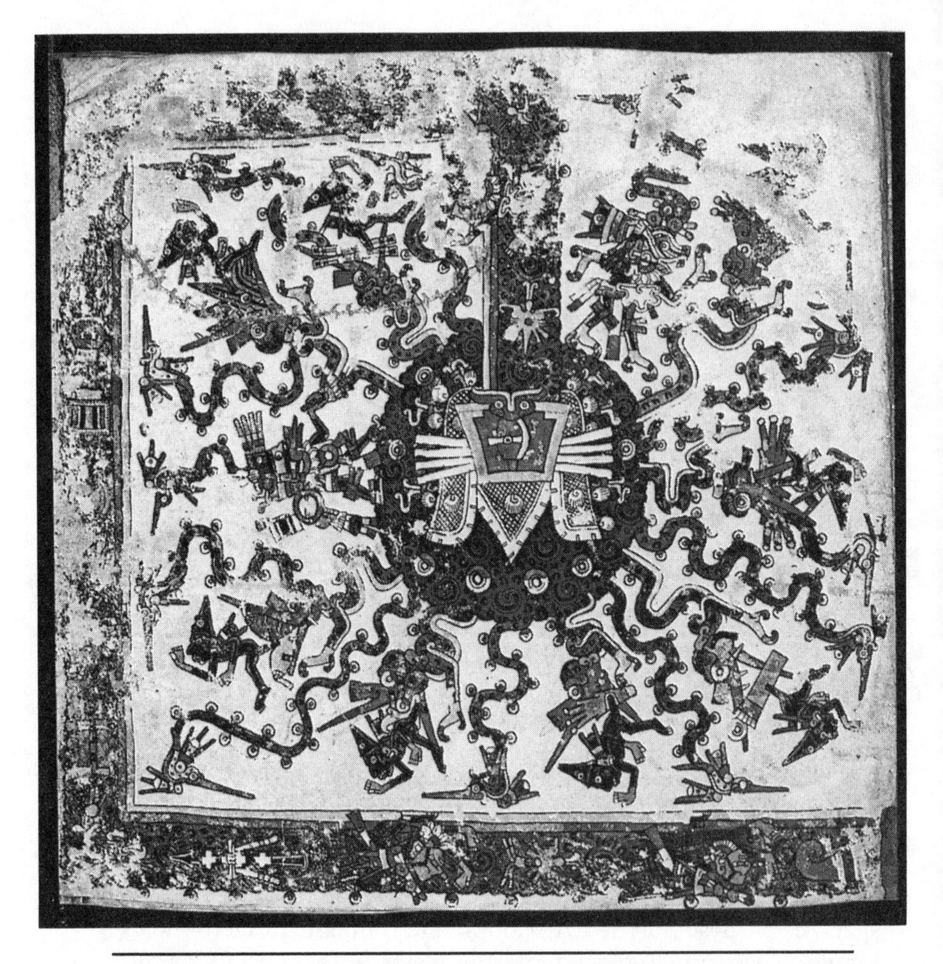

FIGURE 7.2 A dynamic universe. The opening of the ceremonial bundle in the Codex Borgia reveals a living earth from which plants, animals, and spirits emanate almost electrically. This sacred book served as a priestly guide to ritual procedures for one of the highly specialized cultic temples. It was never intended for any other human eyes. SOURCE: Codex Borgia. Codices Selecti, vol. XXXIV (Graz: Akademische Druck-u. Verlagsanstalt, 1976).

depicts the opening of the ceremonial bundle, a device still carried by some shamans. Once the ceremonial bag is unleashed, out springs every conceivable aspect of earthly power: little deity snakes with two heads (some smoking), spiders, and stone knives. Here is a cult that strove to articulate the behavior of real natural beings, not abstract natural laws. The earth explodes with life, and you can feel the electrical discharge as you peer into the ceremonial bag and momentarily glimpse nature's forces. Other pages in the Borgia sequence dutifully give the omens assigned to the different days of the divinatory calendar.

The godly statuary of the Aztec pantheon, which stood in the cultic temples, were called *téotl* (the Spanish translated them as gods or devils); they, too, were not mere representations of nature's forces. As the art historian Richard Townsend characterizes them, the Aztec gods are forces that appear in various forms diffused throughout the universe:

> This force was pre-eminently manifested in the natural forces—earth, air, fire and water—but was also to be found in persons of great distinction, of things and places of unusual or mysterious configuration. *Teotl* expresses the notion of sacred quality, but with the idea that it could be physically manifested in some specific presence—a rainstorm, a mirage, a lake, or a majestic mountain. It was as if the world was perceived as being magically charged, inherently alive in greater or lesser degree with this vital force.[3]

When one of these sacred manifestations or hierophanies took place, it was regarded with great seriousness, and much meaning was attached to it. Like the miracles of the burning bush of the Old Testament, or the healing of the leper in the New Testament, these are not occurrences to be explained away through scientific rationalization—the way we tend to deal with the Bible stories today. For the Aztecs, the wonders of nature were the actions of animate entities who roamed a universe that was vibrantly alive and filled with purpose, a universe we cannot imagine today because we have drastically altered our definition of both what is animate and what is purposeful. We must stretch far to appreciate the way an Aztec priest or citizen knew the natural surroundings.

In many cases in Central Mexico—and this is particularly true among the Aztecs—there were festivals to commemorate each of the 18 months of the seasonal calendar. There were feasts for the dead, for the god of war, for the sun and rain gods, for floral deities, even for the gods of intoxicating beverages. Descriptions of the ceremonies that took place on these occasions come from the Spanish glosses written over the Codex. These show how deeply ingrained agrarian fertility concepts had become in the minds of the people. Take, for example, the feast of the corn god celebrated in the month of Hueytozoztli, which begins at the end of April. At this time, in every one of the temples dedicated to the corn god Cinteotl, worshipers made offerings of cornstalks and leaves, as well as prepared products from the plant, such as tamales and tortillas.

A couple of months later, in the month of Etzalqualiztli, Tlaloc was honored and all activities shifted to his temples. To each of his holy places they brought white and green mats woven out of reeds that grew in the lake. One chronicler listed several Tlaloc shrines among the 76 temples in and about the ceremonial precinct of ancient Tenochtitlan. And so, the tribute list continued through all 18 of the 20-day months of the year.

The World Diagram

The idea of a rotating cycle of celebrations, each one taking place in a different set of temples, suggests that, for the Aztecs, time and place were naturally juxtaposed.

Only within the past century has our culture formally developed a mechanism for tying time and space together. When, early in the century, Einstein proposed the theory of relativity, one of his goals was to find a set of physical laws expressible in terms of mathematical equations that reduced the three dimensions of space and the dimension of time to a single conceptual entity. To give a simple example, suppose I write down the mathematical equation

$$x^2 + y^2 = 1$$

and substitute different sets of paired numbers for values of x and y that satisfy the equation. If I plot the results on a sheet of graph paper (letting x be the horizontal coordinate and y the vertical), the result will be a circle. I can perform the exact analogue of this operation in three dimensions by beginning with the expression

$$x^2 + y^2 + z^2 = 1.$$

If I substitute the correct numbers and graph the results, I will get a sphere, z being the third dimension—the one that gives depth to the circle. One of Einstein's many contributions to modern science was the realization that this process can be extended yet another step, from three to four dimensions. The added term deals with time (t) as a dimension with spacelike qualities if it is multiplied by the speed of light (c). Thus the equation

$$x^2 + y^2 + z^2 + (ct)^2 = 1$$

transforms itself graphically into a four-dimensional sphere. Of course, we cannot see or even imagine such a sphere; yet we know that that equation logically, and by extension from its two- and three-dimensional analogues, belongs to the same family of mathematical curves: circle, sphere, "???." We can think of the "???" as a sphere in four dimensions, wherein the added dimension is (*ct*). For the modern cosmologist, Einstein's contribution is not simply an esoteric mathematical exercise. It can be used to predict the behavior of moving objects in the real, physical world. Indeed, his equations have helped us to unify time and space by thinking of time as the fourth dimension, in effect by subordinating it to space and endowing it with a spacelike quality. The formal and rigorous language of mathematics was the vehicle Einstein employed to achieve this transformation.

So what has all this to do with the Aztecs? Specifically, nothing; but if we think in a general way about the human propensity to create formal systems that enjoin the concepts of space and time, there is an important behavioral similarity between modern physicist and ancient Aztec priest. The Aztecs were just as concerned with seeking order in the universe by trying to find links among seemingly diverse aspects of a chaotic world. The unifying aspects of Einstein's equations in the modern world find a parallel in constructions like the diagram from the first page of the Codex Fejerváry-Mayer (figure 6.6D), a pre-Columbian picture book from Central Mexico.

A wealth of information converges in this all-encompassing space-time diagram, so much that I have found it necessary to construct table 7.1 to guide us through its mazelike configuration. To begin with, forget about the detailed pictures within: just back off and look at the general shape of the diagram from a distance. Its general form is like the Maya quadripartite *kin* sign, which means time itself. The basic feature is a floral design with two sets of four petals: a "Maltese Cross" composed of large, trapezoidal petals and a so-called "St. Andrew's Cross," a floral pattern consisting of four smaller rounded petals positioned at 45-degree angles between those of the Maltese Cross. A square design forms the center of the pattern.

Time enters the picture directly at the border of the design, which is marked with circles whose count totals 260. The ritual day count is divided into cycles of 20 named days, tallied in groups of 13. The first set of 13 commences with 1 Cipactli (alligator), whose teeth are visible just above the upper right-hand corner of the central square. Moving counterclockwise along the border, you proceed to count 12 blue dots

TABLE 7.1

A Scheme for Organizing the Information Contained in the Calendar on Page 1 of the Codex Fejerváry-Mayer (figure 6.6D)

Direction	*Petal*	*Year Bearer*	*Named Days*	*Staff*	*Bodily Part*
East	upper left	Acatl	Atl, Ollin, Coatl, Cipactli	guacamaya plant	hand
North	lower left	Tecpatl	Ehecatl, Itzcuintli, Tecpatl, Miquiztli, Ocelotl	fruit tree	foot
West	lower right	Calli	Cuauhtli, Calli, Ozomatli, Quiahuitl,	cactus plant	throat
South	upper right	Tochtli	Tochtli, Cozcacuauhtli, Malinalli, Cuetzpallin, Xochitl	maize plant	head

Direction	*Tree or Plant*	*Source*	*Bird*	*Color*	*Ritual Subject*
East	blue or turquoise tree	sun	quetzal	red	solar
North	cactus	incense	eagle	yellow	autosacrifice
West	maize	death skull	blue-colored bird	blue	dead woman
South	cacao	serpent Tlaltecuhtli (earth god)	parrot	green	earth

SOURCE: A. Aveni, *Skywatchers of Ancient Mexico* (Austin, Tex.: University of Texas Press, 1980), p. 157, table 16. By permission of the University of Texas Press.

(shown on a dark field in the figure), completing the count of 13 on 1 Ocelotl (jaguar). The third cycle passes across the top of the diagram and ends on 1 Mazatl (deer); and the pattern continues with counts of 13, terminating on Xochitl (flower), Acatl (reed, symbolized on the back of the bird in the upper-left corner), Miquiztli (death), Quiahuitl (rain), Malinalli (grass), Coatl (serpent), Tecpatl (flint knife), Ozomatli (monkey), Cuetzpallin (lizard), Ollin (movement), Itzcuintli (dog), Calli (house), Cozcacuauhtli (buzzard), Atl (water), Ehecatl (wind), Cuauhtli (eagle) and Tochtli (rabbit); finally, 1 Cipactli closes the ritual count. Since all these symbols are pictured at the vertices of the double-cross design, the 260-day cycle seems to encapsulate all other astrological and calendrical matters depicted within the diagram: time surrounds space, and the different parts of time are assigned to different directions.

The regions of the world along with their associated colors are enshrined in the four arms of the Maltese Cross: the four cardinal points (crossarms) and the zenith (center). East (red) is at the top, west (blue) at the bottom, north (yellow) to the left, and south (green) to the right. When the sun rises, he sees the north to his right, the south to his left; straight ahead is the region of the west where he will die each night. The sun appears as a spiked disk in the eastern arm of the Maltese Cross, while the death head of the west hangs below the central square, waiting to swallow the sun each night.

North and south were more to the ancient Mexicans than just a pair of geographic regions on a map. In some parts of Mesoamerica, these areas are the "sides of heaven"; and in hieroglyphic form, north has been taken to symbolize "up" and south "down." We in the West might think of their space as consisting of the cardinal points east-west-up-down, with the center being the "fifth place." But the cosmic hinge in the Fejerváry-Mayer diagram is not north-south; rather, it is the east-west axis. The two flaps of the Maltese Cross that represent these directions are attached to the central square, while the north and south panels seem to swing free.

We tend to think of north and south as the basic cardinal axis, with the east and west secondary crossarms. After all, we conceive of the earth's polar axis in terms of north and south poles and mark noon and midnight by the passage of the sun across the meridian, the imaginary line that passes overhead from north to south. This is why all of our maps are oriented with north on the top. But if we look back far enough into our own past, we find that our ideas about geographic orientation are similar to those of the Aztecs. Many early Old World maps show east at the top. Should you wonder why, you need only proceed to the nearest dictionary. Our word *orientation* is defined in some lexicons as the normal eastward-facing posture worshipers take as well as that part of the church one looks at when viewing the altar from the nave. In many cases, east is the direction along which the rising sun's first light penetrates the church on the day of the patron saint after whom the church was named. As late as the nineteenth century, the British poet William Wordsworth witnessed this orientation process in action and described what took place as the masons attempted to align the base of a structure:

Then to her Patron Saint a previous rite
Resounded with deep swell and solemn close,
Through unremitting vigils of the night,
Till from his couch the wished-for Sun uprose.

He rose, and straight—as by divine command—
They, who had waited for that sign to trace
Their work's foundation, gave with careful hand
To the high altar its determined place.[4]

Our own language betrays the secret. East is aurora (the dawn) or Ush ("to burn," in Sanskrit)—both terms that relate to the sun. Red is the color of the dawn for the Aztecs as well. So there are religious roots to our cardinal directions, and each of the compass points is tied to the pagan art of sun worship buried deep in our past.

What about the figures inside the Maltese Cross? Each cardinal direction has its representative tree, a source for the tree, a bird, directional gods, and a directional ritual (see table 7.1 for details). Thus, the spiked tree on the left, a cactus plant, would literally be the tree of the north, for this is where the largest species of cactus grow in Mexico.

Xiuhtecuhtli, the celestial fire god, is located at the center of the cosmic diagram. He is armed with spears and an *atlatl* (spear thrower) and his blood flows in four streams outward to the corners of the universe. He is the first of the Nine Lords of the Night. The remaining eight Lords are pictured, two to each directional flap of the Maltese Cross, as follows:

2. Iztli (east, right)—flint knife
3. Pilcintecuhtli (east, left)—young maize
4. Cinteotl (south, right)—maize
5. Mictlantecuhtli (south, left)—death
6. Chalchiuhtlicue (west, right)—jade skirt, water (female)
7. Tlazolteotl (west, left)—earth, cleanser of the soul (female)
8. Tepeyollotl (north, right)—heart of the hill
9. Tlaloc (north, left)—rain

Each of them rules for a day in sequential order, and together they make up yet another cycle of Mesoamerican time, not unlike our seven-day week.

Now, look at the second cross in the diagram, the one in the shape of an *X*. The four arms of this St. Andrew's cross signify the four houses of the sun in the sky, two in the east and two in the west. As in the *kin* symbol, they may represent the intercardinal points that stand for the seasonal extremes to which the sun migrates along the horizon over the course of its annual cycle, an idea also expressed in Aztec architecture, as we shall see. Thus, summer solstice sunrise would be at the upper left, winter solstice sunrise at the upper right, winter solstice sunset at the lower left, and summer solstice sunset at the lower right. Elements pertaining to the cardinal directions can be found in the petals of the cross. These include a plant within and a representative bird at the top of the petal.

The year bearer, the one who by virtue of being the name of New Year's Day must carry the burden of the entire year, is borne on the back of each bird. In four of the interstices between the petals appear the five named days tied to that region of space. And in the remaining spaces between the petals, there is a different part of the human anatomy, each body part having its own assigned spatial direction—hand, foot, throat, and head splayed out into the four intercardinal corners of the universe, all joined to the center by streamers of the creator god's vital blood.

The Aztec world diagram is not purely symbolic. It is also a functioning calendar. The Aztecs could use it, for example, to tally the count of the 365-day year as well as the 260-day cycle. Suppose you begin in the east with 1 Acatl, year bearer for a particular year; then counting through the cycle of 20 day names 18 times, you are left with a remainder of 5 days; this takes you 5 days forward and you arrive at the name of the day bearing the second year: Tecpatl. Likewise, New Year's Day of the third year bears the name Calli and that of the fourth year, Tochtli. The fifth year begins on the same day as the first, Acatl (since $365 \times 4 = 1460$, and $1460 \div 20$ gives a remainder of 0), and thus closes the day-counting cycle.

If there is a central theme about order in the universe depicted on this codex page, it is the idea that all things are arranged in categories of four. As we have seen, the quadripartite world view is also one of the guiding structural paradigms in other regions of the Americas (see, for example, some of the artifacts depicted in figure 6.6).

The circular calendar wheel from the chronicle of Diego Durán, which actually looks more like a swastika bent into the shape of a circle, shows, along with four winds and the four directions (again, with east at the top), the division of the 52-year calendar round into its four

13-year segments—*trecenas*, the Spanish called them. The chronicler Sahagun expresses why such a division was conceived and what it meant:

> When (came) the time of the binding of our years, always they gradually neared and approached (the year) Two Reed. This is to say: they then reached and ended (a period of) fifty-two years. For at that time (these years) were piled up, added one to another, and brought together; wherefore the thirteen-year (cycles) and four times made a circle, as hath been made known. Hence was it said that then we tied and bound our years, and that once again the years were newly laid hold of. When it was evident that the years lay ready to burst into life, everyone took hold of them, so that once more would start forth—once again—another (period of) fifty-two years. Then (the two cycles) might proceed to reach one hundred and four years. It was called "One Old Age" when twice they had made the round, when twice the times of binding the years had come together.[5]

Aztec time was a sequence of bundles that needed to be joined together in order to make a circle. Sahagun's quote is yet another reminder of that graphic tying together of two calendar dates with a rope, shown in figure 6.7B.

Unlike the Western wall calendar, which counts only days, the quadripartite calendar in figure 6.6A shows only successive New Year's days. To use it, begin at the top (east) closest to the center with the year bearer 1 Reed, the name day in the 260-day cycle of the first 365-day year. Then pass 90 degrees to the left to 2 Flint Knife, the name day for the next year; then another quarter turn to 3 House at the bottom, the name day of the third year; then to 4 Rabbit at the right; then back up top to complete the first full cycle with 5 Reed; and so on. Rather than following the course of time line by line as with the wall calendar, to read the Aztec New Year's calendar we must journey outward along a space that continually brings us back to the same point and ends where it began, having passed 52 well-marked resting points spaced along the four quarters of the spatial universe. Every native American calendar is cyclic and reusable: like time itself, the calendar has no end.

Being centuries removed from the Aztecs, we might tend to think that unlike us they did not speculate and wonder about the world around them. The idea of joining space and time in world diagrams like the one we have been discussing must have provoked as much discussion among Mexican wise men as did working principles behind the variation of the species among nineteenth-century biologists or hypotheses on how to manipulate genetic materials among those of the twentieth. Take, for example, this series of questions on the nature of the afterlife and in particular the path to be taken by departed souls. Three alternatives are suggested:

> Where do we go, oh! Where do we go?
> Are we dead beyond, or do we yet live?
> Will there be existence again?
> Will the joy of the Giver of Life be there again?[6]

These questions the Mexica were raising about the nature of human existence are as meaningful in their cultural context as those raised by Judeo-Christian theologists on the existence of God, and must not be confused with the separate issue of who controlled knowledge, who used it, and for what outcome. It is all too easy for us to think that their knowledge really had no intrinsic value and that Aztec history depicts only a set of rulers who greedily manipulated a witless citizenry through magic and mysticism. If this is the way we choose to characterize their history, then it tells us nothing about them. Instead, it may tell us something about our own need to feel superior.

An interesting sidelight about the calendar in figure 6.6A is that it represents a clash of ideas. Because it was drawn well after Spanish contact, it shows Old and New World ways of conceiving the universe blended together. The universe may be quartered, but those cherubic wind gods lodged in the corners have very European-looking faces. And what is the sun doing at the center of the world? Is this a reflection of the European Renaissance idea of a sun-centered universe penetrating all the way across the sea to the New World via the Spanish priests who were trying to convert the natives to their way of thinking? As observers who live in another time, we need to be careful about discovering original indigenous customs when we look through the lens of a secondary source like this one.

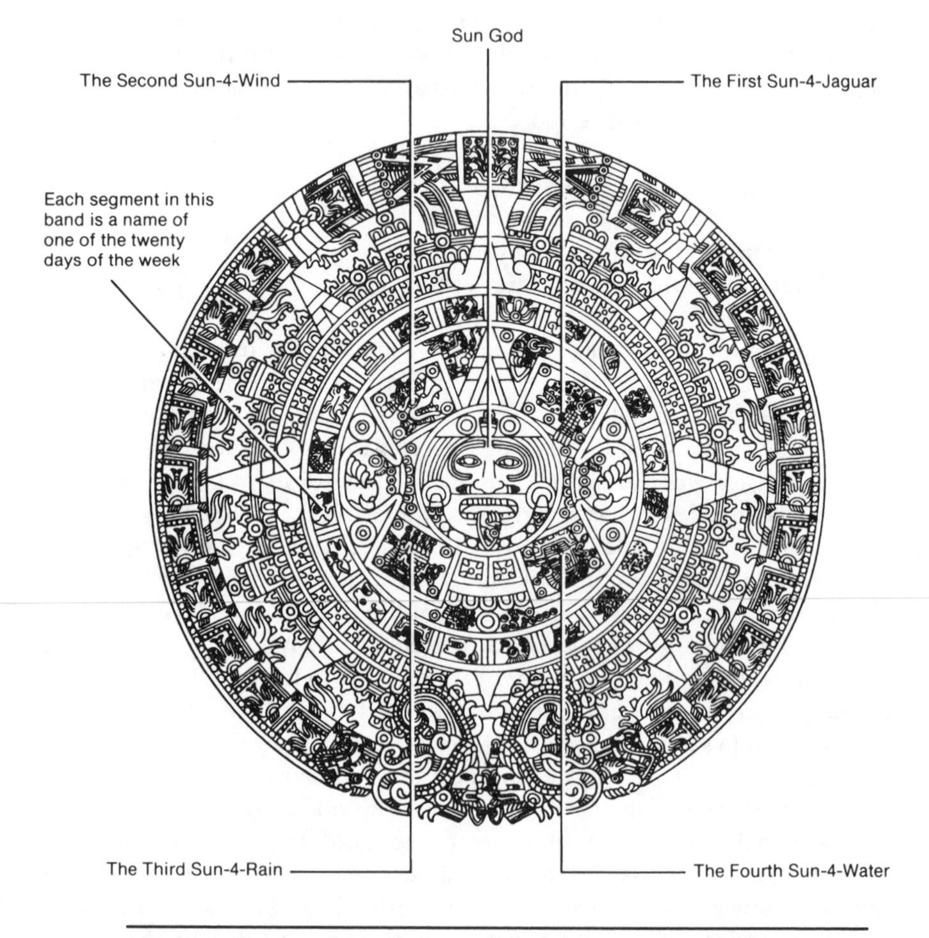

FIGURE 7.3 The Aztec sun stone is not really a calendar, but a symbol commemorating the events of all creations and including every one of time's division as well. The Four Creations are represented by the four blocks designated as "first," "second," "third," and "fourth" sun. SOURCE: National Museum of Anthropology and History, Mexico City. After M. Coe, in A. Aveni, ed., *Archaeoastronomy in Pre-Columbian America* (Austin, Tex.: University of Texas Press, 1975), fig. 10. By permission of the University of Texas Press.

The Creation Story

Why a four-part universe? The Aztec creation story lies at the root of the explanation. Like all good yarns devised to explain how things originate, this version of a people's genesis is dynamic and engaging. Perhaps the story is best told in the graphic central portion of one of the most well-known Aztec works of sculpture. When the famous sun stone was unearthed from the southeastern corner of Mexico City's main square in the middle of the eighteenth century, it probably lay not far from its original location at the base of Tenochtitlan's principal temple, the so-called Templo Mayor. The round stone weighs 25 tons and measures more than 3½ meters in width (figure 7.3). It was carved in commemoration of the fourteenth-century Aztec king Axayacatl, who, like his predecessors, believed he drew his power from the sun.

The story of creation carved on the face of the stone displays the dynamic element in their model of the world. At the center of the stone is the face of Tonatiuh, the sun. His wrinkled countenance is adorned with nose and ear pendants, and his claws firmly grasp the firmament, from which he is suspended. One look at him makes it obvious why the Spaniards, who worshiped a more beneficient-looking god, were repulsed by the unsightly image of such a deity and any form of religious worship he might represent. What god was this whose lolling tongue, a flint knife, beckoned for the blood of human hearts to nourish him on his course?

The four panels surrounding Tonatiuh represent previous ages—or "suns," as the Aztecs called them. The first cosmogonic epoch (upper right) was the "sun of jaguar," named after the day 4 Jaguar in the 260-day cycle on which it terminated. Note the head of the jaguar surrounded by four dots within this panel. During this epoch, the inhabitants of the earth, the result of the gods' first try at a creation, were giants who dwelled in caves. But they did not till the soil as expected and so were destroyed—all of them eaten by jaguars. In the second sun, the "sun of wind," symbolized by the day 4 Wind (upper left), another imperfect human race was blown away by the wind. The gods transformed these creatures into apes, that they might better cling to the world, an act that is said to account for the similarity between apes and men. In the third creation, the "sun of fire-rain" (note the symbol of 4 Rain at the lower left), some men survived by being transformed into birds in order to escape from the destruction of the world by volcanic eruptions. In the fourth creation, the "sun of water," depicted at the lower right, ended with a flood that followed torrential downpours. But this time a transformation from men into fish kept them from perishing. The symbol 4 Water marks this epoch. We live in the "fifth sun," whose symbol houses Tonatiuh and the other four cosmogonic ages. Our age is symbolized by its creation date, Naui-Ollin, or 4 Movement; and the four large dots of this sign's coefficient are easily recognizable on the periphery of the four panels that denote the previous "suns." In each case, the universe was destroyed and re-created anew and each age provides an explanatory temporal framework in which to categorize different forms of life and to relate them to the present human condition.

Two points about Mexican time are worth underlining in this creation story: first, the oscillating repetitive nature of the events taking place. Like the Maya gods' creation attempts described in the *Popul Vuh*, previous "suns" were thought to have been creative ventures that failed to achieve the necessary delicate balance between gods and people. Creation time repeats itself, but it is punctuated by periods of destruction. And, second, the present contains pieces of the past. Each attempt at creation tries to account for the present state of humankind by referring to what remains in the world. Fish and birds are really our kin, the gods' failed children from archaic creations. We were destined not to dominate them, as Old Testament Genesis states; rather, we were meant to respect them.

An uncanny resemblance between ancient Old and New World creation symbols and ideas is revealed in this story. Looked at closely, each of the destructive agents is conceived as a force that can be represented by an *element.* The jaguar is an earth monster; while wind, water, and fire are the other parallel entities in the story. Earth, air, fire, and water—these are the same four basic elements conceived by the Greek philosophers, but with one significant difference. For the Aztec, they were not static and permanent constructs, but violent forces that could erupt at any moment, as they still do today in the fragile ecology of Central Mexico.

Another basic theme of Aztec cosmology seems to be that all of life is a struggle filled with key moments of tension. The cyclic re-creations were the outcome of struggles among the creator gods. In each of the previous suns, one force attained temporary supremacy over the others; but the balance of forces tipped, and further godly combat ensued until another period of accord was reached. To hear them tell it, the present harmony, the age of the fifth sun, the "sun of movement," began not so far from the Aztec homeland. Just as the Maya rulers of Palenque sought to secure their bond to the ancient gods, so, too, the Aztec kings were driven to connect their lineage to deities who surely resided in the nearby colossal ruins of ancient Teotihuacan, where time was created. They pegged the creation of the present epoch to events that took place among the ancient pyramids of that fabulous city. There, in a struggle among the gods, one of their members bravely sacrificed himself by throwing himself into the fire to re-emerge as the new sun:

> And after the god hurled himself into the flame, the other gods looked about but they were unable to guess where he would appear. Some thought he would appear from the north; they stood; they looked to the north. Towards noon, others felt that he might emerge from anywhere; for all about them was the splendor of dawn. Others looked to the east, convinced that from there he would rise; and from the east he did. . . . When it appeared, it was flaming red; it faltered from side to side. No one was able to look at it; its light was brilliant and blinding.[7]

Armed with information from the Aztec sun stone, you can look back at the Fejerváry-Mayer world diagram and begin to sense its dynamic quality. The four panels are the four corners from which the creator gods activated the history of the world. You can begin to make sense of the logic inherent in one of the basic tenets of Aztec religious belief: that only by the act of sacrifice can the sun—and, consequently, life—continue to exist. Only through this act of penitence can the sun keep moving, for his movement is the only action that can sustain the present epoch.

Restructuring Time to Legitimize Aztec Rule

An underlying motive in creating and developing any cosmology is to produce a solid ideological foundation upon which to place society's history. The Aztec imperial state succeeded in integrating elements of its creation theme into a way of life, and thus made the rulership legitimate. In so doing, the Aztecs—as had the Mayas—adopted and modified the beliefs of one of their many creation cults that had already existed among the various tribes that had settled in the basin of Central Mexico. They identified the mission of the state with the task of keeping the sun in motion by fueling it with life's vital energy, thus giving the idea of continuity concrete form. Only by supplying the sun with life's vital fluid could they hold it on its course in the present age. The Aztec people, and particularly the military, played a direct role in extending the outcome of creation; and that role was to see to it that the blood of captives would flow at a sufficient rate to keep the solar deity on course. The Aztecs literally became the "people of the sun," the special race dealt life's most important task: to keep the whole world running smoothly, to keep time marching on. And if they did not succeed, the balance of powers among the cosmic forces might be offset, and the world once again would be destroyed—this time by earthquake—and it might well be the last time!

The solar fuel, supplied by sacrificial victims, was obtained in a kind of ceremonial warfare with the surrounding states, which, one by one, became tributaries of Tenochtitlan. Unlike the Spanish invaders they would ultimately encounter, for the Aztecs the main business of warfare was neither the conquest of territory nor the slaying of the enemy on the battlefield. Rather it was the acquisition of this living raw material to propitiate the gods. The greatest achievement of any male member of Aztec society was to become a warrior of the cult of Huitzilopochtli, god of sun and war, the principal Aztec deity who had been catapulted to the top of a rather large pantheon, the arch representative of Aztec materialism—their patron deity.

The Aztecs were a people of action, and their cosmology was action-packed. All over the capital were signs of the mission of this vigorous expansionist state, and the flow of solar time appears in almost every one of them. It is the power source driving the mill of human sacrifice, the grist being supplied by human hearts from foreign nations. Tizoc's stone, named in honor of one of the Aztec emperors, is another mammoth carved circular monument that illustrates the ideological role of Aztec time. Slightly smaller than the sun stone, it pictures the Aztec empire in the middle of the universe, with their emperor at its center. Around the rim of the cylinder, Tizoc confronts fifteen different foreign chieftains in battle, whom he has just taken as prisoners for sacrifice. Below lies the earth, decorated with masks of the terrestrial deity. Above is the realm of the sky with the stars depicted as precious jade jewels. On top of the cylinder, a huge sun symbol with cardinal and intercardinal points ties cosmos and society together.

Another carved monument—a stone dedicating the passage of rulership from Tizoc, who died in battle prematurely, to his brother Ahuizotl—echoes many of these same themes. It is decorated with the day sign (7 Acatl), which probably corresponds to one of the annual calendrical festivals. The date is carved over a set of instruments for drawing penitential blood located between the figures of the two brothers. Below lies an ultra-large rendition of the year bearer in the year of the accession, 8 Acatl (1487). The scene is reminiscent of the one in the Temple of the Cross at Palenque where the calendar also serves as the framework for the transfer of power from Pacal to his successor. This stone of Ahuizotl also commemorated one of the building phases of Tenochtitlan's Templo Mayor, atop which it was once placed. This monumental art seems to have been spread all over the Aztec capital. The intention behind it was to create a specific public image to communicate the ideology that the forces of nature and the state are really one and the same.

If Tenochtitlan was the center of the world, the Templo Mayor (figure 7.4) was its *"axis mundi."* In the legend concerning the foundation of their city, the Aztecs were guided by the gods to erect their temple there, at the place where the eagle stood perched atop a cactus, on the very spot where the heart of one of their archaic sacrificial martyrs had been hurled. Chroniclers say this is the place where the blood of human sacrifice was drawn: up to ten thousand people were sacrificed in a single year, always by the same process:

> And also at the top of the pyramid were circular stones very large—upon which they slew victims in order to pay honor to their gods. And the blood, the blood of those who died, indeed reached the base; so did it flow off.
>
> They took the victim up the pyramid temple before the devil: the priests first went holding him by his hands. And he who was known as the arranger of victims, this one laid him out on the sacrificial stone. And when he laid him upon it, four men each pulled on his arms, his legs. And already in the hand of the fire priest lay the flint knife with which he was to slash open the breast of this ceremonially bathed one.

> And then, when he had laid upon his breast, he at once seized his heart from him. And he whose breast he laid upon lay quite alive. And when the fire priest had taken his heart from him, he raised it in dedication to the sun.[8]

At the time of Spanish contact, the great Aztec temple was over 40 meters high. They had rebuilt it layer upon identical layer at least seven times, thus architecturally encapsulating the past within the present. Each stage of completion, as in the New Fire Ceremony, constituted an enactment of the renewal of Aztec beliefs and principles.*

The top of the Templo Mayor was crowned by a pair of sanctuaries, one dedicated to Tlaloc, the old rain and fertility god known to all people of Central Mexico all the way back to Teotihuacan. The other temple, slightly taller, honored the Aztec patron deity Huitzilopochtli. The juxtaposition of these gods in a twin-temple structure may reflect a political compromise between a militaristic faction and a theocratically oriented, more traditional element of the Aztec polity. Perhaps the unity of these competing forces had something to do with propelling Aztec leadership to the fore in the basin of Lake Texcoco in the fourteenth century.

The Templo Mayor faced west, toward the setting sun. When the solar deity passed into this region, just before disappearing into the underworld, he would be in the most propitious position to receive the sacrificial blood that would sustain him and keep him on his course. Then when he rose the next day, he would reappear before the public, who could view him from the front of the sacred temple, as he passed over the top of it. At the time of the spring equinox, when the sun's annual cycle began, he would position himself precisely in the notch between the houses of Tlaloc and Huitzilopochtli. Like the Maya builders of Copan, Uxmal, and Chichén Itzá, the Aztec priests engineered their ceremonial space so that time and calendar would be integrated into the performance of the sacrificial ritual.

The technical difficulties posed by trying to make the building face the correct direction so that the sun would fall in the slot must have been considerable. For example, they needed to skew the temple 7 degrees off the east-west line in order to catch the sun when it reached

*There is really no evidence to support the popular notion that the Aztecs built over their structures every 52 years to celebrate formally the completion of a calendar round. More likely, reconstruction was motivated by such events as the royal accession to power or a major victory over the enemy.

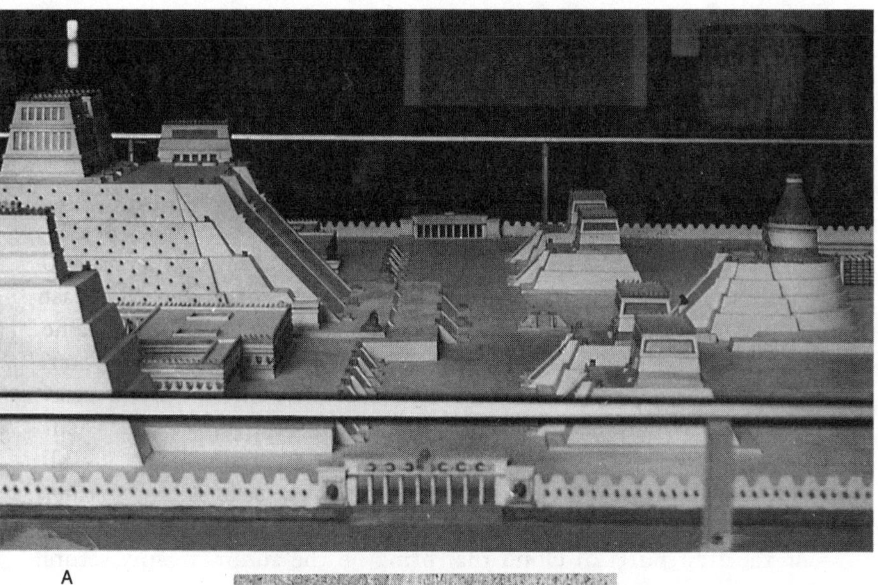

A

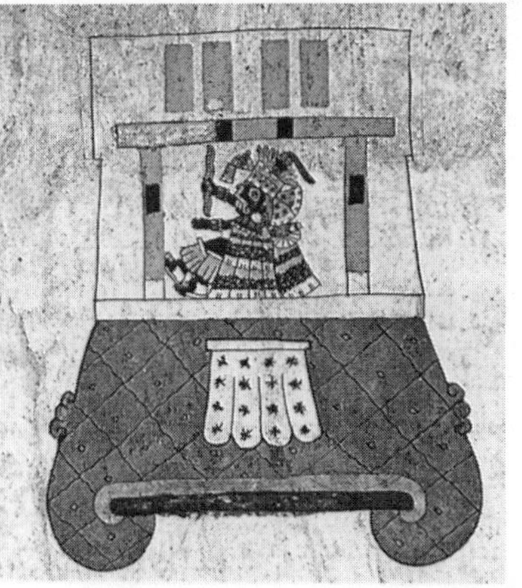

B

FIGURE 7.4 (*A*) The Templo Mayor of Tenochtitlan (tall building on the left), the center of the Aztec empire. To keep the sun on course, captives were sacrificed in front of the twin temples at the top, one dedicated to the god of sun and war, Huitzilopchtli, the other to the god of rain, Tlaloc. Center and periphery of the Aztec Empire were symbolically connected in both space and time: rain was said to come when the sun rose over the rain god's mountain as seen from the temple. (*B*) The Templo Mayor depicted in the Codices after the Spanish contact, with demons flanking the twin sacrificial temples. SOURCE: (*A*) Photo by the author of a model in the Templo Mayor Museum, Mexico City. (*B*) From the Codex Borbonicus, Codices Selecti, vol. XLIV (Graz: Akademische Druck-u. Verlagsanstalt, 1974).

the proper height.* One chronicler tells us that at least one frustrated emperor needed to tear the structure down and rebuild it to get it right.[9]

The Templo Mayor also linked the Aztec center of the world to the rest of the universe, which was represented most immediately by the peripheral mountains and valleys visible from Tenochtitlan and the tribute centers that lay beyond. Thus, the Aztecs extended their solar sight line to the visible horizon, where it passed over Mount Tlaloc, high in the eastern sierra 44 kilometers away. There they built a large enclosed shrine, reached by a 300-meter-long stone-lined causeway. This place, which they called Tlalocan, home of the rain god, was dedicated exclusively to his worship. Inside Tlaloc's mountain lay the source of all water, and only his action could release it. Today, on those rare moments when Mexico City's urban smog permits, one still can see the first puffs of cloud that bring on the summer rainy season forming above Tlalocan. The chronicler Durán tells us that once a year the priests gathered on Mount Tlaloc to sacrifice children to the rain deity, to pay the debt they owed him for letting loose his water from inside the mountain in time for spring planting. From the vantage point of their mountain shrine, the priests could look back across the valley to the Templo Mayor and see the sun set directly over it on the day they made their sacrifices, thus assuring them that all was right in heaven and earth. The time-line between the Tlaloc temple in the ceremonial precinct and the one on the peripheral hill served as yet another concrete mechanism tying the Aztec state to nature's environment.

All the events surrounding the rain-god rituals were programmed to take place at both the right time and in the proper place. For example, Sahagun tells us that in the Aztec month of Atlcahualo the priests brought children for sacrifice to seven different shrines. These children were selected because they possessed favorable day signs, and each was labeled with his or her own particular colored paper streamers—"human paper streamers," they were called. Each group had to be brought to a particular shrine on the valley periphery and offered for payment of their debt to Tlaloc. If we search out the place names given in the chronicles on modern maps of Mexico City, we find that the designated locations not only are time-ordered but also follow a particular course in space, just as the wandering eye follows

*Though it rises precisely in the east, the equinox sun moves slightly southward, in the low northern hemisphere latitude of Mexico City, along a slanted trajectory as it gains altitude.

the circular course of time in the New Year's calendar of figure 6.6A.

The first sacrifices took place in the temple on the mountain Cuauhtepec, a place that shares its name on modern maps with the highest mountain located exactly north of the old ceremonial precinct. Painted potsherds and shattered walls still litter the top of the 10,000-foot-high peak. From there the sacrificial pilgrimage moved like the hands of a clock around the valley rim to Yoaltecatl, then to Tepetzinco and Poyauhtlan, to Cocotitlan and Yiauhqueme.

Studies that attempt literally to map out in real space the locations of the rituals are still in their infancy—probably because up to now no one was really prepared to believe a militaristic state without a system of writing as we know it could preserve and transmit such a wealth of detailed, highly ordered information about the relations between space and time. We might have expected this level of organization from the Maya, who at least had the capacity to write it all down in their books. We have a tendency to view imperial expansionists like the Aztecs as simply too busy with conquest to bother with esoteric details about how to worship their gods. For so long we have been put off by their barbarous acts of human sacrifice. For us they are strictly *post*-Classical. But the rich descriptions of elaborate rites and ceremonies Sahagun gave us in his twelve-volume *General History of the Things of New Spain* should have led us to think otherwise. His work documents Aztec religion, passed down through memory and oral tradition, as the glue that bonded the people together.[10] The sacrificial caches of Olmec and Teotihuacan objects, of seashells and obsidian chips from distant shores recently unearthed from the inner layers of the Templo Mayor, testify to Tenochtitlan's deep and widespread religious roots.

As in Shakespeare, all the world was the stage upon which the Aztecs acted out their rituals. And if the play's the thing, then it must be carefully directed and properly acted out. By linking the performance of their rituals to the regions surrounding the periphery of the Templo Mayor in this programmatic way, the Aztecs had proven that the natural order and the social order were really identical. Their displays of celestial bodies lining up with the *axis mundi* at the right time complete the proof. These spectacles must have made a lasting impression on those who witnessed them. Though we would scarcely think about celestially or environmentally orienting any of our institutional buildings, much less our places of worship, in this peculiar way, this sort of imaginative symbolization is rife in the imperial states and kingships of ancient America.

8

THE INCAS AND THEIR ORIENTATION CALENDAR

> In that Empire, the craft of Cartography attained such Perfection that the Map of a Single province covered the space of an entire City, and the Map of the Empire itself an entire Province. In the course of Time, these Extensive maps were found somehow wanting, and so the College of Cartographers evolved a Map of the Empire that was of the same Scale as the Empire and that coincided with it point for point. Less attentive to the study of Cartography, succeeding Generations came to judge a map of such Magnitude cumbersome, and, not without Irreverence, they abandoned it to the Rigors of sun and Rain. In the western Deserts, tattered Fragments of the Map are still to be found, sheltering an occasional Beast or beggar; in the whole Nation, no other relic is left of the Discipline of Geography.
>
> —JORGE LUIS BORGES
> and ADOLFO BIOY CASARES

THIS surrealistic passage from the short story "A Universal History of Infamy,"[1] about a map so large that it covers the entire landscape, epitomizes the nature of imperial Inca timekeeping, for their calendar was neither a piece of parchment nor a carved stone: their timepiece was the city itself.

Indeed, the ancient Inca, who dominated the Andes and the west coast of South America during the late fourteenth and early fifteenth centuries, developed as sophisticated a timekeeping system as the Maya. Far from their stereotype as New World Romans preoccupied with war and conquest, they, too, turned to nature watching.

Although they possessed no written record as we know it, they managed, like the Aztecs, to rise out of obscurity and, in less than one hundred years, built an empire that stretched from Ecuador to Chile and covered all of Peru. Given their humble beginnings, not even the cagiest Las Vegas gambler would have bet on them as the local tribe in the Cuzco Valley that ultimately would succeed in controlling all the others.

Owing to the latitudinal extent of their empire—over 30 degrees—they needed to regulate time in order to fix civil, agricultural, and religious dates over a wide area and with a host of both different ecologies and skies. Allusions in post-Conquest Spanish chronicles to Inca timekeeping demonstrate how astounding were both the quantity and the quality of careful sky watching and timing by the stars which took place in and around Cuzco, the Inca capital, called Tahuantinsuyu (the four quarters). Unusual, too, was the way the Inca cleverly used time principles to establish political and social order.

As at Tenochtitlan, the rich historical record left us by the Spanish chroniclers of Cuzco, who also were ill disposed toward believing that their subjects were civilized in any sense, offers tantalizing clues that the Inca had built, into the natural landscape surrounding their capital city, an orientation calendar. It consisted of specific sightlines directed to the sun moving at the horizon, and even a scheme for counting the days. Their calendar serves as the epitome of human adaptation to a difficult environment, and played a major role in the Inca's success at empire building—at least until they themselves were abruptly overtaken by Western European imperialism under the sword of Pizarro. Upper Cuzco lay on higher ground upriver; while downriver, or lower Cuzco, denoted the territory through which water flowed as it passed out of the valley. Cuzco also was united by roads: four of them separated the divisions or *suyus* from one another and led out of the Cuzco valley to the remotest places in their empire. Coricancha, called by the Spaniards the Temple of the Sun (figure 8.1), was the focal point of the system, and was situated precisely at the confluence of the two rivers that flow through the valley of Cuzco.

Father Bernabe Cobo's chronicle contains an elaborate description of the radial subdivisions that further make up the unusual organizational plan of the city.[2] He lists 41 invisible radial lines, called *ceques*, that crossed the landscape, each emanating from the Coricancha. These lines were marked out by *huacas*, or sacred places—like beads on a string, or knots on a *quipu* (figure 8.2). Many

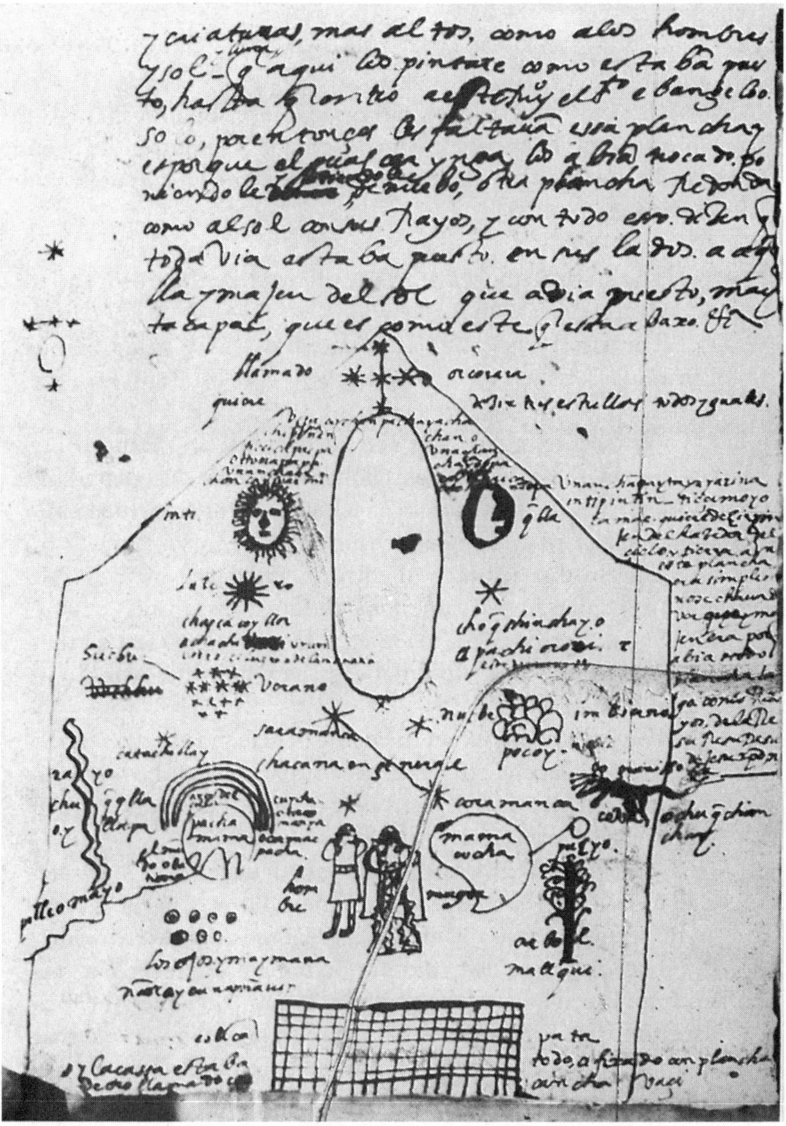

FIGURE 8.1 Coricancha, which the Spaniards called the Temple of the Sun. (*A*) Coricancha was the center of the *ceque* system and the principal temple for Inca ancestor worship. Today the Church of Santo Domingo, symbolic of the Spanish domination, perches atop the original andesite walls of the temple. (*B*) The Coricancha as depicted in the book of the chronicler Santa Cruz Pachacuti Yamqui. Man and woman, sun and moon, evening and morning star, winter and summer, along with a variety of symbolic animals and constellations, all have their place in a dualistic arrangement that consists of the pairing of complementary opposites. SOURCE: (*A*) Photo courtesy of Susan A. Niles. (*B*) After A. Aveni, *Skywatchers of Ancient Mexico* (Austin, Tex.: University of Texas Press, 1980), fig. 111.

of these *huacas* were rocks, springs, valleys, or mountains—even movable objects. These places must have been important to the Inca: witness the detail with which Cobo describes the *huacas* of just one of the 41 *ceques*:

> The sixth ceque was called Collana, like the third, and it had eleven guacas [*huacas*].
>
> The first was called Catonge and was a stone of the Pururaucas, which was in a window next to the Temple of the Sun.
>
> The second guaca was named Pucamarca [*sic*; probably for Quishuarcancha]; it was a house or temple designated for the sacrifices of the Pachayachachic [Creator] in which children were sacrificed and everything else.
>
> The third guaca was called Ñan, which means "road." It was in the plaza where one took the road for Chinchaysuyu. Universal sacrifice was made at it for travelers, and so that the road in question would always be whole and would not crumble and fall.
>
> The fourth guaca had the name of Guayra and was in the doorway of Cajana. At it sacrifice was made to the wind, so that it would not do damage, and a pit had been made there in which the sacrifices were buried.
>
> The fifth guaca was the palace of Huayna Capac named Cajana, within which was a lake named Ticcicocha which was an important shrine and at which great sacrifices were made.
>
> The sixth guaca was a fountain named Capipacchan [*sic*; for Çapi Pacchan], which was in Capi [Çapi], in which the Inca used to bathe. Sacrifices were made at it, and they prayed that the water might not carry away his strength or do him harm.
>
> The seventh guaca was called Capi [Çapi], which means "root." It was a very large quinua [tree] root which the sorcerers said was the root from which Cuzco issued and by means of which it was preserved. They made sacrifices to it for the preservation of the said city.
>
> The eighth was named Quisco; it was on top of the hill of Capi [Çapi], where universal sacrifice was made for the same reason as to the above mentioned root.
>
> The ninth guaca was a hill named Quiangalla which is on the Yucay road. On it were two markers or pillars which they

regarded as indication that, when the sun reached there, it was the beginning of the summer.

The tenth was a small fountain which was called Guarquaillapuquiu, and it is next to this hill. In it they threw the dust which was left over from the sacrifices of the guacas of this ceque.

The eleventh and last guaca was called Illacamarca; it was in a fortress which there was, built on a steep rock on the way to Yucay, and at it the guacas of this ceque ended.[3]

The whole *ceque* system looks strikingly like a *quipu* (figure 8.2), an array of knotted strings bound together on a common cord, long

FIGURE 8.2 An Inca *quipu*, or counting device, consists of arrays of knots on strings; its very structure invokes radial and hierarchical ways of representing space and time used on a grand scale in the *ceque* system. SOURCE: W. Conklin, "The Information System of Middle Horizon Quipus" in A. Aveni and G. Urton, eds., *Archaeoastronomy and Ethnoastronomy in the American Tropics* (New York: New York Academy of Sciences, 1982).

used by Andean people as a tallying device. One chronicler even tells us they kept their history on the *quipu*.[4] But unlike history books and codices, a *quipu* requires more than the eye to be read. You *look* at cords of different colors and *feel* the knots with your hands to acquire the information contained within its text. If one conceives of the *ceque* system as a giant *quipu* that overlay Cuzco, then the cords that reach radially outward from the Coricancha become *ceque* lines, and the knots on the cords represent the *huacas*.

Sun Watching

The key to understanding why the Inca divided their city in this odd radial way is rooted in their attitudes and beliefs about lineage and its connection with ideas about the sun, an association that can be understood only if we delve into their version of creation.

The Inca Creation Story

The Inca say that they descended from the god Viracocha, son of the sun, Inti. He came forth from Lake Titicaca in the highlands of present-day Bolivia, far to the southeast of Cuzco, and created the last race of people—the ones who were built of stone and who dwelled in the ancient city of Tiahuanaco. But Viracocha also created a special pair of people to be his own messengers, and to them he gave explicit instructions: they should come forth from caves surrounding a remote valley where they would establish a new kingdom. All along the way, Viracocha would protect them from both the elements and any attacks by hostile strangers. Making their journey down a river valley, they would know when they had arrived, for one person would split off and found Cuntisuyu, "the shoulders where the sun rises, on the left hand," while the other person would take an alternate route and establish Antisuyu, "which will be on the right." Today anyone who travels from lake Titicaca down the Vilcanota River into the Cuzco valley will find that Viracocha's instructions still fit the landscape. For his "chosen people," Viracocha created the upper half of the moiety, the elevated, upstream section of the city, the part that pertained to things celestial and of a male nature. The complementary, lower half of Cuzco consisted of people and things genealogically related to the earth and feminine aspects.

Prior to all of these developments, there were said to have been four creations (figure 8.3). These were punctuated by destructions wrought by earthquake, flood, fire, and so on, each individual creation having been provoked by struggles among various gods who comprised a competitive pantheon, but all of whom seem to have represented aspects of a single all-purpose creator god, Inti.

Each of these previous creations had been governed by a different kind of time. There had been unbroken durations of light, dark, and "stone-ness" in between the destructions. Out of the alternating cycle of chaos and creation, Viracocha, patron god of the Inca, ultimately triumphed. Today we live in the fourth world, where the principle of activity or change characterizes the passage of time. Ours is the only world in which time, as measured by the periodicities of the changing celestial bodies, has ever existed. It also is the only world in which people matter. Only their action, their activity will be responsible for the ultimate outcome of the present creation—and it may well be the last.

We need not look very far to find remarkable parallels between the Inca and the Aztec creation stories. Just as the Aztecs lay temporal claim to ancient Teotihuacan, imperial Inca history tied the citizens of their empire specifically and directly to established local tradition—and to the nearby ancient ruins of Tiahuanaco, of which they claimed to be the rightful heirs. And like the Maya royal lineage, the Inca fabricated dynastic history, embellishing it all the more the farther back they took it: their blood descendancy goes all the way back to Inti himself. While we are told of previous ages (three rather than four) as in the Aztec case, for some reason those earlier fabrications did not quite make the grade. Creation seems to be a process of becoming in which, by trial and error, things are fashioned with increasing improvement. And, when things do not turn out quite right, a command is issued to go forth and establish a kingdom—the command of a patron god to a special people he has selected to serve his interests and to further carry out his wishes; people whose actions, if followed just as he prescribes them, will ultimately be responsible for the salvation of the world.

The *ceque* system of Cuzco, then, was born out of an ideology of responsible human action. When Cobo describes each *huaca* that ties the *ceques* together, and when he tells us who must worship there and what sacrifices they ought to make, the message is that Cuzco was made up of people who were doing not just what they believed was appropriate, but what, in fact, was necessary and vital to keep the world going. For example, one of the major tasks delegated to the people is to feed Pachamama, the feminine half of the creation principle, which complements the male half represented by the sky god Viracocha. She is "mother earth," and must be nourished at all of the places where she is open or exposed. These open places—like springs, rivers, and wells where water issues forth—become the *huacas* where they make the sacrifices necessary to sustain her.

The Rules of Kinship

Not only was there a purposeful act for everyone in the *ceque* system; there was also a place and a time to act. Thus, each half of the creation principle was identified with the kinship structure of the citizens of the empire. Their ancestors were believed to remain alive within the earth. It was the responsibility of every good citizen to feed his particular *huacas* at the appropriate times, so that the dead, too, could be nourished. In exchange, the ancestors provided their progeny with the most vital commodity of all—water, which would sustain their agricultural economy.

Whether for rich or for poor, this notion of the dead transcending time and interacting with the living gave human action at all social levels a permanent place in the Inca world view. For the wealthy and the regal, the mummified remains of previous generations were lavishly entombed. At the correct time, a deceased uncle or grandfather would be trotted out and paraded around, all adorned with gold offerings. The same for the poor: their mummies were less ornate, bone and shell replacing gold offerings. At all levels of society, a share of ownership of the land continued to remain in perpetuity with the deceased, and only a part of the material goods passed on to the offspring: you *can* take it with you seems to have been their motto. Some archaeologists have suggested that this principle of split inheritance may have been one of the underlying motives behind the drive for Inca territorial expansion, for every ruler needed to continue to provide for the lands, crops, and properties of all of his dead ancestors as well as for himself and his growing family.

The complex rules of connecting kinship and ancestor worship, irrigation of the landscape, the course of the sun and the seasons—all seem to have been tightly woven together in the *ceque* system. At least each of these areas of concern appears in Cobo's detailed description of the *ceque* system.

The anthropologist Tom Zuidema has analyzed Cobo's hierarchi-

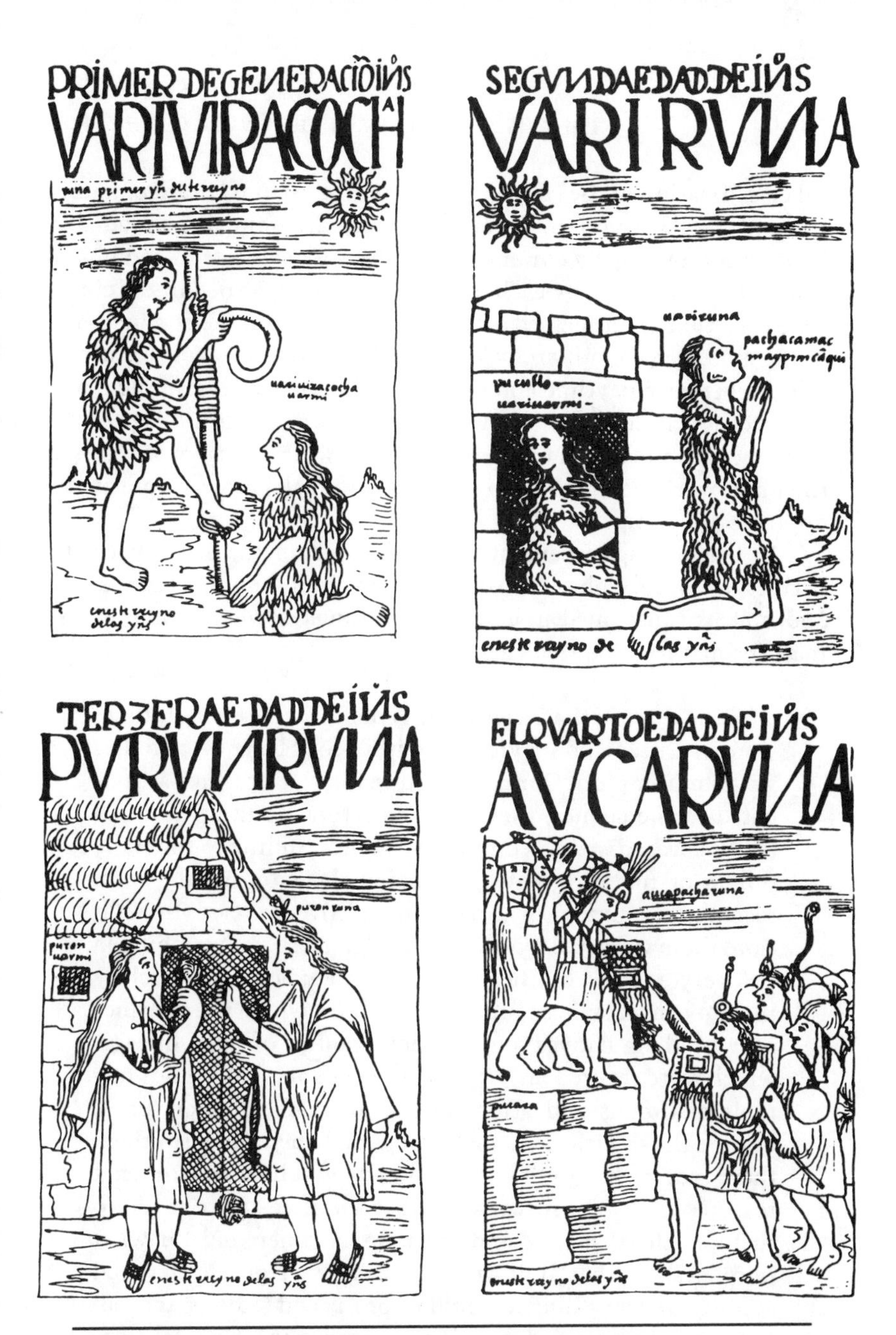

FIGURE 8.3 The four previous ages of the Andean world in the pictorial chronicle of Felipe Guamán Poma de Ayala contains both native and Christian elements. Note the progression from the earthy-appearing humans of primeval times through various unsettled states up to the present uncertain age begun by the great confrontation between Inca and Spaniard. SOURCE: J. Murra and R. Adorno, eds., *El Primer Corónica y Buen Gobierno* (Mexico City: Siglo XXI Editores). By permission of Siglo XXI Editores, Mexico City, from the first unabridged edition of the seventeenth-century manuscript.

cal classification of the *ceques*, which he finds strongly reflect family ties and class distinction. Certain ones were worshiped and tended to by nobility; others, by commoners; some, by people of mixed blood. The order of worship by each of these groups proceeded in the clockwise direction as one passed around the landscape in the northern half of the moiety and counterclockwise in the south, the whole creating a tension between daytime and nighttime. In other words, in the upper (male, celestial) sector, the flow of worship was a mirror image of the movement of the sky in that half of Cuzco; while the flow of water in the underworld, symbolized by the lower (female) moiety, runs in the opposite direction.

In the closing statement of his chapter on the account of the shrines in Cuzco, Cobo offers a number of clues about how the *ceques* and *huacas* functioned in an elaborate timekeeping scheme, whose mainspring was the agricultural year:

> These were the guacas and general shrines which there were in Cuzco and its vicinity within four leagues; together with the Temple of Coricancha and the last four which are not listed in the ceques, they come to a total of three hundred thirty-three, distributed in forty ceques. Adding to them the pillars or markers which indicated the months, the total reaches the number of three hundred fifty at least. In addition, there were many other private [*guacas*], not worshipped by everyone, but by those to whom they belonged, such as those of the provinces subject to the Inca, which were shrines only of their natives and the dead bodies of each lineage, which were revered only by the descendants. Both kinds had their guardians and attendants who, at the proper times, offered the sacrifices . . . that were established. For all of them these Indians had their stories and fables of how and for what reasons they were instituted, what sacrifices were made to them, with what rites and ceremonies, when and for what purposes, so that if it were necessary to give a detailed account of everything it would be prolix and tedious; indeed, I very nearly refrained from listing, even in this brief fashion, the guacas named in these four chapters, and I would have done so, except that I judged it necessary to enumerate them to explain more clearly the gullibility of these people and how the Devil took advantage of it to inflict on them such a harsh

> servitude to so many and such foolish errors with which he had taken possession of them.[5]

The essence of this confusing statement seems to be that there were two kinds of *huacas*, but only one set followed the order of the motion of the sun, thereby indicating the months. At least in Cobo's mind, these constituted an orderly system. But there are other allusions to Inca sun watching and in particular the solar movement about the horizon during the course of the year. Another chronicler, Felipe Guaman Poma de Ayala, himself part Inca, introduces us to a practical motive for following the route of the sun. He tells us that his ancestors employed "observatories with windows" to catch the first and last rays of the sun, and used these observations so that they would know when to shear the llamas, sheep, and alpacas and when to harvest crops—activities that took place during different seasons of the year. In another passage where he describes the duties of office of the state astrologer, he refers explicitly to the changing aspect of the sun in the local environment. The sun "sits in his chair one day and rules from that principal degree (of the December solstice). Then he sits in another chair where he rests and rules from that degree (of the other solstice)." From one seat to the other "he moves each day without resting."[6] During the solstices, he rests for more than a day in his chair, when at those days the motion of the sun from day to day becomes imperceptible, as Guaman Poma says. These seats may be a reference to the chain of rising and setting points of the sun along the horizon over the course of the year.

There is other evidence that the Inca had converted the landscape into a natural, self-operating calendrical device powered by the movement of the sun, a system with no need of formal writing to articulate it—only the celebrated Inca sun pillars. If we turn back to Cobo's chronicle, we discover that certain *huacas* were large stone monuments positioned at key sun positions along the horizon, like the minute markers on a clock dial. One of them is described in the passage quoted on page 282 (the ninth *huaca* of the sixth *ceque*).

The Quiangalla twin pillars probably were placed there to frame the setting sun on the shortest day of the year, the June solstice (in the southern hemisphere). An identical pair marked the sun at the December extreme when it set over the hill Chinchincalla, Cobo tells us. But where would the viewer stand in each case? The most likely observation point would seem to be the Coricancha, the center of the *ceque* system. Indeed, a line drawn from the Coricancha to Chinchincalla on a modern map of Cuzco actually coincides rather well with the direction of the setting sun about 21 December. For the June solstice, the situation is a little more complicated.

Oddly enough, neither the Coricancha nor any other monument near it could possibly have served as the observing station, because Quiangalla is not visible from the center of the city. Could the chronicler have been making a mistake in his description? Not likely, for about 5 kilometers to the northwest of Coricancha is an elaborately carved rock complex, today called Lacco, which consists of sculptured boulders, caves, ruins of buildings, terraces, stairways—even a canalized stream. Adjacent to it lay another *huaca* called Chuquimarca by Cobo. He describes it as a temple of the sun—the one where the Inca king spent his time celebrating the June solstice rituals. A sight line from Chuquimarca to the June solstice sunset position passes directly over the northwestern skyline at Quiangalla.

It may seem a bit unsettling to discover that the Coricancha, the center of the *ceque* system, did not function in all cases as the observing center for events that marked the solar calendar; but if we place the June solstice observation from Chuquimarca in the context of Cuzco's moiety system, we can make sense of what the Inca were trying to accomplish. The June solstice falls in a segment of the year calendar attended to strictly by the people of the northern or upper moiety (called Hanan Cuzco), perhaps because it fell during the time of the year when the sun literally passed through their region of the empire. In effect, they had the special responsibility to give it safe passage by tending to the observations that marked its course.

There is a parallel to the northern-based observation of the June solstice in the southern or lower half of the kingdom. The sun pillars that marked the December solstice sunrise surely must have been mountain peaks themselves, for the southeast horizon, where the sun would first appear at this time, is very distant from Cuzco. The Inca probably marked this sun position from another Temple of the Sun, which Cobo identifies as Puquincancha. Located in the Cuntisuyu quadrant in the southern half of the city (called Hurin Cuzco), Puquincancha, which fits the sight line perfectly, is the most logical site from which the Inca could observe sunrise on the first day of summer. These dual sight lines—a northern one to the June solstice sunset from Hanan Cuzco and a southern one to the December solstice sunrise as viewed from Hurin Cuzco—are paired opposites in time. In Mesoamerica, we became familiar with the notion that different segments of the calendrical cycle can belong to different quarters of the environment; however, we have not really encountered before these dual time directions in the Inca calendar. They reveal a social harmony in which different kinship segments of the community, distributed geographically, acquire local custodianship over certain parts of the year. Later we will see that when the Inca bureaucrats attempted to standardize time in the remotest parts of their empire, they made clever use of the principle of local autonomy as a way of giving their subjects a sense of control over their own regional calendar.

There was much more to Inca sun watching than this pair of astronomical time lines, each cutting across the landscape of its appropriate moiety. The Inca also devised a more ingenious temporal arrangement, one that pivoted the counting of the days about the zenith sunrise/antizenith sunset axis. This sight line sliced horizontally across the *ceque* system and a number of observation stations.

Several chroniclers repeatedly refer to a high hill close to Cuzco on the west as one of the most important locations for marking out the seasonal horizon calendar:

> On the hill of *Carmenga* (today called Cerro Picchu) they have at *definite intervals*, small towers, which serve to keep track of the movement of the sun, which they regard as important . . . and they had a plaza where they say a long time ago there was a swamp or lake at which the founders smoothed over the mortar and stone. . . . From this plaza the four royal roads go out.[7]

Now, Cobo tells us that one of the *ceques* that radiated from the Temple of the Sun passed over this hill. Today we can pinpoint the location of about half the *huacas* on that *ceque* (fortunately for us they still retain their old names). For example, Cobo tells us that the *huaca* Urcoscalla was a place "where those who travel to Chinchaysuyu lose sight of Cuzco."[8] If we take sightings of Cuzco from the hills west of Cerro Picchu, we can climb up to the place where the city disappears from view along this direction. From this point, if we walk radially outward from Coricancha, the ruins of an old mill built by the Spaniards can still be seen near the modern village of Poroy. This must be Cobo's Poroypuquio ("the well of Poroy") which he lists as another *huaca* of this *ceque;* he even tells us that the Spanish built a water mill there.

By identifying places in the landscape that correspond to well-documented references in Cobo, Zuidema and I have traced out the *ceque* line containing the pillars. Though we have been able to stand on the very spot on Cerro Picchu where the sun pillars once stood, today we can see no remains of them. They were torn down by the Spaniards to quell the native habit of worshiping idols; but Garcilaso de la Vega, who wrote his chronicle a whole generation after the Conquest, claims to have seen them when he was a little boy; and colonial paintings hanging in Cuzco's churches and museums exaggeratedly depict the hill of Picchu studded with so many tall slabs like a high-altitude graveyard.

An anonymous chronicler writes that there were not two but rather a total of four pillars in the little array on the hill. As the sun passed each of the pillars, it signaled the different times to plant in the environment of Cuzco. The most important time of the agricultural year occurred

> when the sun stood fitting in the middle between the two pillars they had another pillar in the middle of the plaza, a pillar of well worked stone about one estado [about 6 feet] high, called the *Ushnu*, from which they viewed it. This was the general time to plant in the valleys of Cuzco and surrounding it.[9]

Evidently, the northernmost pillar warned of the coming of the planting season in the Cuzco valley, for the chronicler explicitly states that people who cultivated crops at the higher altitudes, where growth occurred at a slower pace, would be allowed sufficient additional time to sow their seeds before planting commenced in the valley.

This early witness gives us two additional items of importance in sketching out the structure of the Inca calendar. First, he says that the main activity happened in the middle of August—still today the most important time of the year, for this is when the rains come and farmers commence their first planting. Secondly, he tells us where the observer stood to view the sun enter in between the pillars—at the *ushnu* stone in the middle of the plaza.

Some chroniclers say the *ushnu* was a low rounded pillar, but that it also incorporated an altar and a throne. Sometimes it is represented as a basin, an aperture in the earth or a place that sucks rain water down into the earth—a detail that may mean that the *ushnu* was correlated with the time when, as one chronicler says, the earth opens up; or when the sun drinks from it in the plaza so that he can fertilize the earth. This is the time in August marked by the disappearance of the sun over the middle of the pair of horizon pillars on Cerro Picchu.

Like the pillars, the *ushnu*, too, has disappeared; but at the spot where it must have stood, a busy street corner adjacent to modern Cuzcos's Plaza de Armas, you still can look northwest over the top of Picchu. The sun is still bright and relatively high in the sky when, at about four o'clock on a mid-August afternoon, it continues to keep its appointment by dropping down behind the place where the pillars once stood. Now, the middle of August (specifically 17–18 August) is also the time of year when the sun passes through the nadir, or the point opposite to the zenith. The importance of the passage of the sun across the overhead point is mentioned frequently in the Spanish chronicles all over the tropics. If we think of the nadir or "antizenith" not as a position in space but rather a point the sun arrives at in time—the date directly opposite the day when the sun passes overhead—we perhaps come close to the way the Inca understood it. Given their penchant for thinking in terms of paired opposites, they probably would have had little trouble recognizing that the polar opposition of these two times of the annual solar cycle was related to paired complementary directions in space. Standing on top of Cerro Picchu and looking in an easterly direction, they would have sighted the place where the sun rose on the day it passes overhead. Going to that remote spot in the east and looking west six months later, they would then see the sun set over Cerro Picchu. Just as the Aztecs sought to connect the rain god's mountain with their Templo Mayor, the Inca tried to establish a concrete physical relation between Coricancha at the center of their city and natural events that took place at the mountain periphery of Cuzco. This relationship helped structure an urban plan that would assure perfect harmony between their works and those manifest in nature, and most notably the course of the sun god.

The Zenith Principle

But why did the Inca choose this particular visual line as the mainspring of their solar timepiece? Why, of all the directions in space, did it become so symbolically laden? What is special about the zenith-antizenith direction? To convey the importance of the zenith in the

Andean world, I must digress for a moment to some of the differences between tropical- and temperate-zone celestial environments.

If you watch the stars over the course of the night, you will discover that all celestial motion seems to pivot around the Pole Star. Stars in the north turn in circles around it, while those farther away follow daily paths in the shape of even larger circles. As seen from tropical latitudes, the polar pivot lies close to the horizon. In the tropics, the sun, moon, and stars follow more or less vertical paths; they move straight up from the eastern horizon, over the top of the sky, and then plunge straight down in the west.

The celestial framework seems to be centered on the ground-based observer because whatever happens in the northern part of the sky is pretty much mirrored in the southern half. However, as you begin to move out of the tropics, the sky begins to take on a radically different appearance. In higher latitudes celestial motion becomes a combination of both vertical *and* horizontal. Stars rise and set along oblique tracks, and the angle a star path makes with the horizon gets smaller and smaller as you move farther north or south from the equator.

As you approach the higher middle latitudes, the motion about the celestial poles becomes dominant and the asymmetry of celestial motions in a framework centered upon the observer begins to increase. As the celestial pole rises higher and higher in the sky, the center of action seems farther removed from the earthly observer (see figure 8.4).

Some sky events happen in the tropics that simply cannot be seen from anywhere else on earth. The overhead passage of the sun is one of them. Between the latitudes 23½° N and S, the sun stands in the zenith at noon on the longest day, the first day of summer (21 June in the northern, and 21 December in the southern, hemisphere). A shadowless moment occurs when the sun arcs over the zenith and returns to the southern part of the celestial hemisphere.

In the tropics, a pair of dates, which depend upon latitude, will mark the sun's zenithal crossing. These dates divide the year into two parts; the first when the sun passes north, and the second when it passes south of the overhead point at noon. No surprise, then, that all over the world within the tropical band are evidence of calendar and celestial alignment schemes based on the times of solar zenith passage. It is as logical a reference point in the annual cycle as an equinox or a solstice.

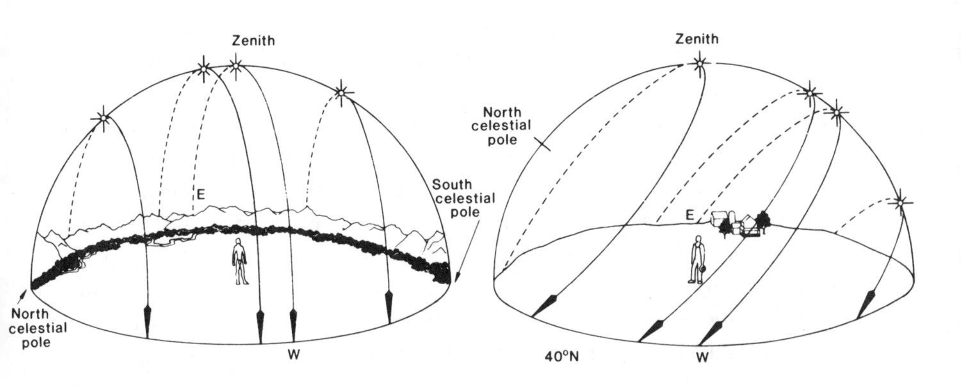

FIGURE 8.4 Why the heavens do not look the same everywhere on earth. Hemispheric diagrams of the sky show how things move as seen from the tropics (*left*) as compared with the high latitudes (*right*): in one case, stars pass "straight-up-and-over"; while in the other, they turn "round-and-round." SOURCE: A. Aveni, "Tropical Archaeoastronomy," *Science* 213 (1981): 161–71. Copyright © 1981 by the American Association for the Advancement of Science.

For example, the people of ancient Java (7°S latitude) developed a shadow-casting device (figure 3.2B) to partition the year in an unusual way. They reckoned the start of the year from the June solstice, the first day of winter in the southern hemisphere; but their so-called rustic year was segmented into 12 months of unequal duration, ranging from 43 days (the first, sixth, seventh, and twelfth months) to 23 days for certain months in between. Early anthropologists thought this peculiar array of month lengths meant that the Javanese were generally uninterested in or slovenly about keeping time. However, these unequal time intervals turn out to be precisely measurable if one uses as the guiding principle not the equal units of time we try to create in our calendar, but instead equal units of length traveled by the shadow cast by the tip of a gnomon at noon; the passage of the sun through the zenith really serves as the pivotal point for dividing up the year in this way.

The people who lived in the latitude of central Java (7°S) latched onto a special circumstance that occurs in the geometry of shadow casting there. When the sun stands on the meridian at noon north of the zenith on the June solstice, the length of a shadow marked off to the south of the base of a vertical pole happens to be exactly double the length that occurs when the noonday sun lies south of the zenith at the December solstice, when the shadow is projected to the north. Javanese chronologists halved the shorter segment and quartered the longer one, thereby producing their ingenious 12-month calendar. As in most cultures, they probably chose the number 12 so that they could roughly integrate full moons into the yearly cycle. Because the sun approaches the solstices slowly, we would expect the first and last months in the sequence, along with the sixth and seventh months, when the shadow approaches its southerly and northerly extensions, respectively, to be those of greatest duration; whereas months defined by equal shadow lengths (when the sun is near the equinox) might tend to be brief. This is precisely what happens in the Javanese year-reckoning scheme. Later calendar keepers needed to alter certain months by 1 or 2 days either way in order that they might better correspond to the sequence of agricultural activities they were designed to represent.

The Javanese calendar is a good example of environmental determinism. The celestial and terrestrial aspects of nature conspired together to give rise to a unique framework for the solar cycle. Using the zenith principle, the architects of this curious tropical calendar had transformed space into time. What was regarded by earlier visitors coming down from the higher latitudes to Java as calendrical anarchy was, in fact, a rational timekeeping system they could hardly recognize because they did not know the celestial environment.

In Mexico, we find the zenith principle in the form of zenith tubes housed within specialized ceremonial structures. These time markers were used to capture the rays of the sun when it crossed overhead. Recall that 16 July, one of the solar zenith passage dates throughout northern Yucatan was commonly regarded there as the first day of the year. Shortly after the conquest, King Philip II of Spain, recognizing the importance his tropical subjects gave to the passage of the sun across the zenith, ordered colonial officials to give him reports of the latitudes of those places, together with their solar zenith dates, so that he would be able to foresee any activities pertaining to the ancient solar cult that might take place.

This astronomical diversion should demonstrate what an obvious sky phenomenon the event of solar zenith passage is to somebody who lives in the tropics—even more obvious than the first days of summer or winter, which often have few meteorological correlates, and certainly easier to notice than the passage of the sun by the equinox, which the Greeks used to start their year. In the tropics the time between overhead passes by the sun is one of nature's fundamental beats waiting to be recognized; and just as it became a chronological hinge in the calendars that evolved all over the tropics, so, too, it was pivotal for the Inca in their attempts to regulate the passage of time.

A Dualistic World View

The Inca fixed attention on the zenith sunrise–antizenith sunset axis not only because that sector of the celestial environment was so suggestive, but also because they were deeply influenced by the terrestrial half of the environment as well. Imagine a mountainous land in which faraway places are reckoned not by the distance east, west, north, or south of the major population centers, but rather by how far above or below them one is situated. The Inca lived in a vertical world, a space in which the time for human action—for planting potatoes, burning off the scrub, worshiping the gods—depended critically upon where a person was positioned in a vertically based ecology, each tier of which was dependent upon every other one.

If we take nature's verticality and couple it with the opposing, complementary dualities that make up the Andean world view as a whole, we begin to move closer to what time meant to the Incas. The Inca classified natural phenomena not by the divisions into phylum-order-genus familiar to us from Western biology, but rather according to the principle of sexual complementarity. The male-sky-Viracocha interacts with the female-earth-Pachamama in a cosmic dialogue that promotes and symbolizes the regeneration of life. Modern Andean villages are still organized according to the moiety system. Plowing, harvesting, even the work preceding the organization of festal days is doled out according to the dualistic principle.

The seasons are no different. Unlike Vivaldi's Four in the Old World (see page 340) in most tropic zone cultures there are only two: the wet and the dry. Viewed as opposite sides of a coin, they contrast and oppose one another; but taken together, they form a whole made up of complementary halves. In the dry season, the Inca conducted the activities of the state: they took the census, they distributed land, they made war. April, the month just before the harvest when the storehouses lay at their lowest ebb, was the time of transition. It was then, when the sun on its northerly passage arrived at the antizenith, that the temporal coin flipped over. For the next few months, the fields lay fallow. Agriculturally speaking, there *was* no time. Not until August, when they planted and made sacrifices to the *huacas* of the poor and the commoners, did nature suddenly reawaken. This was the start of the wet season, a time when Pachamama opened herself and became fertile. Only then could she be penetrated by the nurturing rays of the sun. Once humans began to plow the land, the union of the principles from above and below could regenerate life.

But mid-August is also one of the times of year when the sun reaches the antizenith at midnight in the latitude of Cuzco. Having touched its solstitial marking post on the hill of Quiangalla late in June, it already has started back on a southerly course. In Cuzco, 25–26 April and 17–18 August—agriculturally the two most important dates of the year—also are the calendrical reciprocals of 30–31 October and 12–13 February—the days when the sun crosses overhead. The sight line in horizontal space that connects sunrises and sunsets on these pivotal days can be linked to the cardinal dates in the calendar, the very days when the sun is situated on the vertical axis of Cuzco.

The Inca had discovered the quintessence of vertical complementarity: they recognized in the celestial landscape a set of events that made perfect sense when cast in terms of their ideology. Discovering the significance of the zenith-antizenith principle was as profound for the Inca as the dawning in the Maya mentality that 260 days was the time cycle par excellence upon which all of life's important events and processes converged: the rhythm of Venus, of eclipses, of conception and birth. In the Inca world, the pulse beat of the sun's cycle and the planting cycle coalesce about the vertical solar axis. This coalescence becomes imaged in the form of a horizontal line that cuts across the *ceque* system from the mountains ringing one end of the valley of Cuzco to the other. The Inca harmony of the world consists of a rhythm we can begin to listen to only by putting ourselves into their shoes, only by walking the sides of their mountains under that strange and different tropical sky.

We have good reason to think the Inca employed the *ceque* system and its zenith-antizenith time axis as a principle of calendrical and social organization throughout their vast empire. For example, the ruins of their colony of Huanuco Pampa, far to the west, are built up around a central plaza from the corners of which four roads depart. Each quarter-like *suyu* of the city is split into three distinct segments, reminiscent of the division of Cuzco's *ceques* into three kinship classes. A series of city gates betrays the principal orientation of Huanuco's buildings, which is generally unlike that in Cuzco. A line through the gates passes directly over the main temple at the center of the open plaza, however it points in the direction of sunrise at the equinox rather than at zenith or antizenith passage. Were the Inca giving those subjects who lived on the periphery of the empire some degree of flexibility in the regulation of their temporal affairs—the way our federal government allows each state of the union to decide whether or not to keep daylight savings time? A careful look at the architecture of Huanuco reveals the hidden imperial presence. Two large stone buildings in the eastern quarter, the only two buildings aside from the main temple made completely of stone, are skewed 10 degrees out of line in relation to all the other structures. They align with the zenith-antizenith direction—Cuzco standard time. This paramount axis also turns up in the architecture of other Inca sites along the coast of Peru, like Inkawasi and Tambo Colorado, where the seasonal cycle and climate are totally different from that in the highlands and where the imperial calendar has no more environmental significance than New England harvest time does in Arizona or the salmon run up the Columbia River for a resident of Miami Beach.

Counting the Days

While *ceques* and *huacas* are parts of an orientation calendar that graphically follows the solar cycle, they also functioned as a calendar in the way we are acquainted with that term—as a mnemonic scheme for counting the days. Begin by playing with a few numbers: take the number of *huacas* in the *ceque* system, 328, This is a reasonable representation of a year length—37 days short to be exact. We already have seen that many year calendars, quite unlike our own, keep a count well short of 365. Their designers simply do not count the time when nothing is happening; or if they keep a calendar by the moon, as the Trobrianders did with their 10-month count, they tally a number of months beginning with the moon that coincides with the first major event in eco-time; then they correlate the rest of the months with human activity, ending the count when the activity ceases.

Now, 328 does not break down very well into whole lunar months as reckoned by the phases (29½ days), but it *is* a perfect number for integrating sidereal lunar months. A *sidereal month* (27⅓ days) is the time it takes the moon to move from a given constellation all the way around the zodiac back to that same background pattern of stars. The number 328 is exactly twelve such cycles.

Next, take the number 41, which represents the count of the *ceques;* it too fits perfectly into a lunar sidereal counting scheme, for 8 times 41 equals 328. Because some chroniclers speak of the Andean people having used an eight-day week, it may be that 328 was both a day count consisting of 41 eight-day weeks as well as 12 sidereal months—just as our 365-day year can be broken down into 52 seven-day weeks as well as 12 lunar synodic months.

A sidereal month cannot be detected by marking out successive moons. Because of the ⅓-day leftover interval after the 27-day count, the moon's reappearance in the same constellation of the zodiac happens about eight hours (⅓ day) later. Suppose I begin my first sidereal month by witnessing the moon in a constellation of the zodiac at midnight. The next time around this same event will occur at eight o'clock in the morning—that is, during daylight, when the stars are not even visible. But if I were to reckon sidereal intervals in groups of three, then I could see a visible moon against the same star pattern at the same time of night, because the third multiple of the sidereal month is a whole number: that is, $3 \times 27⅓ = 82$—or twice the count of radial "spokes" in the *ceque* system. Are these unusual coincidences that emanate from juggling numbers, or did such time intervals really play some role in the creation of the 41 *ceques* and 328 *huacas*?

Zuidema suggests that the chroniclers offer a clue. Recall the way Cobo grouped the *ceques.* Taken in kinship-based units of three, the *huacas* add up to numbers generally in the high twenties or low thirties. These quantities, like the number units that make up the Javanese year I discussed earlier, begin to take on the appearance of distorted month intervals. As in Java, there are 12 such triplets, 6 belonging to the region of Hanan Cuzco and the other 6 spanning Hurin Cuzco. The Inca seem to have organized the days into paired year-halves so that each half of the moiety system would pay tribute to or manage its own part of the cycle.

But where to begin and end the counting in such a calendar? One suggestion is provided by some statements in the chronicles that peg the start of the year to the heliacal rising of the Pleiades, an event that occurred in Cuzco about 9 June during pre-Conquest times. Andean people still call the Pleiades *collca*, or a storehouse in which to keep the goods they harvest. Once the Pleiades appear in Cuzco, the length of time they spend in the sky each night is correlated with the duration of the sun in the sky as well as with the accompanying change of climate. In other words, the longer they appear, the longer the days become and the warmer the climate; then, as they recede toward heliacal setting, the days grow shorter and the nights colder. Plants also follow the pattern, growing as the rains increase and ripening when the rain diminishes.

Finally, the sidereal lunar cycle and the Pleiades cycle fit together neatly into the *ceque* system. Suppose you begin the calendar count at the eastern boundary between Hanan and Hurin Cuzco and then count the days clockwise, allowing each *huaca* to stand for a day. Let the first day in the count be 9 June in our calendar, so that it matches the date of the Pleiades's first appearance just before sunrise. Following the count of the *huacas* all around the horizon, you end at a point in early May (37 days before 9 June), or pretty close to the last day in the year on which the Pleiades can be seen in the west after sunset. This 37-day period, the inactive time the Inca did not count formally, now finds its logical celestial correlate and the disappearance and reappearance of the Pleiades makes counting these days unnecessary.

If this hypothetical time-reckoning scheme is laid over the *ceque* system, not only does it reveal the significance of the number of days in their year; it also maps out reasonably well the distribution of some of the subordinate numbers in the counting of the *huacas*. For example, the first three clusters (26, 30, and 29), which are assigned to the southern half of Cuzco, add up to almost exactly three full lunar-phase cycles. In fact, the alternation of lunar synodic months of 29 and 30 days would be an excellent approximation to the "southern half of the year." The northern half of the year seems more erratic by contrast. Hanan and Hurin Cuzco may have been conceived as opposing dualities in the sphere of lunar as well as solar timekeeping, the former following a sidereal and the latter a synodic reckoning system.

To understand better the role the moon played in the Inca calendar as well as other celestial timings they employed, let us return to the spatial point of origin of all of Cuzco's *ceque* lines. Chroniclers said of the Coricancha that it had gold sheathing, and that sheets of gold were removed from it by the Spaniards—over 700 plates averaging 500 pesos (4½ pounds) apiece! There was a seat in its western wall where the king sat; and when the sun struck him as it rose, the precious stones and emeralds set in his throne and the wall around it glittered.

This was the "golden enclosure," as its Quechua name implies; today its ruined andesite-block walls (clearly visible in figure 8.1) still stand at the junction of the Huatanay and Tullumayo rivers beneath the Church of Santo Domingo the Spaniards had erected on top of it. The Coricancha must have been on a par with the Church of the Holy Sepulchre in Jerusalem, remarked one of the archaeologists who excavated it.[10]

Like the Templo Mayor at the center of the Aztec empire, Cuzco's most important building is oriented in a way that ties it logically to the *ceque* calendar. A long hallway in its interior western wall, near the splendid throne, faces sunrise on 25 May; but the Pleiades also rose in that same direction about the time the Inca built the temple. Furthermore, a *huaca* named after the Pleiades is located at the end of a *ceque* running out in this direction. What relation might exist between the Pleiades and the 25 May solar date? The heliacal rise of the Pleiades in Inca times would have taken place about half a lunar synodic month before the June solstice—about 9 June. If the Inca ruler observed the first visible crescent moon falling after 25 May, then he would have noted that, on the average, the first full moon after that date would fall on or near the June solstice. A general rule, in the form of a timing mechanism with built-in crosschecks, is: the first full moon after the heliacal rise of the Pleiades always defines that month which includes the June solstice. Thus, the first glimpse of the Pleiades rising would serve as a warning that the solstice and its attending festival were imminent, while the first crescent would have announced the arrival of both the Pleiades and the sun in the middle of the passageway in between the western walls of Coricancha. All of this complex time reckoning that joined sun, moon, and stars together could occur with neither counting nor writing.

But here I begin to approach the limits of our understanding of how the *ceque* calendar works. What about the minute subdivisions of the *ceque* system that glare out at us from some portions of Cobo's description? Perhaps some days were prescribed by rules that came from noncelestial considerations; perhaps a principle of kinship or a fragment of mythic history is tucked away in different segments of the all-encompassing cosmographic map of Cuzco. There is much we do not understand about the *ceque* system, but we do know enough to comprehend what the Inca bureaucracy had in mind, at least in a general way, when they devised it. They sought to cover all activity in the universe under a single ideological umbrella made up of the dimensions of space and time. The *ceque* system was neither image nor representation: it was Tahuantinsuyu itself.

Articulated through the *ceque* system, the Inca calendar emerges truly as a calendar for the people. Timekeeping for them was a participatory enterprise. Every person of the populace was ordained to act at the proper time by feeding the *huacas* of his kin group. Only then could they keep the world going, for their present creation was a world of action and tension, and its future depended on the acts committed by all the people—ordinary human beings as well as royalty.

In ancient Cuzco, the people's welfare was the welfare of the state. A good citizen could even work up the social ladder by exhibiting military prowess. One could achieve material rewards and earn a place in heaven. As one scholar put it, "In feeding the celestial spirits at earth shrines (*huacas*) at properly discerned times and places, individuals properly related in space and time unite the dual cosmos and nourish the circulation of its parts in order to perpetuate its generative cycle of birth and death."[11] This mission closely parallels that of the Aztecs in relation to their philosophy of time, except that for the Aztecs it was the warrior class that engaged in the paramount human activity—obtaining souls for sacrifice. In both cultures, time was kept moving by a nurturing or feeding principle. Incidentally, Cobo tells us in his description of the *huacas* that the Inca were not above sacrificing human lives: they offered children at approximately one third of all the *huacas*, in addition to sheep and llamas, gold, silver, ground shells, textiles—each item carefully prescribed in the time sequence: "Each ceque was the *responsibility* of the partialities and families of the city of Cuzco, from within which came the attendants and servants who cared for the huacas of their ceque and saw to offering the established sacrifices at the proper times."[12] Moral overtones come through in Cobo's description of the *ceque* system: there is a right way to behave, a moral, as well as cosmic and social, order in the organization of Cuzco.

We do not know where the Inca got their ideas about the *ceque* system—perhaps from the empire of Tiahuanaco to the south, from where they claimed their origins. Except for the archaeological record, pre-Inca history is practically nonexistent: their empire was short-lived; and once the Spaniards discovered it, they plundered Cuzco beyond recognition. In fact, the devastation wrought by Pizarro makes Cortés's conquest of Mexico look mild by comparison.

When these vulgar representatives of high European culture

climbed down from the galleons moored off the west coast of South America and ascended the Andes, like their counterparts in Mexico in search of their own materialistic betterment, they saw not a sinew of this vibrant culture for which they could muster any curiosity or desire for understanding. Even the wise men, the missionaries, who followed them, praised only the heroic deeds of their brothers, the conquistadors, who won the empire of Peru for Roman Catholicism. They consigned all native ways and customs to devil worship, a condition only conversion to the true cross could cure. One of them wrote:

> It is held to be certain that in the shrines and temples where conversation was held with [the devil], he let it be known that it was to his service for certain youths to be attached to the temples from childhood, so that at the time of sacrifice and solemn feasts, the chieftains and other men of rank could indulge in the cursed sin of sodomy.[13]

And another:

> The Indian who does not learn the Law of God is like a beast who neither knows nor wants anything but to eat and to drink; he has no other gratification but to graze on grass.[14]

In 1992, the Western world will celebrate the anniversary of the first chapter in one of the great stories of human misunderstanding. These statements—and hundreds like them—written by early European historians, those upon whom we have grown dependent for much of the information we have acquired about the strange, exotic *others* who populated pre-Columbian America, reveal how short of the mark we have fallen in understanding the New World legacy—a legacy our forebears almost snuffed out of existence before we really had a chance to see what made it tick. Here we stand, five centuries later, trying to pick up the few pieces left, and ask, What can we learn from this? But there is a far deeper question: What moves us?

9

EASTERN STANDARD: TIME RECKONING IN CHINA

> Development of Western Science is based on two great achievements, the invention of the formal logical system (in Euclidean geometry) by the Greek philosophers, and the discovery of the possibility to find out causal relationships by systematic experiment (Renaissance). In my opinion one has not to be astonished that the Chinese sages have not made these steps. The astonishing thing is that these discoveries were made at all.
>
> —ALBERT EINSTEIN

IN ANCIENT CHINA, all the elements that might have led up to our scientific way of reckoning time were already in place. Their astronomers were precise and meticulous observers, they used sophisticated mathematics, and they even developed an advanced technology fairly early in their history. In fact, gunpowder, the magnetic compass, and printing—three of the most important mechanical inventions that altered the course of our own history—were gifts of the Chinese. Why, then, did they fail to develop that special synthesis between higher mathematics and the observation of nature we have come to call science? Why was there no Chinese Kepler, Newton, or Einstein? One answer may be that a spectacular accident of history, rather than a deliberate set of carefully considered options, led us to our modern scientific way of comprehending the world. As Einstein put it in the epigraph to this chapter, "the astonishing thing is that these discoveries were made at all."[1]

Einstein was actually referring to two discoveries. First, the arithmetical quantitative methods inherited from the Babylonians were combined with the logical, geometrical, and pictorial way of thinking developed by the Greeks in the fifth century B.C. The second accident, which took place nearly two thousand years later during the Renaissance, was, as Einstein characterizes it, the "discovery of the possibility to find out causal relationships by systematic experiment."[2] Together these steps led to the miracle we call science. As a result of quantitative logic and experimentation, we have come to believe in an objective truth—the real truth about the order of nature considered as an entity separate from the human mind and soul. Scientists share a faith that truth can be determined by making a series of one-to-one match-ups between mathematically expressed ideas conceived in the mind (we call them theories) and the experimental outcome of tests concerned with how a specific isolated part of this separate natural world might be expected to behave if the idea actually were true. We formulate directed, highly focused questions from our ideas and then manipulate and carefully observe nature's tangible entities. We repeat our test questions several times in exactly the same way, conducting each test objectively and without bias. The experimenter's feelings, emotions, moral, or religious beliefs have no bearing on the results of the test. If the anticipated outcome does, indeed, take place in the laboratory, then we give strong marks to the idea; if not, we change at least one component of it, make another prediction, and carry out another test. Though we do not operate in quite this orderly way on a daily basis, it is by this painstaking stepwise process that our knowledge and understanding of the world outside of ourselves accumulates and progresses.

Take a simple example: if the earth really revolves about the sun in a vast sea of stars (idea), then we would anticipate that, as a reflection of the earth's motion, all nearby stars lying perpendicular to the plane of the earth's orbit would trace out tiny circular orbits against the background of more distant stars over the course of a year (prediction). We can look through our telescopes (experimental test) to see whether we can detect these annual star shifts, or parallaxes. If we do not witness them, then we must amend our statement of the idea. Could it be that the nearby stars are so far away that, even though they move in tiny circles, our telescopic technology is not powerful enough to detect the motion? Or perhaps all the stars actually lie at the same distance; and, therefore, there is no relative motion of one with respect to another to be detected. Maybe the earth is fixed and the sun does the revolving. Or might phantom forces exist that push the stars in the opposite direction by just the correct amount to cancel out the reflection of the earth's motion that otherwise would be visible? The amendments can seem ever more disparate, but regardless of which alternative we opt for as a cause to explain the experimental effects, new information is added to our stockpile of knowledge about the nature of the universe.

It took nearly three centuries to realize that the first of the preceding causes is to blame for the disagreement between theory and fact. Today we have adjusted our theory to read: the earth revolves around the sun, but the nearest stars are so distant that it takes a telescope of relatively large aperture to detect any annual circular motion of the nearest stars relative to the more distant ones that might be produced by the earthly movement. Since the time of the prediction, we have taken the appropriate experimental route and built bigger and bigger telescopes, until finally, by the middle of the nineteenth century, we were able to detect the stellar parallaxes.

In a nutshell, this is basically how science works. We cast ourselves in the role of careful, patient, dispassionate, skeptical, and logical inquirers of nature. Idea is our driving force and precision, rigor, and repeatability, our credo. This is how to get at the truth of the matter; and because it works, we are satisfied. Indeed, often it works so well that we become intoxicated with its results. We begin to believe that our way of proceeding constitutes the only road to admissible truth. And so we ask: Why don't the Chinese look at the world this way? Where have they failed? The answer most often given is that although they had all the raw materials of science at hand, for one reason or another they simply failed to connect them together. Maybe they did not realize how important it would be to mathematize all of nature, to trust only what the eye could see, to be aware of the power inherent in stripping the material world of its anthropic and godly qualities. Perhaps worshipers of Buddha were not materialistic enough to develop our notion of historical time. Their search for a spiritual, rather than a material or social, order is often said to be responsible for an absence of historiography.

In giving each of these answers, it seem that we have firm hold of the yardstick, and the Chinese come up short. We made the correct choices; they did not. Rarely do any of our answers take any cognizance of political conditions in the East and West, or the deeply rooted principles that might exist among their philosophies that could lead to other equally satisfying ways of dealing with the natural world.

Compared with the rest of the world, one peculiar aspect of Chinese society is that it always has been bureaucratically organized; there never was a merchant-trade class like the one that developed independently in Europe during the Middle Ages, and made great demands on clock and compass in order to develop a more efficient and profitable economy. In Europe, the merchant, rather than the landed gentry, charted the direction of the state.

Equally different in outlook were our Greek philosophers compared with those of early China: Plato and Aristotle lived, thought, and taught in a basically democratic city-state, while the Chinese sage was part of a tightly organized and rigidly controlled feudal bureaucracy, an organization that revered history and drew deep moral meaning from it. The Greeks, by contrast, exhibited little reverence for their own past. Their focus, much like our own today, lay in the present. Furthermore, the Greek mode of discourse focused on public argument, on proving a point by logic in its most rigorous form. In China, one proved a point by history, not by logic. When a scholar-philosopher cited a historical example to his ruler, it made a significant and lasting impression. Furthermore, unlike Plato who taught in the academy and took his rest breaks in the marketplace of the agora, the ruler was the only ear to which the Chinese sage had access. No wonder the Chinese developed very different views of nature and the world from ours.

The Chinese World View and Sense of History

One can no more talk about all of China than about all of Native America—Hopi, Inuit, and Iroquois taken together; for China is a blend of cultures spread over half a continent and spanning millennia of time. To make matters even more difficult, we really do not know how long China was isolated from the West. Early contacts likely resulted in a mixture of ideas among the several European and Asiatic cultures. These societies were not as isolated as the New World Maya or Inca from the rest of the outside world. Still, there does emerge, from the millennia of available historical records, a certain characteristic peculiar way of thinking about nature, space, and time that can be termed basically Chinese.

This way of thinking is founded upon a general belief about how the world is controlled. Chinese cyclic time is embedded in certain natural resonances that operate on the principle of complementary dualism. Most important of all, deeply rooted moral fibers lace the whole system together.

The celestial gods are powerful spirits who control nature's behavior, and we human beings are at their mercy. Will the lord command rainfall in this month? Will he send down drought in the 47th year of this 60-year cycle? Will he give protection to the king when he tours the land this spring? As far back as 2000 B.C., these were the sorts of question celestial diviners asked the sun, moon, and stars that ruled over them. The chief timekeepers of the state were delegated to observe carefully the signs in the sky so that they could infer and anticipate every subtle dictum in the will of Heaven. Everything that took place in the celestial environment, from the top of the sky down to the bottom of the atmosphere—whether a bolt of lightning, a shooting star, or an eclipse of the moon—was a sign, a revelation about the course and assessment of human conduct.

These officials were neither priests as among the Maya nor quasi-independent scholars as in the Western Renaissance: they were bureaucrats. And the consequences of an astronomer-bureaucrat not paying attention could be serious. The epitaph on the gravestones of a pair of early astronomers is said to read:

> Here lie the bodies of Ho and Hi
> Whose fate tho' sad is visible,
> Being hanged for they could not spy
> The eclipse which was invisible.[3]

Whether these scientific sages of the royal court failed in their perspicacity we don't really know, but the Chinese written record makes it clear that even though celestial knowledge was directed solely toward the needs of the state, the astronomers were noting with great care and extreme precision the changing positions of sun, moon, and stars. From meticulous observations such as these, they developed the earliest cycles in their time-reckoning machine.

Calendrics was not a free enterprise in East Asia, and its makers and practitioners were not allowed the luxury of scholarly communication that took place at various times in the West, albeit at a much later date. So secret and inwardly directed was the bureaucratic business of timekeeping in ancient China that one order issued in the astronomy department of the ninth-century Tan dynasty even forbade communication between official timekeepers and their subordinates, as well as other civil servants and the public in general.

In China, nature is process—specifically, the process of transformation; and nature's laws are manifested in the changes visible in the sky. But humans have little advance knowledge of change; long-range developments are nature's secret alone—a secret housed in an intangible essence called the "spirit." One must be content with what one can grasp through careful observation of nature.

Change in human society takes place in the form of a tension between two opposing polarities that, under the most ideal moral conditions for humans, will come into balance. The celebrated yin and yang, the twofold way of knowing, are not real substances but rather essences that suggest the potentialities of things; they are master symbols for organizing the world, and around them all other symbols gather in hierarchical order. Yin and yang constitute a principle of interplay that duplicates the harmonic beat of the cosmos in the life and makeup of each individual as well as in the state. *Yin* is the female, the passive, dark part of any entity; and *yang*, the active, male, light; the balance process consists of a waxing and a waning, like so many waves. Actually, however, no element of good and evil that one might attempt to attach to this sort of terminology exists within the yin-yang formalism.

In addition to the twofold, the Chinese also conceive of a fivefold makeup of all things. These polarities combine to generate the five elemental forms: metal, wood, water, fire, and earth—all of which sounds something like the Greek degeneration of the world into the first forms (provided we eliminate metal and wood from the Chinese scheme and substitute air). Out of this fivefold way of organization come the five flavors (sour, salty, acrid, bitter, and sweet) and the five colors (green, yellow, scarlet, white, and black). Also, there are five kinds of wild creatures, five virtues, even five seasons. This fivefoldedness, like the five cardinal directions in Aztec cosmology, constitutes nature's format for the revelation of basic qualities of the world as well as aspects of human behavior. It is reminiscent of the way we use mechanical models in our culture to classify and explain how nature and human society work—the nervous system or DNA as a computer, the brain as a sponge, atomic structure as a planetary system, moneyed economy as water in a hydraulic pump.

Just as the Greeks commemorated the Golden Age of Kronos, the Chinese also seem to have conceived of the past as a time when the world was populated with figures of heroic proportion from whom they descended. Harmony reigned in that primal world, a kind of collective cooperation among all of its tribes; it was something to revere, something from which sound moral lessons could be extracted. Their historical foundations were built upon it.

The Chinese think about the past in terms of three ages: the mythological, the ancient, and the modern. The first two are of special concern. Creation was an event that, by most accounts, took place close to half a million years ago. It began just as the sages in many other cultures tell it—by the opening of heaven and earth. It was not a creation *ex nihilo*—that is, out of nothing—but rather a stepwise fabrication fired up by the Tao, an unknowable principle that resulted in the sorting out of an originally undifferentiated chaos.

First there was a One, and out of that One, the Two was produced, the yin-yang principle that constitutes the makeup of everything conceivable. As in biblical Genesis, there is also an anthropic version of the fabrication of the universe in which Pwan-Ku, a creator god, going at his work with hammer and chisel as he sculpts masses of chaotic matter into the correct shape. His labors lasted eighteen thousand years; and day by day as he worked along, he increased in stature and the heavens rose and the earth expanded around him. Once he made the stage ready, he died. But even his death benefited humans directly, his body parts becoming the entities that fill up earth's basin: his head became the mountains; his left eye, the sun; his right eye, the moon; his beard, the stars; his limbs, the four quarters; his blood, the rivers; his flesh, the soil; his breath, the wind and clouds; his voice, the thunder. His limbs were the four directions; metals, rocks, and precious stones were made from his teeth, bones, and marrow. His sweat became the rain, and the insects that stuck to his body: they were the people (see figure 9.1).

Pwan-Ku was followed by three sovereigns: the Celestial, the Terrestrial, and the Human—all of whom were giants whose reigns also lasted eighteen thousand years. They invented good government and the union of sexes; it was then that humans learned to eat, drink, and sleep. Other godlike sovereigns brought fire from heaven so that man could cook. People gradually became more civilized. They climbed down from their tree nests and began to live in groups. They made clothing and they began to farm. The story begins to sound like our own version of social evolution two thousand years before its time.

The ancient period was marked by Fuh-hi and his four successors,

FIGURE 9.1 According to an ancient Chinese creation myth, the first man was the creator Pwan Ku ("to secure the basin"), who chiseled the sky out of rock. When the great architect of the universe breathed, he caused the wind; when he opened his eyes, he created daylight; and when he spoke, his voice made thunder. SOURCE: C. Williams, *Outlines of Chinese Symbolism and Art Motives* (Shanghai: Kelly & Walsh, 1941).

together called the "Five Sovereigns." They are the Founders of the Empire (around 3000 B.C.), and with them came the invention of marriage and the expansion of the kingdom to the eastern sea. With these founders, Chinese history takes a step from the purely mythical to the quasi-mythical. Fuh-hi built a new capital and divided his territory of the empire into ten provinces, each into ten departments, each department into ten districts, and each district was split into ten towns. He regulated the calendar by introducing the sixty-year cycle in the sixty-first year of his reign (2637 B.C.).

A succession of rulers then blends quasi-mythical with real time, and the rest of the written record from about 2200 B.C. forward reads, like Palenque's Tablet of the Cross, as pure dynastic history. It is extremely lengthy, quite detailed, but basically factual, as "in the 12th month of the 5th year of the Khai-Cheng reign period," and so on. The record is complete through twenty-five successive dynasties dating from slightly before the Christian era. The Chinese have not only names but also detailed biographies, treatises on law, geography, taxation, and engineering—much more detail than in Western history books.

Clearly this bureaucratic account of historical time from the mythical imagery of Pwan-ku to the king lists of more recent times has linear, progressive undertones. It was pieced together by state officials probably with the same drive and motivation of Bishop Ussher or one of the chronologists of the Eastern Holy Roman Empire. The idea was to implant firmly the organized establishment into a traditional past consisting of godlike heroic figures.

In the Chinese account of creation, time emerges as an accumulation of eras, epochs, and seasons, just as space becomes a summation of a group of locations, domains, and orientations. Chinese space is distributed according to a pattern of hierarchy; and time, according to a number of limited cycles. The temporal cycles and the spatial hierarchy pivot about the emperor, the Son of Heaven, and their powers acquire maximum concentration in the capital city of the world when the great celebrations and festivities are held. This Son of Heaven was required to travel around his domain every five years to regulate the progress of his empire. He would go to the east during the spring, to the south in summer, to the west in autumn, and to the north in winter, each appropriate time being assigned its proper space.

As in other societies, there was a density to Chinese time. The emperor condensed it with the help of official celebrations. In the dead summer months of hard work and little social life, time's density was slight, but it increased in autumn and achieved its maximum during ritual jubilees.

Time's order was manifested in a succession of closed cycles, each being a multiplication of phenomena that emanated, like space, from a center. When a cycle—say, a dynastic cycle—ended, its separation from the preceding one needed to be rigorously defined, but the revolved cycle must not come to a bad end. Like the Yucatec Maya, who could not accept Christianity from the Spaniards until the right time, the scions of a fallen dynasty often were endowed with an enclosed estate wherein they were obligated to continue the harmonious completion of their microcosmic cycle. Time-past reverberated in time-present, and the fully revolved cycle became artificially preserved, its ghost unable to interfere with the harmonious development of the contemporary cycle in progress.

The Chinese Calendar

Another manifestation of this organic, moralistic cosmogony, and the central role the emperor played within it, can be seen in the way the Chinese developed their calendars. They produced over one hundred calendars and calendar changes between the fourth century B.C. and the eighteenth century A.D., most of them dealing with the determination of the length of the month and year with ever-increasing accuracy as well as with eclipses and planetary conjunctions. Practically all of the celestial observations to which we trace the Chinese calendar are couched in a framework foreign to Western astronomy.

While the passage of time in Greek astronomy is regulated by the ecliptic, Chinese astronomy is decidedly equatorial in character. The north celestial pole, today marked by Polaris, rides high in the sky at latitude 30°N where China is situated. My discussion in the last chapter of the differences in the sky environment in temperate relative to tropical latitudes makes it clear why the Chinese favored a polar-based time system. The visible fixity of the pole and the perpetual visibility of the circumpolar stars (see figure 8.4) were suggestive natural symbols for the unflagging, eternal power of heaven. (The polar pivot held the same fascination for the Egyptians and the Chumash Indians of California, who called the polar region the Sky Coyote, a guardian deity of the stable condition of the earth and its people.)

In their celestial coordinate system, the microcosm was a mirror reflection of the macrocosm. Thus, just as the bureaucratic agrarian state pivoted about the Son of Heaven, so, too, stars revolved around the fixed celestial pole. Lines that marked the hours radiated from the pole, just as the power of the Son of Heaven fanned out in all directions from his central city into each of the four quarters of his empire.

The emperor, thought to be the terrestrial complementary half of the celestial pole, built his city with its main axis aligned along the precise north-south direction; thus did he bring humans' most important earthly work into perfect symmetry and harmony with the sky (figure 9.2).

In much the same way one would quarter an apple using the core as the axis of symmetry, the sky was split at the pole into four segments, or "palaces." In accord with the sacred fiveness of Chinese cosmology, the circumpolar region acquired special importance as a fifth zone because it surrounded the imperial pole star. Like the Central Mexican world model in figure 6.6D, each of the cardinal directions had its symbolic bird, color, and season.

At right angles to the pole lies the celestial equator; one can think of it as an extension of the earth's equator onto the sky and the principal circle by which we measure the day. The Chinese divided this equatorial band into twenty-eight star groups, or "lunar mansions":

> The moon in her waxing and waning is never at fault
> Her 28 stewards escort her and never go straying
> Here at last is a trustworthy mirror on earth
> To show us the skies never-hasting and never-delaying.[4]

This inscription, carved by an emperor on his armillary sphere, a bronze-iron working spherical model of the sky with its principal circles marked out, was written down about A.D. 750; it reflects that everpresent oriental preoccupation with keeping careful track of celestial motion.

Another inscription on a similar instrument from approximately the same period typifies the prophetic side of Chinese timekeeping and harbors as well the theme of the balance of natural forces affecting a society that never once seems to have harbored a thought about stripping the spiritual element from chronology as we have done in the Western world:

> This metal mirror has the virtue of the Evening Star
> And the essence of the White Tiger of the West,
> The natural elements of Yin and Yang are present in it
> And the mysterious spirituality of mountains and rivers.

The passage goes on to name the parts of the instrument, and closes with a kind of prayer-prophecy:

> Let none of the myriad things withhold their reflection from it
> Whoever possesses this mirror and treasures it
> Will meet with good fortune and achieve exalted rank.[5]

Keeping Track of Time

Dividing the sky into 28 parts, like segments of an orange, easily enabled Chinese astronomers to keep track of the days of the lunar month, as the first of the preceding two passages suggests. Since it takes 27⅓ days to make a complete circuit among the stars, the moon would change its position by one lunar mansion per day. But the system also gives the observer a ready reference to the solar position among the stars, so that Chinese timekeepers also could determine the time of the year as well. To do so, the careful sky watcher would need to make a nighttime observation of the passage by the meridian (the north-south line that passes overhead) of any of the perpetually visible circumpolar stars. Knowing the equatorial constellations that lay along the extension of a line from the celestial pole through the given star group, and knowing that the full moon would always lie in the mansion directly opposite the one housing the sun, a timekeeper could thus keep track of the position of the sun on their equatorially based zodiac.

Technology is one element of Chinese time reckoning that particularly fascinates Westerners, mainly because many of the Chinese physical celestial models antedate our own. They used the gnomonic sundial as early as 1500 B.C. (a full thousand years before the Greeks). Three thousand years later, though the basic design, like much Chinese astronomical instrumentation, remained unchanged, they developed sundial clocks of notable proportions (figure 9.2). They also kept time by water clocks. Figure 3.2C shows a thirteenth-century European water clock whose design was likely influenced by the Chinese. Their armillary spheres are even more complex: some were celestial analogue devices like the ancient Greek simulacra, powered by waterwheels and designed to move precisely in tune with the sky above. By the eighth century A.D., they even had invented and employed a clock escapement.

In building such immense time-measuring machines and keeping meticulous records of the observations made with them, the goal of astronomy in the Chinese state was to remove the ominous threat that accompanies any unforeseen event—to bureaucratize the heavens. In practical astronomical terms, this goal translates into the replacement of any uncertainty inherent in the sky with a set of mathematically computable cycles.

The linear nature of Chinese bureaucratic timekeeping is the

FIGURE 9.2 Sundial on front court of Hall of Supreme Harmony, Imperial Palace, Beijing (compare figure 3.2A and B). SOURCE: Photo courtesy of Robin Rector Krupp.

product of a historically minded people very much like ourselves. Yet these people were also concerned with temporal events that repeat themselves. Chinese cyclic time had a very early origin that was likely astral, though we cannot prove it. The basic time unit, like the Babylonian system, is based upon multiples of the number 6, the largest definable unit being a 60-year cycle composed of 10 written characters of one set, known as the heavenly stem, that were joined to 12 characters of another set, called the 12 earthly branches. Since 60 is the lowest common multiple of 10 and 12, no consecutively paired set of characters will repeat until 60 years have elapsed. This principle of commensuration is quite like the harmonic pairing of the cycles of 13 and 20 days of the Maya *tzolkin* to yield 260 uniquely definable date pairs.

The 12 earthly branches, or *ti-chih*, like the 28 lunar houses, employ yet another way of dividing up the equatorial zodiac, which is familiar to us in the naming of Chinese New Year. Each branch bears the name of an animal: Rat, Ox, Tiger, Hare, Dragon, Snake, Horse,

Sheep, Monkey, Fowl, Dog, and Pig. These 12 assigned names resemble the Maya year bearers in their 52-year cycle, except that they circulate five times in each 60-year cycle instead of four times in 52. Reflecting the principle of complementarity, six of the animals are wild and six are domestic. Each belongs to either the yang (active, wild, male) element or the yin (passive, domestic, female); for example, the obedient dog who guards the house at night, and the slow pig who bends his eyes earthward, partake of yin, while the yang element dominates in the swift horse and terrible tiger.

Thus, the flow of time through the year passes in waves from one phase to the opposite and back again, the totality always achieving perfect balance. No one really knows where this 60-year cycle originated, but astronomical roots have been suggested. We know that the 60-year period is a perfect fit to the period of conjunction of Jupiter and Saturn, the two slowest-moving planets. Actually close conjunctions of Jupiter and Saturn happen every 20 years, but the third of each set takes place in the same constellation of the zodiac. These constitute the trigon of conjunctions, also recognized and developed in Western astronomy.

The ancient Chinese watched Jupiter as avidly as the Maya tracked Venus. As a result of their early attention to a moon-based calendar and the importance of the number 12 in the lunar zodiacal cycle, they may have become fixed upon the twelveness inherent in Jovian cycles. They related the 12-year Jupiter cycle to the 12 earthly branches, each station of the zodiac corresponding to one of the feudal states. Thus, by observing the position of Jupiter in the zodiac, the astrologer acquired a prediction or omen concerning the political turf in whose celestial station the planet resided.

About the same length as the Mesoamerican Calendar Round of 52 years that unites the 260- and 365-day cycles, every part of the Chinese 60-year cycle likewise can be experienced perhaps but once in a lifetime. Coincidentally, the Dogon people of Africa also measure a 60-year time base; the three 20-year stages of life that comprise it were said to represent the phases of life. At twenty, one attains adulthood; at forty, elderhood; and at sixty, seniority.

The Lunar Month and Other Cycles

Like most other cultures of the non-Western world, the lunar month, and not the seasonal year, served as the earliest short-term baseline in Chinese calendars. Months of 29 and 30 days alternated in groups of 12 to make up a 354-day year. But the battle of year-month commensuration was fought in the East as well as in the West; and, as in the Mediterranean, the Chinese shifted from the lunar to the solar year base when they devised their own month intercalation scheme. At first, they simply added months when needed; but then they adopted a formal scheme. Some sinologists have argued that the discoveries of the 76- and 19-year intercalation cycles, which I discussed in respect to the Roman calendar, were made independently of contact with the West. The 19-year cycle may have been used by the Chinese well before the Greek astronomer Meton's recognition of it in 432 B.C.

As the year base gained a foothold in the calendar, the Chinese proceeded to determine its length with greater precision. Such careful observers were their astronomers that, by the twelfth century, they had recognized minute short-period variations in the length of the year that were not known in the West until well after the Renaissance. Each succeeding emperor acquired the appetite for the quest for precision as a way of outdoing his predecessor. Each sought to become a little more familiar with subtle hints about the future course of the state that could be extracted by peeking around nature's corner.

The Chinese invented other ways of representing time cyclically. To monitor changes in the agricultural year, they split the year into 24 two-week periods of 15 days, and gave them names like Corn Rain (about 20 April, when it was said to be time to sow), Ripening Grain (about 21 May, when winter corn sown the previous autumn begins to fill out its ears), and small heat (about 7 July, which usually falls in one of the heat-wave periods). These divisions were said to constitute the Joints and Breaths of the year cycle.

This concept of jointed time has a remarkable parallel in the contemporary East Indonesian community of Kedang. The British ethnologist R. H. Barnes informs us that the terms *laru* and *wog*, which refer to interval and event, are named, respectively, after the long sections between joints in a human body or piece of bamboo and the joints themselves.[6]

The joints become the critical moments of transformation, the balance points between alternating intervals of tension and relaxation, when nature's forces inject themselves into the flow of time. These contact points between long periods of duration are attended by elaborate rites and festivals.

The Chinese also had a 10-day "week" cycle at least by the first century B.C., probably even earlier; but like our unbroken interval of 7 days, its origin is obscure, though it does not seem to be based on any recognizable celestial period.

Inevitably, societies organized at the state level always seem to build small cycles into larger ones. The Chinese wove the fabric of recurrent time well beyond the 60-year Jupiter-Saturn conjunction period. For example, the 135-month period, the so-called Triple Concordance, brought together the Jupiter period, the seasonal year, and the lunar-eclipse cycle.

But their longer time cycles seem to consist of a curious alternation of astronomically based intervals and pure numbers. In one eleventh-century version of a long-term calendar, an Epoch Cycle is said to amount to 129,000 years expressed as $12 \times 30 \times 12 \times 30$; it is made up of 12 Conjunction Cycles, each possessing its own 30 cycles, each of these cycles being 12 generations long, and each generation 30 years long. The 12 and 30 alternation might have originated in the 12 hours in the day cycle and the 30 days of the monthly round. These long ages are said to be governed by alternating conditions of chaos and differentiation, one following continuously and endlessly upon the other.

The multiplication of large into still larger and more abstract, complex cycles may have been an influence of Indian Buddhism, at least by 200 B.C. Their "day in the life of Brahma" (4,320,000 of our years) was the basic temporal unit of currency used in meting out the *yugas* or cosmic-age cycles of worldly existence. Each of these world ages was tied to a virtue, and each alternated with equivalent periods of darkness in degenerating durations in the ratio of the pure numbers 4:3:2:1. Thus a *krtayuga* marks 4,800 divine years of self-discipline; a *tretayuga*, 3,600 divine years of wisdom; a *dvaparayuga*, 2,400 divine years of sacrifice; and a *kaliyuga*, 1,200 divine years of generosity.

These four world ages or epochs taken together constitute 12,000 divine years, or a *mahayuga*, each made up of 360 days in the life of Brahma. These intervals, imperceptibly vast, make no real sense to us. They are intended to span only the domain of the gods. Upon their completion, the world is dissolved, reabsorbed, molded, and reconstituted. Cosmic creations and destructions occur in endless succession. There is no first cause to plague the Hindu. Brahma goes on and on forever alternating between states of being and nonbeing.

Borrowed or not, the Chinese habit of thinking in cycles transcends both time and cosmos. It spills over into the water cycle in meteorology and reflects the circulation of bodily fluids in humans and animals in Chinese medicine. Medieval Chinese physiologists were rather more circulation-minded than their Western counterparts. They believed that blood flowed in vessels while "pneuma" circulated in "tracts" or interstitial tissues. The flow, like the movement of time, was continuous and uninterrupted; it had neither beginning nor end. Since each organ contributed its specific influence to these fluids, in effect it had contact with and influence upon every other organ in the body. Thus, Chinese medicine is geared toward more holistic, systemic qualities than that which developed in the West.

Accepting nature as it is and finding the hidden complementary principle in all things is characteristically Chinese. Whether the naming of a house of the zodiac, the length of a time cycle, or even the reason for the six-sided shape of a snow crystal, this principle always rises to the surface. To paraphrase one naturalist-philosopher of the twelfth century: "Six is a yin number. It is generated from earth and is the perfected number of water and because snow is water condensed into flowers, these are always six-pointed, like the crystals of gypsum that originate in the earth. Everything comes from the numbers inherent in nature."[7]

For the commonsensical West, this kind of explanation comes across as metaphysical, mystical, and without foundation. We do not see the connection between snow and gypsum, and surely we cannot entertain the notion of personified minerals in any explanatory theory of nature. Physical objects simply do not possess gender. The sun is a ball of hydrogen and helium: there is nothing male about it. And the moon is a solid subplanetary body with mountains and craters: it is not a female. A recognizable connection between snow and gypsum in Western meteorology does, however, offer us an opportunity to glimpse what the Chinese naturalist may have been getting at. Gypsum, precisely because of its crystal shape, turns out to be a highly effective ice-nucleating agent used in cloud seeding, and particles of clay from the earth's surface are known to be one of the most likely condensates that cause precipitation.

The Chinese culture has a tradition of both kinds of basic temporal models: cyclic and linear. Chinese historical-mindedness was at least as obsessive as the Judeo-Christian world. Like our ancestors, the Chinese also tried to fasten one-way, irreversible time to events of universal significance, such as the unification of the empire in 220 B.C. by Emperor Chhin Shih Huang or the birth of Confucius in 551 B.C.

Enframed by open-ended linear time are the eternal returns of cyclic time—possibly an influence from India. The cycles are replete with the same complementary dualism and tension between polar opposites we have encountered elsewhere. They tell of how historical events, timed by planetary conjunctions, eclipses, and sometimes large intervals dictated by abstract number alone, repeat themselves. For the ancient Chinese, Nature was destined to control both kinds of time, while we humans must learn our limits and not combat her. Still, we are allowed to pry open a tiny crack in the celestial compartment housing nature's secrets, if only to fit our human concerns into the temporal framework ordained by heaven.

PART IV

A World of Time

Enigma of the Hour. Giorgio De Chirico. Milan, Private Collection, Marburg/ Art Resource, New York.

10

BUILDING ON THE BASIC RHYTHMS

> **I do not want to call an astronomical theory scientific until it gives us control over all the irregularities within each period and thus frees us from constant consultation of observational records.**
>
> —ASGER AABOE

AT LEAST the process of acquiring information about the nature of time starts out the same: it comes from human experience. The seventeenth-century British philosopher John Locke likened the human mind to an empty blackboard, a *tabula rasa* on which experience is written.[1] (Today we might substitute the more familiar computer screen.) Human sensation is the contact point between chalk and blackboard. Gradually our slate accumulates perceptions of sounds, tastes, and smells—stimuli that fly across space into our sharply tuned receivers: we're all ears.

How we decipher the writing on the blackboard—how we interpret what it means—is another matter. We think, we reflect, we form ideas. We are forever dipping back into our reservoir of acquired experience, recollecting the information we have chosen to make note of, as we try to cast it in some order that makes sense of our particular circumstances and for our particular aims. Time begins in human consciousness with the determination of a sense of order. As the philosopher-physicist Jacob Bronowski has remarked:

> For order does not display itself of itself; if it can be said to be there at all, it is not there for the mere looking. There is no way of pointing a finger or a camera at it; order must be discovered, and in a deep sense, it must be created. What we see, as we see it, is mere disorder.[2]

Even if this sounds simple, how can we come to know an idea about time's order that resides in somebody else's brain? Because we speak the same language and come from the same culture, if I am successful at expressing myself, by using words and pictures, by giving examples, by designing metaphors and models you can relate to, then I might be able to set off a ringing alarm of familiarity in your head. Consequently you might be able to get an inkling of what is rattling around in mine.

Concrete human expression is the raw material out of which we build up our impressions of what made our predecessors tick. Their material remains become a testament. They have left us mementos: books, carved bones, standing stones, buildings, sculpture, and paintings—which we eagerly photograph and measure, or collect and cram into boxes and museum cases. The stuff they once possessed gives legitimacy to our present thoughts about them; tangible expressions of human knowledge are all we have to go on.

The process of expressing our thoughts in concrete form may have begun when humans first started counting on their fingers to divvy up the kill after a day's hunting, to bundle together the sequence of days and nights—all the intricate components of the harmonic sequence of events that makes up the biorhythmic and astrorhythmic cycles I have discussed. Each beat in a given time sequence came to be identified with one of the human appendages. Native tribes of North America witnessed and identified each day of the month with a part of the body. Day 1, the first crescent, was the baby finger of the right hand; day 2, the next finger; and so on. After the thumb (the fifth day), the count passed to the wrist, to the bone between wrist and elbow, then to the elbow. The lunar count waxed upward, going from joint to bone over the shoulder to the head, which represented the fourteenth day or full moon; then it waned down the left side of the body, each named body part corresponding to its lunar likeness in reverse.

Time counting by the body is not just a practical convenience or a mnemonic, like tying a string around your finger to remember something. It is part of a process of creating meaning by the act of associating seemingly different objects with one another. When I move over the top of my body from fingertip to fingertip, I display and express a hidden likeness between the makeup of my body and the composition of time; for a moment the tips of the baby fingers of my extended arms are transformed into opposing crescent moons, and the flow of the count up one side of my body and down the other becomes a manifestation of the reversible phase cycle. My head, like the full moon, is the only singular essence in the chain. Similarly, the flow of blood among the organs becomes a body metaphor, one that can be likened to the passage of the moon around the zodiac, each of its component constellations being associated with one of the diverse bodily parts (figure 10.1).

Recurring Cycles

This way of expressing what it means to be human by linking the parts of the universe within to those that lie outside of ourselves seems fundamental. Yet as we have seen, time is not conceived in the same way by all human minds, nor is it reckoned by all peoples of the world in precisely the same manner, though at the most general level, there are certainly parallels. In many parts of the world both past and present, for example, there is an overwhelming tendency to think of time in the form of recurring cycles, a continuous sequence of events with neither beginning nor end, the past forever repeating itself.

All around us, as I discussed in chapter 1, nature broadcasts its rhythm—a decidedly cyclic rhythm. We do not know how the living world caught its rhythmic sense: whether the beat of life is driven by a ticking from without rather than a vibration from within. But we do know that all of us in the community of life, from oysters to emperors, mimic celestial and seasonal movements and changes. To me it does not really matter where to draw the line that divides who behaves pensively and reflectively as opposed to instinctually and innately, be it gorilla or lemur—maybe even my cat. (Since I have lived with her for fifteen years, I know the difference between instinct and calculated mischief!)

The domination of cyclic time in human thought and action is the result of a conditioning process we have all been through because we live in an environment filled with interminable repetitive alternations—day and night, the seasons, the phases of the moon.

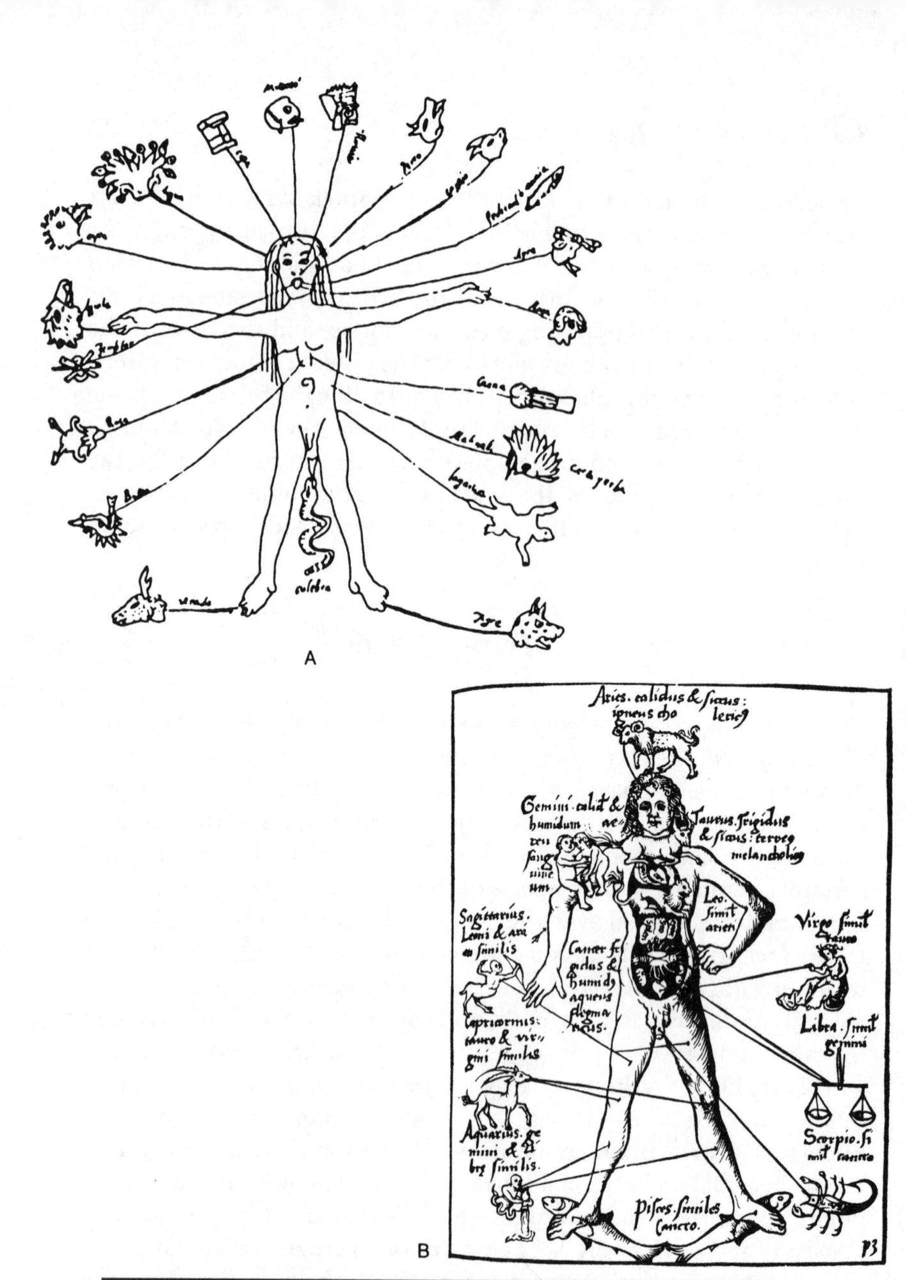

FIGURE 10.1 Time's clock in the parts of the body. Reminiscent of the biorhythms of chapter 1, these diagrams, (*A*) from Central Mexico and (*B*) from Central Europe, depict the association between various body parts and time units. For other cross-cultural similarities in the graphic expression of the meaning of time, see figure 4.2. (*A*) Twenty day signs in the Mexican Calendar from the Codex Vaticano Latino, pl. LXXIII in A. Lopez Austin, *The Human Body and Ideology*, vol. I (Salt Lake City, Utah: University of Utah Press, 1988), p. 348. Copyright © 1988 by the University of Utah Press. (*B*) Organs, limbs, and joints controlled by the zodiac. SOURCE: G. Reisch, *Margarita Philosophica*, a late medieval encyclopedia (Freiburg, 1503). The Huntington Library, San Marino, Calif.

Every visceral flip-flop in the biological world marches in time with these metronomic beats: a plant raises and lowers its leaves, a groundhog mates and hibernates, corals and clams lay down growth ridges. If we all really believe our insides are part of the outside world, then isn't it normal for us to develop cyclic forms of expression of human action and interaction? And isn't it natural, by extension and extrapolation, to suppose other cycles exist that tie into the orderly framework of our world, cycles simply too long to be easily experienced—life and death, the royal succession of kings, the occurrence of floods, earthquakes and volcanic eruptions, destructions and re-creations of the entire universe? Common sense is the logic of cyclic time reckoning.

The cosmogonic creation myths in the *Popul Vuh* and the Bible are not mystical spinoffs of overimaginative, primitive fantasizers bent on anthropomorphizing the universe. The myths I have talked about can be understood as expressions of some of the basic ways our ancient ancestors thought about time. There is nothing illusory in their celebration of the harmony and rhythm we can all feel in our bones, nothing deceptive about back-pedaling cyclic time into the domain of the gods.

It takes many cycles to make a god—beyond a lifetime, several generations, centuries. People begin to draw the line between human and suprahuman worlds when they feel crushed by a time dimension they cannot really perceive—a stretch so vast that they are compelled to gear it to some form like themselves that resides outside of time as we live it. World ages, mega-units of time, become the boundary line where the multiplication of cycle upon cycle, *baktun* upon *katun*, *mahayuga* upon *yuga* become less natural and more abstract, less environmentally related and more purely numerical. In old Western European cultures, the 26,000-year cycle of the precession of the equinoxes emerges as one of nature's long-distance signposts on the cyclic road of time.

The wheels of time roll on level ground and uphill as well as down. They can be perceived as static, like those of ancient Hinduism; progressive, as revealed in the Maya and Inca origin cycles; or even degenerative, as the pre-Classical Greeks interpreted the world and—indeed, the way modern cosmologists still tend to view it today—as a system tending toward calculated disorder. Still, modern cosmology does have something in common with the Maya and Aztec views of creation and re-creation, for we retain vestiges of this cyclic view of time. Many cosmologists continue to believe that previous universes may have existed before the Big Bang and that other universes may yet rise out of the ashes of our own.

Our modern scientific creation myth seems not so out of step with many of the others. Modern astrophysics places the beginning of time approximately fifteen billion years in the past—but a few days in the life of Buddha and only a thousand times the length of the longest time units in the Maya Dresden Codex. Like the creation stories of other cultures, ours began with a drastic transformation—a great separation of one form into many, followed by a number of rapid shuffles in the states of matter and energy—all of it happening in but a few seconds of time. The early action was played out in a relatively small sector of the domain of space and time—a domain of colossal temperature and near-infinite density—existence in its most abstract, mathematically extrapolated form. Creation's place and time is both forever impenetrable by our probing scientific instruments and incapable of revealing any empirical evidence about itself, at least within a time radius of 10^{-43} seconds after the creation event, at which time the laws of physics we know today first became applicable. Opponents of this view, some of them retreaded outcasts of the largely discredited steady-state theory of cosmology, see the Big Bang as a local effect, a tiny ripple in an infinite, eternal space-time manifold. Some visionaries have hypothesized a space-time manifold perforated by countless microcosmic creation points.

The numbers are staggering and the scales awesome, but the basic themes of our creation myth—cataclysmic transition, the separation of entities, and the rapid-fire alternation between states of turbulence and quiescence, between anxiety and relaxation—are not terribly different from the series of creation battles in *Enuma Elish*, from the sequence of violent changes that brought about the emergence of the Inca Viracocha's dominant state of being, even from the flashy, hierarchical creation by word of mouth in Genesis. The flow of the rhythm is the same.

Stripped of its detail and attendant mathematical jargon, however, there is one cardinal difference in the creation myth of the twenty-first century with respect to all those of the past: it is purposeless. We humans are totally left out of it (recall the epigraph to chapter 4). There is no body of Pwan Ku to decay and give birth to us, nor does Yahweh speak to us of human direction, action, and purpose. Whether the galaxies expand apart infinitely leaving us in our own dark little corner of space-time, or whether we are caught in the instantaneous crush of an impending collapse of the universal fireball, the cosmic myth told by modern science offers humanity as indifferent an ending as beginning. We emerge as the great nonparticipants who have no influence on the outcome. There is no Tonatiuh or *huaca* for us to feed, no year and day bearers we must help to carry the burden of time. We have not the slightest cosmic duty to perform. We pay a dear price for the freedom we gained in our struggle for cosmic independence when we willfully separated consciousness from its god-bound fetters. And we are only just beginning to deal with the consequences of venturing outside a participatory universe.

Modern cosmology also has managed to iron out the segment of the time circle on which we run—to put off the future so far that it matters little to think about, much less to act upon it. Besides, we have enough short-period kinks in our time machine to contemplate, such as the depletion of the ozone layer, global climatic change, and colliding tectonic plates.

Cyclic time dominates, but it is not all the same. It comes in different flavors: dream time and agri-time, seasonal time and mythic time. Hesiod had both Works *and* Days. No one thinks past and future cycles are exactly the same, for none of our days is truly identical to another, and one winter can differ markedly from the next.

Small cycles imitate big cycles. Contemporary Maya say the day is like the year in each of its parts. The *Popol Vuh* follows cyclic creation, the sowing and the dawning, by paralleling the cycle of Venus. And in the ancient version of our calendar, the twelve days of Christmas signal what will come to pass in the forthcoming twelve lunar months. Pick up any farmer's almanac and discover next year's weather forecast. Micro-time structures macro-time.

Despite the danger of model building (once you make a model, you can be tempted to force-fit everything into it), we nevertheless have tried to fashion various mental loops and twists that incorporate all of these complex properties of cyclic time. We clipped our circular wire and made it into a coiled spring that always returns us to the same side of our circular course but places us slightly higher up as we travel around it.

The Linearization of Time

But how does such a model incorporate the Maya understanding of the past, not only as the prologue to the present but also the prophecy for the future—time past and time future all rolled into one, as the historian Nancy Farriss has characterized it?[3] And how do we devise a cyclic model so that it emphasizes the importance of transition in every reckoning system I have discussed: the signal event of the dawning of Venus that follows the sowing in the Maya calendar, the drastic shifts in Nuer behavior occurring in that sharp borderland between wet and dry seasons in eco-time, or the Chinese and Balinese joints in time? Major changes are always accompanied by festivals, the rites of passage that break up the monotonous flow of ordinary time, the way our weekend calls an end to the duration of the workday week.

Perhaps the best model for time in the modern world is a series of interconnected hoops rolling up a great hill of progress. While we live cyclically from sleep to wakefulness, day to day and week to week, in the long run our existence seems to be a string of endless *nows*, a sequence of completed successive stages, each unique and nonrecurrent. Our language of time is filled with terms and ideas that denote linear succession. We use lines of evidence to find out who committed a crime. We follow a line of thought, bridge a gap, draw a conclusion, take a sequence of courses in school. Our means have ends; we set a career path, pursue our goals, look forward to a new millennium, some of us to a second coming. Effect follows cause, response follows stimulus, and reward comes from hard work. We glide forever upward on our endless time line. Our evolutionary tree sprouts skyward from ages past into the present; its branches grow upward and outward into the future. Our chain of events is said to progress and develop. TV newspeople continually update us, and we are forever trying to catch up with what's happening in the sequence.

Modern Westerners stand in stark contrast to those Trobriand gardeners whose language possesses no tenses, no way of distinguishing by the use of words between past and present. One ethnologist tells us that they despise directive behavior; that they perceive reality atemporally, in terms of fixed patterned wholes.[4] A tree is supported by its trunk, a house by posts, an incantation by its opening line, a hunting expedition by its leader. In each case, the latter term has the same meaning and is identified by the same word: *u'ula*. Pattern is the source of truth in this culture. Linear connections, causality, and sequentiality are no concerns of theirs. They place no value on the future as we do when we use the word *progress* to describe which way we shall go to get to it. For them, the present is not the road to the future, and the future is neither good nor bad, better nor worse than the past. Even their means are not climactic. There is no "just dessert" to look forward to. Instead, the relish, the most savory part of the dinner, is consumed with the rest of the staple food. The persistent sameness that seems so irksome to us is perfectly normal to them.

The linearization of time that dominates Western thought is often regarded as a property unique to our culture. But I have shown that other highly organized, bureaucratic cultures, like the Chinese and the Maya, also linearize time to some degree. I do not mean, however, that linear time is a property of advanced civilized societies that have real history, while the cyclic perception of eternal returns is conceived exclusively in primitive or savage minds. Anthropologists also have learned that advancement is a value-laden term. While we might say that Spanish technology was more advanced than Aztec, we could also say the Nuer system of kinship is more advanced than ours. The anthropologist Johannes Fabian has suggested that we invent such dichotomies as a way of distancing the unfamiliar.[5] We magnify the differences between us and them and append human values to each side. Even after five hundred years of exploring new lands and strange people, our Western minds continue to resist the essential humanness in the Other that we see in ourselves. Yet, paradoxically, we leapfrog backward over the present-other and search the pages of prehistory for people like ourselves in the Stonehenges of the past, people upon whose shoulders we might stand. Are they so far away from our own space and time that they can pose us no threat?

Our obsession with linear progression conditions us to look downward to past time as well as outward, away from ourselves in space, at other contemporary cultures. We label them all "traditional" as if to imply that they are all the same—all static, unformed, undeveloped, not yet having a history of their own.

One historian has said of the early Greeks that their astronomical knowledge was primitive, "a farmer's or shepherd's astronomy [which one] should not be at all surprised at finding in any settled community, whether literate or not"; it lay below the advanced or scientific level.[6] And another characterizes most foreign calendars in general as laden with superstition and unemancipated from the fetters of the religious cult.[7]

Superstition is another value-laden term that gives us license to pass judgment upon a culture without ever examining or experiencing it. Literally, it means "to stand in awe and amazement"; but Webster's also describes it as a "system of beliefs by which religious veneration or regard is shown toward objects which *deserve none*," or "the assignment of such a degree or such a kind of veneration or regard toward an object, as such *does not deserve*," and "a faith or article of faith based on insufficient evidence [italics mine]."[8] When we use the word *faith* to characterize a belief held by some other culture, we take it as an article of our *own* faith that the evidence for such a statement is insufficient, and therefore that the belief is not deserving. Yet we reach this conclusion not by examining any evidence but, instead, by invoking the principle that there can be no causal connection between two such entities. The connection, therefore, makes no sense under the only valid set of terms by which *we* understand the world.

Controlling the Rhythms

By comparing our time with theirs, we have also discovered that whether a civilization develops a long-term methodology for categorizing events depends more upon who is in charge of time and what purpose it will serve, than upon any measure of the native intelligence (whatever that is?) of time's user. And while simple basic categories like *cyclic* and *linear* are useful concepts, the way people invent and use time also has much to do with the control of activities in the society in which they live. Man-made year cycles are fastened to the agrarian economy at points of high tension. They begin and end when consumer goods are most readily available and in their greatest abundance. The farmer's crisis always occurs at planting time and is never exactly the same. Neither is the anxious tension when harvesters measure out the store of grain to feed the populace, nor the sigh of relief when they finally can assure themselves there will be enough for everyone. Festivals for invoking or paying debt to the gods are positioned strategically in relation to periods of work.

Business calendars, social calendars, and religious calendars—all are components of time in an organized society, each one serving different interests. In state and imperial societies like Rome and Cuzco, bureaucratic rulerships worked toward merging cycles and schedules into a single, all-encompassing calendar, one capable of addressing all interests.

In our case, the cultural taproot of calendrical organization grew out of economic determinism and materialism, though, as we have seen, the Judeo-Christian religion played a major role. Babylonian counting and timekeeping began with an economic motive, and the idea that time is money grew out of medieval mercantilism. Today the rigid control of human time is still powered largely by the business of making money in a highly industrialized, technological world.

When speaking of control, we usually think of empire; but cultural control has both material and ideological aspects. Territorially, it can mean direct subjugation by tribute and taxation, the way the Aztecs dominated the basin of Lake Texcoco; but it also can imply stylistic and ideological influence without direct occupation or intervention, the way ancient Teotihuacan captivated the interest of parts of the Maya world. As cultures expand, they become more segmented socially, authority becomes more centralized, agriculture intensified. Political, military, and religious institutions proliferate.

The *etic* approach to cultural evolution explains these changes by applying the Darwinian principle of adaptive change to societies immersed in a complex gearbox of interacting systems. All cultural institutions respond to a combination of biological and social needs. Human technological and subsistence systems interact with environmental systems to generate all of society's institutions.

Only the ruling powers can introduce new festivals and rites or eliminate the old ones. To control society, they must cultivate roots with the past. The further back they trace their connections, the more safe and secure any act they perform or deed they commit. They are forced constantly to write and rewrite their own history; and to do it, they are compelled to devise and sanction explicit official ways of doling it out.

Since time must be controlled minute by minute, second by second, in our complicated world, we universalize it so all can march in lockstep. Just as, to survive, we develop systems of population control, crop control, and control in the distribution of wealth, so, too, we have perfected time control. Geoffrey Conrad and Arthur Demarest are among the modern anthropologists who have trouble with this functionalist approach to cultural evolution.[9] It leaves too little room for free will. Ideology only justifies; it does not promote. It becomes the device rulers invent or modify to legitimize their rulership. People's values and beliefs always seem to follow in the wake of developments within the sphere of economic subsistence. Perhaps we persist in placing an ancient person's stomach before the head because we have not emerged from the mental rinse cycle that followed the great wash of deterministic ideas about time and evolution let loose in the middle of the last century.

How does pure function explain the temporal extremism of the Maya written calendar or account for anti-Semitism as a major controlling element in the Gregorian reform? As Conrad and Demarest suggest, we must not think of economics and ideology as cause and effect. Rather, we need to view them together in a more holistic way if we are to have any hope of understanding cultural change.

Time control began when somebody drove a stick into the ground and began to use the varying lengths of its shadow as a means of signifying the quiescent duration that separates one event from the next. Longer periods, like days and months of the year, could be marked out by timing the heliacal risings of bright stars or by notching the phases of the moon onto a piece of a bone—as our ancestors may have done twenty thousand years ago. The events they chose were not the comets and supernovae that suddenly emerge from the celestial scenery to dazzle our modern cataclysmically oriented minds. These people were looking instead for phenomena that repeat in a dependable way and occur in the right time at the right place—the passing of the sun overhead in Java or Yucatan, the emergence of the Pleiades from behind the sun at planting time in highland Peru, the first daggers of Venus's returning morning light at the onset of the rainy season in Yucatan.

But there also lay in every Javanese and Yucatecan, if not in us, a recognition of the parallels between natural and sacral periodicity. The spiritual side of humankind molded cosmogonic myth and its enactment through ritual to re-create the creation, to bring time-past into time-present, to regenerate the world. Why bother to debate whether sacred time preceded utilitarian time or whether the two are the same or different? What emerges as a constant in each calendar I have discussed is that time captivates not as a pure fact of nature but instead as a dimension of life that ultimately can be submitted to cultural control. The material embodiment of time evident in the book, the codex, the carved door lintel, the ceremonial stela, as well as the grandfather clock—all have the effect of making past and future concrete and official. Even the oral folk tale, the Andean *quipu*, the Inca sun pillars, and the skewed axis of Uxmal's House of the Governor: all ways of expressing time, without the use of writing in the traditional sense, give structure to the calendar.

Time systems became more complex and ornate as an economy and its attending bureaucracy grew and diversified. In China and

Europe, mechanical clocks replaced sundials. We slowly began to manipulate nature's direct input into the timekeeping process for our own benefit. Intercalation was one of the first steps toward human intervention, an insertion of society's time into celestial time. Thus, we make the year complete by improving upon nature where we believe it has failed.

In a sense, the Maya did to the Venus cycle what medieval Christendom did to the sun cycle. The Venus table in the Dresden Codex tampers with time and reduces it to a cultural creation based on minor variations in nature's harmonic heartbeat which can be detected only by careful listening and close observation. In bureaucratic societies, human actors take over both nature's script writing and directing.

The modern mass production of timepieces—with their artificial hours, minutes, and seconds—symbolizes the extent of our single-minded struggle to exercise control over that ghostly mechanical entity we imagine to be jogging alongside us, as close as a shadow but uninfluenced by the way we behave. When you say you are strapped for time, perhaps you are only expressing your frustration at the way you have become enslaved to that oscillating chip you carry about on your wrist.

Human culture emerges as the great processor of time. Like the rest of the biological world, our ancestors began by sensing the orderly biorhythms of natural time—the beat of the tides, the coming of the rains, the on-and-off stroboscopic flickering of the full moon's light, the comings and goings of swallows, locusts, and the red tide. Unlike the New Haven oysters that relocated in Evanston, somewhere back in the distant past we became impatient and dissatisfied. We grabbed hold of the controls; we changed the order. We manipulated time, developed and enhanced it, processed, compressed, and packaged it into a crazy-quilt patchwork to conform to our perceived needs: greater efficiency in dividing up the day means more earning power for both the corporate head and his workers; greater precision in Olympic timing makes for a better Reebok sneaker; and strategic positioning of daylight-saving time gives us more rest and recreation, and that leads to a longer personal time line.

The sociologist Michael Young uses the terms *habit* and *custom* as ways of giving humanness to the process of cyclic entrainment, that behavioral locking-in to nature's cycles in which every segment of the living world participates.[10] Driven to conform to the monotonous strains of linear, techno-time, we do great harm to both our health and our psyche. We work both night and day, we keep our environment at constant time and temperature, and we think of the future only in terms of prolonging individual human life. Young's three-part agenda for reversing social evolution calls upon us to change our attitude toward death by emphasizing its role in a cyclic process rather than its finality, to reawaken human awareness to the seasonal cycle, and to break the metronomic gridlock of our daily round. His regimen seems to say, "let's all loosen up and raise a little real-time consciousness." While Young's consciousness-raising scheme may seem a bit naive, I applaud him for recognizing that our subjugation to the yoke of fatalism is indeed conscious and that it has the effect of removing us from any real participation in the rhythm of the universe. We desperately want to take up an instrument to play, but our ambition to conduct the whole orchestra forbids us from doing so. Evans-Pritchard envied the Nuer because they seemed content just to play along.

While the cultural filter, that great processor through which nature's rhythms pass, has produced many different versions of time, as a fifth millennium of recorded history completes its cycle all the ways of reckoning this elusive medium have slowly begun to merge toward one. Human diversity dissolves into unity as strange languages and customs become extinct. As we evolve slowly and inexorably into one world, we also move toward reckoning change and activity by a single clock. We are united by time zones, an international dateline, and a universal second. Modern secular time transcends both nation and religion, both environment and demography. Time becomes more objective, less spiritual and unemotional. The World Calendar may have failed on its last go-round; but like the metric system, it will ultimately triumph.

Why bother, then, to tell stories about circular, linear, and in-between time-reckoning systems long buried in the past? Because a spark of human desire still burns within us to reconnect ourselves to the rest of the universe, to renegotiate the dialogue about human limitations. We all seek contentment in what we do—felicity, as Aristotle called it.[11] And the common denominator underpinning all of our stories from the *Theogony* to daylight-savings time has been the deep-seated human desire to live in an orderly universe. Temporally speaking, we desire the capacity to anticipate where things are going, to relieve our anxiety by peeking around nature's corner as far as it will allow.

Paradoxically, our mechanical way of repatterning time has led to a way of knowing it that is totally divorced from the real world. We have reduced time to pure number, Aristotle's measure of all things. With a push of the button and a quick glance at a series of registers, an array of blinking numerals triggers my anxiety or allays my fears about the future, and it does so instantaneously and with Olympic accuracy.

Or, perhaps technology only masks my fears about what the future may hold. Since there is great insecurity in direct observation, we always try, as the chapter epigraph suggests,[12] to reduce empirical data to the barest minimum possible. If we bureaucratize the cosmos, we need never interact with it. We can become insensitive to it.

Such a futuristic scenario seems an ironic capstone to a pyramid whose foundation is the ancient connection between human beings and the rest of nature. It was not so long ago that we were all behaving like tide-bound oysters and barometric potatoes. The human race began as a part of the harmony of nature: we were written into the score. We accepted our role and responded to nature's sheet music, sometimes being surprised by it but always going with its flow. What changed it all may have taken place in that dark cave in France over a thousand generations ago when an exhausted hunter, running calloused fingers over a piece of bone on which he had long been making deliberate notches, first began to recognize—a pattern.

Score of *The Four Seasons* by Antonio Vivaldi. E. Eulenberg Ltd., London. Photo composition by Warren Wheeler.

EPILOGUE

> Every man is born, lives and dies in historic time. As he runs through the life-cycle characteristic of our species, each phase of it joins in the events of the world.
>
> —Everett Hughes

This statement[1] about the generations by the historian Everett Hughes reminds me that I am a passage between two resting points. The events in my life are shaped and given meaning by the world with which I interact. My grandfather could not have written a book any more than I could have spent my efforts tilling a field in southern Europe to feed and support a family. I came with different equipment.

And I can only speculate on the encounter between history and my grandson, for between the making of my introduction and my epilogue, he has made his initial contact point with the flow of human history. I entered time's river at a different point from either grandfather or grandson, and know that my thoughts and actions respond to different conditions. The voyage I take today may seem less meaningful to navigators downstream from me, where the shoals may be shallower, the rapids more dangerous, the flow more rapid. Because I cannot see around the tortuous bends in the stream, I can never know how those navigators will interpret my ship's log should they ever pick it up and try to decipher it. But the good sailor widens his field of vision by climbing up into the crow's nest and carefully scanning the horizon in all directions. Then, by using a combination of the record in his journal together with what he sees, he sorts out the myriad possibilities and chooses the best course.

If anyone in my bend of the river is helped by reading my log, then I am gratified.

NOTES

Introduction: Our Time—and Theirs

1. *Webster's New Twentieth Century Dictionary of the English Language Unabridged* (New York: Standard Reference Works Publishing, 1956), p. 1793.
2. Ibid. This definition is attributed to the physicist James Clerke Maxwell.
3. C. D. Broad, "Time." *Encyclopedia of Religion & Ethics*, ed. J. Hastings, vol. 12 (Edinburgh: Clark, 1921), p. 334.
4. B. Malinowski, *Argonauts of the Western Pacific* (New York: Dutton, 1961), p. 517.
5. Ibid., p. 518.

Chapter 1. The Basic Rhythms

The subject of biorhythms, particularly in lower evolutionary forms, is treated at a popular level in Ritchie Ward's *The Living Clocks* (New York: Alfred A. Knopf, 1971). Frank Brown, archproponent of the exogenous hypothesis of circadian biochronometry (the idea that time sensing is controlled from the outside), explains his view particularly well in *Biological Clocks* (Englewood, N.J.: Heath, 1962); while E. Bünning, *The Physiological Clock* (New York: Springer-Verlag, 1973) may be consulted for the other (endogenous) side of the issue. These opposing viewpoints are juxtaposed in a pair of long essays in *The Biological Clock*, edited by F. Brown, J. Hastings, and J. Palmer (New York: Academic, 1970). See also D. Saunders, *An Introduction to Biological Rhythms* (New York: John Wiley, 1977). For particular reference to circadian timing in human beings, an excellent source book is provided by M. Moore-Ede, F. Sulzman, and C. Fuller in *The Clocks That Time Us* (Cambridge, Mass.: Harvard University Press, 1982). Or, at a more popular level, see J. Campbell, *Winston Churchill's Afternoon Nap* (New York: Simon & Schuster, 1986). K. von Frisch's treatise on bees is still the classic work in its field: *The Dancing Bees* (New York: Harcourt Brace 1953).

1. See F. A. Brown, Jr. "Life's Mysterious Clocks," in R. Thruelson and J. Kobler, eds., *Adventures of the Mind* (New York: Random House, 1962), p. 159.

2. See F. A. Brown, Jr., "The Rhythmic Nature of Animals and Plants," *American Scientist* 47 (1959): 164–86.

3. K. von Frisch, *The Dancing Bees* (New York: Harcourt Brace, 1953), esp. chap. 11.

4. Ibid., p. 143.

5. J. J. de Mairan, "Observation Botanique," *Histoire de l'Académie Royale des Sciences* (Paris, 1729), p. 35.

6. M. Moore-Ede et al., *The Clocks That Time Us* (Cambridge, Mass.: Harvard University Press, 1982).

7. E. Bünning, *The Physiological Clock*, 2nd ed., revised (Berlin: Springer, 1967).

8. R. Ward, *The Living Clocks* (New York: Alfred A. Knopf, 1971), p. 330.

9. D. Saunders, *An Introduction to Biological Rhythms* (New York: John Wiley, 1977), pp. 106–7. See also Moore-Ede et al., *The Clocks*, pp. 278–84.

Chapter 2. Early Time Reckoning

Early Greek astronomy is detailed in D. R. Dick's thorough *Early Greek Astronomy to Aristotle* (London: Thames & Hudson, 1970). For a broader perspective on early Greek mythology, see J. Harrison, *Epilegomena and Themis* (Cambridge: Cambridge University Press, 1912). M. Eliade's ideas on cyclic time appear in *The Myth of the Eternal Return* (Princeton: Princeton University Press, 1954), and *The Sacred and the Profane* (New York: Harcourt Brace, 1957).

Some of the interpretations of the Works and Days calendar are explained further in A. F. Aveni and A. Ammerman, "Early Greek Astronomy in the Oral Tradition: The Search for Archeological Correlates" (paper presented at conference on "Symbol and Text: Archeology in Lettered and Unlettered Societies," Colgate University, 1983).

For an appreciation of the quantity of information that actually can be communicated strictly by oral means, and for an understanding of the formulaic and mnemonic quality of the oral mode of transmission, see F. Yates, *The Art of Memory* (Chicago: University of Chicago Press, 1966); or A. B. Lord, *The Singer of Tales* (New York: Atheneum, 1978).

For a discussion of evidence on the earliest notations about the calendar, see A. Marshack, *The Roots of Civilization* (New York: McGraw-Hill, 1972); and W. B. Murray, "Calendrical Petroglyphs of Northern Mexico," in A. F. Aveni, ed., *Archaeoastronomy in the New World* (Cambridge: Cambridge University Press, 1982). The foundations of cuneiform writing are traced in D. Schmandt-Besserat, "An Ancient Token System: The Precursor to Numerals and Writing," *Archaeology* 39 (6 [1986]): 32–39. The classic work on the modern decoding of Stonehenge is G. S. Hawkins, *Stonehenge Decoded* (New York: Delta Dell, 1965). For excellent counterpoint on the Stonehenge debate, readers should consult C. Chippindale, *Stonehenge Complete* (Ithaca: Cornell University Press, 1983).

Among the interesting material on the origins of human communication, I recommend *Human Communication: Language and Its Psychobiological Bases,* introduction by W. Wang (no editor), Readings from *Scientific American* (San Francisco: W. H. Freeman, 1982). This excellent reader consists of eighteen articles from *Scientific American* on subjects ranging from the origin of speech and writing in different theaters of human experience to the development of various languages, including Chinese and Bantu (African), and includes work on the language of bees, birds, and apes. See also I. Davidson and W. Noble, "The Archeology of Perception," *Current Anthropology* 30 (1989): 125–55; and *Toward a New Understanding of Literacy,* ed. M. Wrolstad and D. Fisher (New York: Praeger, 1985).

An encyclopedic work on the foundations of technology is volume I of *A History of Technology,* ed. C. Singer, E. Holmyard, and A. Hall (Oxford: Clarendon Press, 1954).

Finally, some ideas in this chapter on the development of literacy, writing, and the earliest records of calendrical knowledge were culled from readings of J. Goody, *Domestication of the Savage Mind* (Cambridge: Cambridge University Press, 1977); and E. R. Leach, "Rethinking Anthropology," London School of Economy Monographs in Social Anthropology, no. 22 (1971).

The quotation in the chapter epigraph is from *Hesiod,* trans. R. Lattimore, (Ann Arbor: University of Michigan Press, 1972), lines 561–64.

1. For different viewpoints on this issue, see *Human Communication: Language and Its Psychological Bases,* Readings from *Scientific American,* (San Francisco: W. H. Freeman, 1982); and D. Griffin, *Animal Thinking* (Cambridge: Harvard University Press, 1984).

2. See D. Premack, "Language in Chimpanzee?," *Science* 172 (1971): 808–22.

3. A. Lord, *The Singer of Tales* (New York: Atheneum, 1978), pp. 26–27.

4. This and subsequent quotations from *Works and Days* in this chapter are from *Hesiod,* trans. R. Lattimore (Ann Arbor: University of Michigan Press, 1972).

5. This and the following quotations from Genesis are from *The New Oxford Annotated Bible with the Apocrypha,* ed. H. May and B. Metzger (New York: Oxford University Press, 1973).

6. This and subsequent quotations from the *Theogony* in this chapter are from *The Poems of Hesiod,* ed. R. M. Frazer (Norman: University of Oklahoma Press, 1983).

7. This and subsequent quotations from *Enuma Elish* in this chapter are from *Enuma Elish* in A. Heidel, *The Babylonian Genesis* (Chicago: University of Chicago Press, 1942).

8. E. R. Leach, "Rethinking Anthropology," London School of Economics, Monographs in Social Anthropology, no. 22 (1971).

9. M. Eliade, *The Sacred and the Profane* (New York: Harcourt Brace, 1959).

10. Heidel, *Babylonian Genesis,* tablet VII, lines 132–33.

11. A Marshack, *The Roots of Civilization* (New York: McGraw-Hill, 1972).

12. R. White, "The Manipulation and Use of Burins in Incision and Notation," *Canadian Journal of Anthropology* 2 (129): 1963; and F. d'Errico, "Paleolithic Lunar Calendars: A Case of Wishful Thinking?" *Current Anthropology* 30 (117): 1989.
13. W. B. Murray, "Calendrical Petroglyphs of Northern Mexico," in A. F. Aveni, ed., *Archaeoastronomy in the New World* (Cambridge: Cambridge University Press, 1982), pp. 195–204.
14. G. S. Hawkins, *Stonehenge Decoded* (New York: Delta, 1965); original paper, "Stonehenge Decoded," *Nature* 200 (306): 1963.
15. G. S. Hawkins, "Stonehenge: A Neolithic Computer," *Nature* 202 (1964): 1258–61; F. Hoyle, *On Stonehenge* (San Francisco: W. H. Freeman, 1977), esp. chaps. 3–5.
16. Hawkins, *Stonehenge Decoded*, p. vii.
17. J. Hawkes, "God in the Machine," *Antiquity* 41 (1967): 174–80. For another critical assessment see R. J. C. Atkinson, "Moonshine on Stonehenge," *Antiquity* 40 (1966): 212–16.
18. D. Schmandt-Besserat, "An Ancient Token System, The Precursor to Numerals and Writing," *Archaeology* 39 (6 [1986]): 32–39.

Chapter 3. The Western Calendar

One of the most thorough, rigorous, detailed discussions of the calendar reform is presented in the *Encyclopædia Britannica* (New York: J. J. Little & Ives, 1910). M. Nilsson's *Primitive Time Reckoning* (Lund: Gleerup, 1920), though somewhat outdated in its analytical methodology, is nevertheless an excellent source book of information on the divisions of the calendars kept by non-Western societies. On the origin of the week, see F. H. Colson's delightful little book *The Week* (Cambridge: Cambridge University Press, 1926); and the more extensive E. Zerubavel, *The Seven Day Circle: The History and Meaning of the Week* (New York: Free Press, 1985). Information on the biological week cycle was taken from J. Campbell, *Winston Churchill's Afternoon Nap* (New York: Simon & Schuster, 1986). On time standards, see the many articles in *The Voices of Time* ed. J. T. Fraser (Amherst: University of Massachusetts Press, 1981); and on the history of clocks, the highly readable *Revolution in Time* by D. Landes (Cambridge, Mass.: Harvard Belknap, 1983). Derek de Solla Price, *Science Since Babylon* (New Haven: Yale University Press, 1975) offers insight into the general role Western scientific thinking has played in devising timekeeping systems.

1. "A Timepiece For the Tasteful: Merely $3 Million," *New York Times*, 15 March 1989.
2. D. DeSolla Price, *Science Since Babylon* (New Haven: Yale University Press, 1975), p. 53.
3. E. E. Evans-Pritchard, *The Nuer* (New York: Oxford University Press, 1969), p. 103.
4. A. Heidel, *The Babylonian Genesis* (Chicago: University of Chicago Press, 1942), tablet V, lines 15–17.

5. C. Lévi-Strauss, *The Raw and the Cooked* (New York: Harper & Row, 1969), p. 9.

6. C. W. Jones, *Bedae Opera de Temporibus* (Cambridge, Mass.: Medieval Archives of America, 1943), p. 9.

7. For an exhaustive compendium on the reform, see *Gregorian Reform of the Calendar, Proceedings of the Vatican Conference to Commemorate Its 400th Anniversary, 1582–1982,* ed. G. Coyne et al. (Vatican City: Pontificia Academia Scientarum, Specola Vaticana, 1983). For the Kepler citation, see G. Moyer, "The Gregorian Calendar," *Scientific American* 246 (5 [1982]): 147.

Chapter 4. The Year and Its Accumulation in History

A number of ideas for this chapter, particularly the portion that details the historical development of our thinking about time (the history of history, so to speak), were developed as a result of my early and sustained contact with S. Toulmin and J. Goodfield, *The Discovery of Time,* (Chicago: University of Chicago Press, 1965). I taught a survey course in the natural sciences out of this book more than twenty years ago and recently resurrected it in a seminar on comparative cosmologies. Anyone interested in how we developed our idea of Western history should read it. E. R. Dodds has written an interesting essay on the idea of progress in ancient Greece in *The Ancient Concept of Progress* (Oxford: Clarendon, 1973), which provides further background on the concept of evolutionary history.

Ideas on Western European time and the philosophy of time in general are further discussed in J. Whitrow, *The Natural Philosophy of Time* (Oxford: Clarendon, 1980); C. M. Sherover, *The Human Experience of Time* (New York: New York University Press, 1975); and the compendium *The Study of Time,* edited by J. T. Fraser et al. (New York: Springer-Verlag), in 3 volumes, beginning 1972. See also J. T. Fraser, *Time, the Familiar Stranger* (Amherst: University of Massachusetts Press, 1987); and *Genesis and Evolution of Time* (Brighton, Mass.: Harvester, 1982).

Jack Goody, *Domestication of the Savage Mind* (Cambridge: Cambridge University Press, 1977), is useful in the philosophical underpinning of my discussion about timekeeping systems; note in particular his discussion on associative thinking. Environmental time as viewed from the modern perspective of a city dweller is discussed in K. Lynch, *What Time Is This Place?* (Cambridge, Mass.: MIT Press, 1972). Michael Young, *The Metronomic Society* (Cambridge, Mass.: Harvard University Press, 1988), is another interesting treatise on modern social time.

There are several outstanding resources on time as perceived in modern everyday life. Among them, Lewis Mumford's prophetic *Technics and Civilization* (New York: Harcourt Brace, 1934), deals with the effects of the high-tech revolution at the time of its inception. *The Beer Can by the Highway* by J. A. Kouwenhoven (New York: Doubleday, 1961), and M. Lerner, *American Civilization: Life and Thought in the US Today* (New York: Simon & Schuster,

1957), offer illuminating perspectives on how modern Americans think about time as well as how they use it at the practical level.

An account of "The French Republican Calendar: A Case Study in the Sociology of Time" is given by E. Zerubavel in *The American Sociological Review* 42 (1977): 868–76; while the failed world calendar is outlined in a tidy little book by Elizabeth Achelis, one of the most ardent crusaders for a lost cause: *Of Time and the Calendar* (New York: Hermitage, 1955).

George Gamow's article "Modern Cosmology" in *Scientific American* 190 (3 [1954]):55–63, remains a standard popular work on Big Bang cosmology. In the same journal, one can find George Gale's "The Anthropic Principle," *Scientific American* 245 (1981):154–71. For greater detail, see J. Barrow and F. Tipler, *The Anthropic Cosmological Principle* (Oxford: Oxford University Press, 1988). The biological and geological inquiries into the nature of time's arrow are lucidly discussed in S. J. Gould, *Time's Arrow, Time's Cycle* (Cambridge, Mass.: Harvard University Press, 1987).

As I mention early in the chapter, the 1980s have produced a spate of popular works that profess to deal with the history of time but actually deal only with recent scientific approaches to knowledge. I can recommend as among the most lucid and effective: David Park, *The Image of Eternity* (Amherst: University of Massachusetts Press, 1980); Steven Weinberg, *The First Three Minutes* (New York: Basic Books, 1988); Steven Hawking, *A Brief History of Time* (New York: Bantam, 1988); and Heinz Pagels, *The Cosmic Code* (New York: Simon & Schuster, 1982) and *Perfect Symmetry* (New York: Simon & Schuster, 1985).

1. J. Harrison, *Epilegomena to the Study of Greek Religion and Themis, A Study of the Social Origins of Greek Religion* (New Hyde Park: University Books, 1962), p. 497.
2. M. Eliade, *The Sacred and the Profane* (New York: Harcourt Brace, 1959), p. 68. See also E. Lyle, "The Dark Days and the Light Month," *Folklore* 95 (ii [1984]):221–23; A. Rubel, "Prognosticative Calendar Systems," *American Anthropologist* 67 (1965):107–9.
3. St. Augustine, *The City of God* (New York: Fathers of the Church, 1950), book XII, chap. 14.
4. Aristotle, *Physics*, trans. H. G. Apostle (Bloomington: Indiana University Press, 1969), book IV, chap. 12, p. 221.
5. Ibid.
6. *The Confessions of St. Augustine*, trans. Edward B. Pussey (New York: Collier Books, 1961), book XI, chap. 14.
7. R. Descartes, *Principles of Philosophy*, 1644, as quoted in S. Toulmin and J. Goodfield, *The Discovery of Time* (Chicago: University of Chicago Press, 1965), pp. 80–81.
8. C. Darwin, *The Origin of Species* (1859) (New York: Avenel, 1959), p. 130.
9. Ibid., p. 133.
10. G. Gamow, *The Creation of the Universe* (New York: Viking, 1952), p. 57.
11. G. Gamow, "Evolutionary Cosmology," *Scientific American* 9 (1956):148.

12. Oscillating models are summarized in M. Rowan-Robinson, *The Cosmological Distance Ladder* (New York: W. H. Freeman, 1985), chap. 5.
13. S. Weinberg, *The First Three Minutes* (New York: Basic Books, 1988), chap. 2.
14. Augustine, *City of God*, book XXI, chap. 10.
15. J. Barrow and F. Tipler, *The Anthropic Cosmological Principle* (Oxford: Oxford University Press, 1988).
16. H. Pagels, *Perfect Symmetry* (New York: Simon & Schuster, 1985), pp. 376–77.
17. S. J. Gould, "Asymmetry of Lineages and the Direction of Evolutionary Time," *Science* 236 (1987):1437–41.
18. T. Rothman, "The Seven Arrows of Time," *Discover* 8 (2 [1987]):70.
19. L. Mumford, *Technics and Civilization* (New York: Harcourt Brace, 1934), p. 15.
20. M. Lerner, *America as a Civilization* (New York: Simon & Schuster, 1957), p. 618.
21. Weinberg, *First Three Minutes*; S. Hawking, *A Brief History of Time* (New York: Bantam, 1988).
22. J. T. Fraser, *Time, The Familiar Stranger* (Amherst: University of Massachusetts Press, 1987), p. 272.
23. E. Achelis, *Of Time and the Calendar* (New York: Hermitage, 1955), p. 27.
24. Ibid., p. 13.
25. Ibid., p. 15.
26. Ibid., pp. 81–82.

Chapter 5. Tribal Societies and Lunar-Social Time

Whether you agree with him or not, the ideas of the British anthropologist E. R. Leach are central to any comparison between Western and non-Western timekeeping systems. Above all others, he seems to be acutely aware that the problem intercalation has posed for us is viewed differently by other people of the world. See especially his "Rethinking Anthropology," London School of Economy Monographs in Social Anthropology, no. 22 (1971); "Primitive Calendars," *Oceania* 20 (1950):245–62, where the Trobriand calendar is discussed; and "Primitive Time Reckoning," in C. Singer, E. Holmyard, and A. Hall, eds., *History of Technology* (Oxford: Clarendon Press, 1954), vol. I, pp. 110–27. A cognitive psychologist's approach to Trobriand notions of time can be found in D. Lee's "Lineal and Nonlineal Codifications of Reality," *Psychosomatic Medicine* 12 (1950): 89–97. The life cycle of the Trobriand palolo worm is described in detail in D. Saunders, *An Introduction to Biological Rhythms* (New York: John Wiley, 1977).

Much of the material on the Nuer is derived from E. E. Evans-Pritchard's *The Nuer* (Oxford: Oxford University Press, 1940). Other sources of ethnographic information referred to in this chapter include S. Fabian, "Eastern Bororo Space Time: Structure, Process, and Precise Knowledge Among a Native Brazilian People" (Ph.D. dissertation, University of Illinois, 1987);

and, on the Mursi, D. Turton and C. Ruggles, "Agreeing to Disagree: The Measurement of Duration in a Southwestern Ethiopian community," *Current Anthropology* 19 (1978):585–600. Body counting in native American cultures is discussed in several articles in *Native American Mathematics*, ed. M. Closs (Austin: University of Texas Press, 1986).

On the social determination of temporal knowledge concerning time, see R. H. Barnes's excellent chapter on timekeeping among the contemporary Kedang of Indonesia in *Kedang: A Study of the Collective Thought of an Eastern Indonesian People* (Oxford: Clarendon Press, 1974). This subject is also dealt with by M. Bloch, "The Past and the Present," *Man* (n.s.) 12 (1977):278–292. For critiques of Bloch's view, see M. Bourdillon, "Knowing the World or Hiding It: A Response to Maurice Bloch," *Man* (n.s.) 13 (1979):591–99; and L. Howe, "The Sound Determination of Knowledge: Maurice Bloch and Balinese Time," *Man* (n.s.) 16 (1979):220–34.

1. E. E. Evans-Pritchard, *The Nuer* (Oxford: Oxford University Press, 1940), p. 103.
2. Ibid., p. 100.
3. For some recent studies that address this issue, see A. Weiner, *Women of Value, Men of Renown* (Austin: University of Texas Press, 1976), a re-analysis of Trobriand exchange systems, funerary and other rituals, and so on, with women as the focus; also S. Rogers "Woman's Place, A Critical Review of Anthropological Theory," *Comparative Studies in Society and History* 2 (1 [1978]):123–62.
4. S. Fabian, "Eastern Bororo Space-Time: Structure, Process and Precise Knowledge Among a Native Brazilian People" (Ph.D. dissertation, University of Illinois, 1987), p. 96.

Chapter 6. The Interlocking Calendars of the Maya

Many exciting breakthroughs have recently taken place in the study of Maya writing. One of the most significant new texts that places Maya timekeeping in a sound historical context is L. Schele and M. Miller, *The Blood of Kings, Dynasty and Ritual in Maya Art* (Fort Worth, Tex.: Kimbell Museum, 1986). Other recent contributions include the collected papers presented at several interdisciplinary conferences on the art, iconography, and dynastic history of Palenque, held at the ruins of Palenque: *The Palenque Round Table Series*, under the general editorship of Merle Greene Robertson (Austin: University of Texas Press, 1978). The paper by D. Earle and D. Snow, "The Origin of the 260-Day Calendar: The Gestation Hypothesis Reconsidered in Light of Its Uses among the Quiché-Maya," appears in this series as well. On the historical theme, see also M. Miller, *The Murals of Bonampak* (Princeton: Princeton University Press, 1986), wherein the Bonampak wall murals are presented and discussed; and L. Schele and D. Freidel, "The Maya Message: Time, Text and Image," a paper presented at the "Symposium on Art and Communication," Israel Museum, Jerusalem, 1985.

Standard works on Maya hieroglyphic writing include J. E. S. Thompson,

Maya Hieroglyphic Writing (Norman: University of Oklahoma Press, 1960), and *A Commentary on the Dresden Codex* (Philadelphia: American Philosophical Society, 1972). For those interested in the earliest archaeological records of Maya and pre-Maya calendrical symbols, I recommend Joyce Marcus's brief discussion, "The Origins of Mesoamerican Writing," *Annual Review of Anthropology* 5 (1976):35–47.

For detailed discussions of the operation of the calendar, with examples of how the calculations referred to in the present chapter actually were performed, consult A. Aveni, *Skywatchers of Ancient Mexico* (Austin: University of Texas Press, 1980). Details on the Venus calendar appear in Thompson's *Commentary* (previously cited); in A. Aveni, "Venus and the Maya," *American Scientist* 67 (1979):274–85 (where the Venus symbols and orientations in the architecture are discussed); and in F. Lounsbury, "The Base of the Venus Table of the Dresden Table and Its Significance for the Calendar Correlation Problem," in A. Aveni and G. Brotherston, eds., *Calendars in Mesoamerica and Peru, Native American Computation of Time* (Oxford: British Archaeological Reports no. S 174, 1983), pp. 1–26 (it is there that the Venus table correction scheme discussed in this chapter originally appeared).

Other articles consulted in preparation of the section on Maya numeration include M. Macri, "Phoneticism in Maya Head Variant Numerals" (thesis, University of California, 1982), where the idea of the expansion of 13 into 20 is presented; and several publications by H. Neuenswander, of the Summer Institute of Linguistics, Guatemala.

General references to Maya astronomical calculations as they pertain to interdisciplinary studies in archaeoastronomy appear in A. Aveni, ed., *Archaeoastronomy in Pre-Columbian America* (Austin: University of Texas Press, 1975); A. Aveni, ed., *Native American Astronomy* (Austin: University of Texas Press, 1977); and A. Aveni, ed., *Archaeoastronomy in the New World* (Cambridge: Cambridge University Press, 1982) (see the companion volume to the latter—D. Heggie, ed., *Archaeoastronomy in the Old World*); and finally, for interesting cross-cultural comparisons in archeoastronomical studies, see A. F. Aveni, ed., *World Archaeoastronomy* (Cambridge: Cambridge University Press, 1989); and R. Williamson, ed., *Archaeoastronomy in the Americas* (Monterey: Ballena and College Park, Md.; Center for Archaeoastronomy, 1981). On the building alignments at Uxmal and other Puuc sites, see A. F. Aveni and H. Hartung, "Maya City Planning and the Calendar," *Transactions of the American Philosophical Society* 76 (part 7 [1986]).

Among the ethnohistoric references employed in this chapter are Diego de Landa, *Relación de las Cosas de Yucatan*, edited by A. Tozzer (Cambridge: Papers of the Peabody Museum of Archaeology and Ethnology, vol. 18, 1941), originally written in the Yucatan in the 1550s; and the other post-Conquest documents contained in M. Edmonson, ed., *The Ancient Future of the Itzá, The Book of Chilam Balam of Tizimin* (Austin: University of Texas Press, 1982); and *The Codex Peréz* and *The Book of Chilam Balam of Mani*, edited by E. Craine and R. Reindorp (Norman: University of Oklahoma Press, 1979).

In my discussion of Maya time after the Spanish contact, I have drawn insight and information from several excellent current ethnological and ethnohistorical studies; among them: E. Hunt, *Transformation of the Hummingbird* (Ithaca: Cornell University Press, 1977); N. Farriss, "Remembering the Future, Anticipating the Past," *Comparative Studies in Society and History* 29 (1987):566–93; B. Tedlock, *Time and the Highland Maya* (Albuquerque: University of New Mexico Press, 1982); and G. Gossen, *Chamulas in the World of the Sun* (Cambridge: Harvard University Press, 1974). I have tried to be faithful throughout to D. Tedlock's translation of the *Popul Vuh, The Mayan Book of the Dawn of Life* (New York: Simon & Schuster, 1985).

J. Watanabe (*Man* [n.s.] 18 [1983]: 710–28) has written an excellent short article entitled "In the World of the Sun: A Cognitive Model of Mayan Cosmology," which offers a real sense of the Maya *kin* concept, as does M. León-Portilla's *Time and Reality in the Thought of the Maya* (Boston: Beacon Press, 1973).

One of the most illuminating general discussions on the first Spanish contact with mainland Mesoamerica is that of T. Todorov, *The Conquest of the Americas* (New York: Harper & Row, 1982). For other general reading on the Maya civilization, see M. Coe, *The Maya* (New York: Praeger, 1973); and G. and G. Stuart, *The Mysterious Maya* (Washington, D.C.: National Geographic Society, 1971).

1. *The Bernal Díaz Chronicles*, trans. and ed. A. Idell (New York: Doubleday, 1956; originally published 1632), p. 243.
2. D. Durán, *Book of the Gods, Rites, and the Ancient Calendar*, trans. F. Horcasitas and D. Heyden (Norman: University of Oklahoma Press, 1971; originally published *c.* 1588), p. 240.
3. J. E. S. Thompson, *The Rise and Fall of Maya Civilization* (Norman: University of Oklahoma Press, 1954), p. 162.
4. *Book of Chilam Balam of Mani*, trans. E. Craine and R. Reindorp (Norman: University of Oklahoma Press, 1979), p. 47.
5. M. Edmonson, *The Ancient Future of the Itzá* (Norman: University of Oklahoma Press, 1982), p. 32.
6. D. Earle and D. Snow, "The Origin of the 260-day Calendar: The Gestation Hypothesis Reconsidered in Light of Its Use Among the Quiché-Maya," in V. Fields, ed., *Fifth Palenque Round Table* (San Francisco: Pre-Columbian Art Research Institute, 1985), pp. 241–44.
7. J. E. S. Thompson, *A Commentary on the Dresden Codex* (Philadelphia: American Philosophical Society, 1972), pp. 75–77.
8. M. Edmonson, *Ancient Future*, p. 97.
9. D. de Landa, *Relación de las Cosas de Yucatán*, trans. and ed. A. Tozzer (Cambridge: Peabody Museum, 1941), pp. 27–28.
10. B. Tedlock, *Time and the Highland Maya* (Albuquerque: University of New Mexico Press, 1982), chap. 6.
11. Landa, *Relación*, pp. 28–29.
12. P. Velazquez, trans., *Codice Chimalpopoca (Anales de Cuauhtitlan y Leyenda de los Soles)* (Mexico: Imprenta Universitaria, 1945), appendix to vol. III. Translated in E. Seler, "Venus Period in the Picture Writings of the Borgia Group," *Bureau of American Ethnology Bulletin* 28 (1904 [Washington, D.C.: Government Printing Office]): pp. 359–60.

13. F. Lounsbury, "The Base of the Venus Table of the Dresden Codex and its Significance for the Calendar Correlation Problem," In A. Aveni and G. Brotherston, eds., *Calendars in Mesoamerica and Peru: Native American Computations of Time* (Oxford: British Archaeological Reports International Series S174, 1983), pp. 1–26.
14. J. E. S. Thompson, "Maya Astronomy," in F. R. Hodson, ed., *The Place of Astronomy in the Ancient World* (London: Philosophical Transactions of the Royal Society, 1974), A276, p. 94.
15. Juan Pío Perez, in J. L. Stephens, *Incidents of Travel in Yucatan* (New York: Dover, 1961; originally published 1841), vol. I, p. 280.
16. D. Tedlock, trans., *Popol Vuh* (New York: Simon & Schuster, 1985).
17. Ibid., pp. 197–98.
18. Landa, *Relación*, p. 37.
19. Edmonson, *Ancient Future*, pp. 8 and 6–7, respectively.

Chapter 7. The Aztecs and the Sun

There are a few reliable Spanish chroniclers (missionaries, mostly) who describe the way of life in ancient Aztec Tenochtitlan. One of the best, whom I have mentioned repeatedly, is Bernardino de Sahagun, whose authoritative 12-volume *Florentine Codex, General History of the Things of New Spain* ("Florentine" as it was rediscovered in Florence, Italy) appears in English translation by A. J. O. Anderson and C. E. Dibble (Santa Fe: School of American Research; and Salt Lake City: University of Utah, 1953–81). Another is Diego Durán, *Book of the Gods, Rites and the Ancient Calendar,* translated by F. Horcasitas and D. Heyden (Norman: University of Oklahoma Press, 1971).

Among the most expressive pictorial codices from Central Mexico are the Codex Nuttall; see the popular edition edited by A. G. Miller (New York: Dover, 1975); and the Codex Borgia (Graz, Austria: Akademische Druck-u. Verlagsanstalt, 1976). See also the Codex Magliabecchiano and the Codex Nuttall, edited by E. Boone (Berkeley: University of California Press, 1983).

A clear exposition on the use of monumental art as an expression of cosmic and social order in the Aztec capital is given in R. Townsend's monograph, "State and Cosmos in the Art of Tenochtitlan" (Washington, D.C.: Dumbarton Oaks Center for Pre-Columbian Studies, 1979). And the work of M. León-Portilla more than adequately explains the Aztec philosophy of time; see, for example, his *Aztec Thought and Culture* (Norman: University of Oklahoma Press, 1963). I have drawn extensively from the latter two foundation sources in the preparation of this chapter.

For a diagrammatic analysis of the Aztec sun stone, see R. S. Flandes's pamphlet "The Sun Stone or Aztec Calendar," originally published in 1936 and printed by Lito Seldes, Mexico City.

For general reading on the history of the Aztecs, I recommend G. Conrad and A. Demarest, *Religion and Empire* (Cambridge: Cambridge University Press, 1984); these authors also make interesting contrasts between the Aztec and Inca imperial states. See also the eminently readable books of N. Davies, such as *The Aztecs* (London: Abacus, 1977) or *The Ancient Kingdoms of Mexico* (New York: Penguin, 1982). The exciting story of the archaeological excavations in downtown Mexico City that have unearthed the magnificent caches taken from the Templo Mayor, now on display in the museum adjacent to the site, are chronicled in *The Great Temple of the Aztecs* by the chief archaeologist on the project, Eduardo Matos Moctezuma (London: Thames & Hudson, 1988); and their implications are discussed further in J. Broda, D. Carrasco, and E. Matos Moctezuma, *The Great Temple of Tenochtitlan* (Los Angeles: University of California Press, 1987).

1. *The Bernal Díaz Chronicles,* trans. and ed. A. Idell (New York: Doubleday, 1956), p. 139.
2. B. de Sahagun, *General History of the Things of New Spain,* trans., A. Anderson and C. Dibble (Santa Fe: School of American Research; and Salt Lake City: University of Utah Press, 1953), book 7, p. 25.
3. R. Townsend, "State and Cosmos in the Art of Tenochtitlan" (Washington, D.C.: Dumbarton Oaks Center for Pre-Columbian Studies, 1979), p. 28.
4. W. Wordsworth, "On the Same Occasion [that is, On Seeing the Foundation Preparing for the Erection of Rydal Chapel, Westmoreland]," in *Wordsworth: Poetical Works,* E. de Selincourt, ed. (Oxford and New York: Oxford University Press, 1969, p. 417; poem originally published, 1823).
5. Sahagun, *General History,* book 7, p. 25.
6. Cantares Mexicanos, fol. 61, in M. León-Portilla, *Aztec Thought and Culture* (Norman: University of Oklahoma Press, 1963), p. 128.
7. Sahagun, *General History,* book 2 (1981), pp. 14–15.
8. Ibid., appendix, p. 197.
9. Quoted in A. Maudslay, "A Note on the Position and Extent of the Great Temple Enclosure of Tenochtitlan," XVIII Acts of the International Congress of Americanists (London, 1912), part 2, pp. 173–75.
10. Sahagun, *General History.*

Chapter 8. The Incas and Their Orientation Calendar

The works of relatively few Andean chroniclers have been translated into English; and, worse still, many descriptions are not reliable. Among the better ones are F. Guaman Poma de Ayala; see R. Adorno's *Guaman Poma, Writing and Resistance in Colonial Peru* (Austin: University of Texas Press, 1986). The chronicle of Garcilaso de la Vega, called *Royal Commentaries of the Inca,* 2 vols. (Austin: University of Texas Press, 1966), is somewhat less trustworthy although widely available. One of the most authoritative histories of the Inca conquest is J. Hemming, *The Conquest of the Incas* (New York: Harcourt Brace, 1970).

There are several excellent ethnologies that can be examined as a way of contrasting present-day life and customs in the Andes with that pertaining to the pre-contact period. I recommend B. J. Isbell, *To Defend Ourselves: Ecology and Ritual in an Andean Village* (Austin: University of Texas Press, 1978), which deals with the village of Chuschi in southern Peru; and G. Urton, *At the Crossroads of the Earth and the Sky* (Austin: University of Texas Press, 1981), which centers on Misminay, a modern town close to Cuzco. The latter deals specifically with concepts of space, time, and cosmology.

Many of the ideas about the structure of the *ceque* system expressed in this chapter are drawn from a long series of articles by R.T. Zuidema, who has conducted research in Cuzco for the past thirty-five years. Among the most recent articles by him are "Catachillay: The Role of the Pleiades and of the Southern Cross and Alpha and Beta Centauri in the Calendar of the Incas," in "Ethnoastronomy and Archaeoastronomy in the American Tropics," ed. A. Aveni and G. Urton, *Annals* of the New York Academy of Sciences, vol. 385, (1982): 203–30; a more general article, "The Inca Calendar," in *Native American Astronomy*, ed. A. Aveni (Austin: University of Texas Press, 1977), pp. 220–259; and, finally, "Inca Observations of the Solar and Lunar Passages Through the Zenith and Nadir," in *Archaeoastronomy in the Americas*, ed. R. Williamson (Los Altos and College Park, Md.: Ballena Press and Center for Archaeoastronomy, 1981), pp. 319–42. In the latter volume can also be found my article, "Horizon Astronomy in Incaic Cuzco" (pp. 305–18), which explains the technique of tracing the *ceque* lines and astronomical alignments in the physical environment of Cuzco and also summarizes the results.

Concepts of religious and moral order present in the *ceque* system are explicated in an excellent article by L. Sullivan entitled "Above, Below, or Far Away: Andean Cosmogony and Ethical Order," which appears in the edited text of R. Lovin and F. Reynolds, *Cosmogony and Ethical Order* (Chicago: University of Chicago Press, 1985). Some of the other articles in this text (on the modern-day Maya by K. Warren and on the ancient Greeks by W. H. Adkins) are relevant to subject matter I discussed earlier. On this theme of South American cosmologies in general, see also Sullivan's *Icanchu's Drum* (New York: Macmillan, 1988).

And I ought to recommend, last of all, a couple of standard references on Andean archaeology in general: A. Kendall, *Everyday Life of the Incas* (London: Batsford 1973); and L. Lumbreras, *The People and Cultures of Ancient Peru* (Washington, D.C.: Smithsonian Institution, 1974).

1. J. L. Borges and A. B. Casares, "On Exactitude in Science," in *Fantastic Tales* (New York: Herder & Herder, 1971), p. 23. In their anthology, they attribute the quotation to S. Miranda (*Viajes de Varones Prudentes*, libro cuarto, cap. XIV [Mérida, 1658]).

2. B. Cobo, *Historia del Nuevo Mundo*, 4 vols., ed. D. Marcos Jiménez de la Espada (Seville: Sociedad de Bibliofilos Andaluces, 1890–93; originally published 1653), especially vol. IV, pp. 9–47.

3. Translation by J. Rowe, "An Account of the Shrines of Cuzco," *Nawpa Pacha* 17 (1979):25–26.

4. Garcilaso de la Vega, *Royal Commentaries of the Inca*, 2 vols., trans. and ed. H. Livermore (Austin: University of Texas Press, 1966; originally published 1616), pp. 764–65.

5. Rowe, "An Account," p. 61.

6. P. Guaman Poma de Ayala, *El Primer Nueva Corónica y Buen*

Gobierno, ed. J. Murra and R. Adorno; trans. J. Urioste (Mexico: Siglo XXI, 1980; originally published 1614) folios 353–54 and 883–84.

7. P. Cieza de León, *La Crónica del Peru* (Lima, 1973; originally published 1550), chap. 44.

8. Rowe, "An Account," p. 27.

9. V. Maurtua, anonymous chronicler, in *Juício de Limites entre el Perú y Bolivia* 8 (1906):149–65 (Madrid). For a recent discussion of the problem of the location of the pillars, see R. T. Zuidema, "The Pillars of Cuzco: Which Two Dates of Sunset Did They Define?," in A. Aveni, ed., *New Directions in American Archaeoastronomy* (Oxford: British Archaeological Reports International Series 454S, 1988), pp. 143–70.

10. J. Rowe, "An Introduction to the Archaeology of Cuzco," *Papers of the Peabody Museum on American Archaeology and Ethnology* 27 (2 [1944]):26.

11. L. Sullivan, "Above, Below, or Far Away: Andean Cosmogony and Ethical Order," in *Cosmogony and Ethical Order*, ed. R. Lovin and F. Reynolds (Chicago: University of Chicago Press, 1985), p. 119.

12. Rowe, "An Account," p. 15.

13. *The Incas of Pedro Cieza de León*, trans. H. de Onis (Norman: University of Oklahoma Press, 1959), chap. 107.

14. Quoted in R. Adorno, *Guaman Poma, Writing and Resistance in Colonial Peru* (Austin: University of Texas Press, 1986), p. 67.

Chapter 9. Eastern Standard: Time Reckoning in China

The Chinese cosmology and world view is a vast topic still not sufficiently explored and consequently not well understood. Without doubt, the acknowledged master of dissemination of Chinese science to the modern Western world is J. Needham, whose 7-volume *Science and Civilization in China* (Cambridge: Cambridge University Press, 1954), is the best resource to consult on astronomy. An excellent comparative study by Needham appears in *Clerks and Craftsmen in China and the West* (Cambridge: Cambridge University Press, 1970); and *The Voices of Time*, edited by J. T. Fraser (Amherst: University of Massachusetts Press, 1966), contains Needham's "Time and Knowledge in China and the West" (pp. 92–135), which deals specifically with the Chinese historiographic view. In *The Place of Astronomy in the Ancient World*, ed. F. R. Hodson (London: Procedings of the Royal Society, vol. A276 [1974], pp. 67–82), Needham deals specifically with astronomical and time-measuring instruments and methods in a paper entitled "Astronomy in Ancient and Medieval China."

Chinese Science: Exploration of an Ancient Tradition, ed. S. Nakayama and N. Sivin (Cambridge: MIT Press, 1973), is a collection of works on aspects of different branches of Chinese science. More recent is N. Sivin, "Chinese Archaeostronomy: Between Two Worlds," in A. Aveni, ed., *World Archaeoastronomy* (Cambridge: Cambridge University Press, 1989), which deals with the possibility and potential for exploration in that interdisciplinary area. D. de Solla Price, *Science Since Babylon* (New Haven: Yale University Press,

1975), reflects on the issue of how like the West the Chinese really were in their astronomical practice.

For an elementary, well-illustrated discussion of how the Chinese practiced astronomy, see the series of articles by E. C. Krupp printed in the June, July, and August 1982 issues of *The Griffith Observer*, a publication of Griffith Observatory and Planetarium, Los Angeles. Specific details about the various Chinese calendars and their festivals and rites can be found in J. Bredon, *The Moon Year* (Shanghai: Kelly & Walsh, 1927). More general accounts of Chinese civilization that deal in some respect with Chinese ideas about time include A. de Reincourt, *The Soul of China* (New York: Coward McCann, 1958).

1. Albert Einstein, in a letter dated 23 April 1953, quoted in D. de Solla Price, *Science Since Babylon* (New Haven: Yale University Press, 1975), p. 15.
2. Ibid.
3. Anonymous quote in S. M. Russell, *The Observatory* 18 (1895):323.
4. J. Needham, *Clerks and Craftsmen in China and the West* (Cambridge: Cambridge University Press, 1970), p. 225.
5. Ibid., p. 81.
6. R. H. Barnes, *Kedang* (Oxford: Clarendon Press, 1974), pp. 141–42.
7. Needham, *Clerks and Craftsmen*, p. 101.

Chapter 10. Building on the Basic Rhythms

1. R. Aaron, *John Locke* (Oxford: Clarendon Press, 1955), pp. 32, 35, and 114.
2. J. Bronowski, *Science and Human Values* (New York: Harper & Row, 1975), pp. 13–14.
3. N. Farriss, "Remembering the Future, Anticipating the Past: History, Time and Cosmology Among the Maya of Yucatan," *Comparative Studies in Society and History* 29 (1987):566–93.
4. D. Lee, "Lineal and Non-Lineal Codifications of Reality," *Psychosomatic Medicine* 12 (1950):89–97.
5. J. Fabian, *Time and the Other* (New York: Columbia University Press, 1983).
6. A. Aaboe, "Scientific Astronomy in Antiquity," in F. R. Hodson, ed., *The Place of Astronomy in the Ancient World* (London: Proceedings of the Royal Society, 1974), vol. A276, p. 22.
7. M. Nilsson, *Primitive Time Reckoning* (Lund: Gleerup, 1920).
8. *Webster's New Twentieth Century Dictionary of the English Language Unabridged* (New York: Standard Reference Works Publishing, 1956).
9. G. Conrad and A. Demarest, *Religion and Empire, the Dynamics of Aztec and Inca Expansionism* (Cambridge: Cambridge University Press, 1984).
10. M. Young, *The Metronomic Society* (Cambridge, Mass.: Harvard University Press, 1988).

11. *The Politics of Aristotle*, trans. and ed. E. Barker (Oxford: Oxford University Press, 1956), p. 118.
12. Aaboe, "Scientific Astronomy," p. 23.

Epilogue

1. E. C. Hughes, "Cycles, Turning Points and Careers," in E. C. Hughes, ed., *The Sociological Eye*, vol. I (Chicago: Aldine, 1984), p. 124.

INDEX